T0313168

A Revised Handbook
to the
FLORA OF CEYLON

VOLUME XV, PART B

Aspleniaceae	Hymenophyllaceae	Parkeriaceae
Azollaceae	Isoetaceae	Polypodiaceae
Blechnaceae	Lomariopsidaceae	Psilotaceae
Cyatheaceae	Loxogrammaceae	Pteridaceae
Davalliaceae	Lycopodiaceae	Salviniaceae
Dennstaedtiaceae	Marattiaceae	Schizaeaceae
Dryopteridaceae	Marsileaceae	Selaginellaceae
Equisetaceae	Oleandraceae	Thelypteridaceae
Gleicheniaceae	Ophioglossaceae	Vittariaceae
Grammitidaceae	Osmundaceae	Woodsiaceae

A Revised Handbook
to the
FLORA OF CEYLON

VOLUME XV, PART B
FERNS AND FERN-ALLIES

Edited by
MONIKA SHAFFER-FEHRE

Sponsored jointly by the

University of Peradeniya,
Department of Agriculture, Peradeniya, Sri Lanka,
National Science Foundation of Sri Lanka
and the Overseas Development Administration,
United Kingdom

General Editor
M.D. DASSANAYAKE

Editorial Board
M.D. DASSANAYAKE
W.D. CLAYTON

CRC Press
Taylor & Francis Group
Boca Raton London New York

CRC Press is an imprint of the
Taylor & Francis Group, an **informa** business
A SCIENCE PUBLISHERS BOOK

CIP data will be provided on request.

CRC Press
Taylor & Francis Group
6000 Broken Sound Parkway NW, Suite 300
Boca Raton, FL 33487-2742

First issued in hardback 2019

© 2006 by Taylor & Francis Group, LLC
CRC Press is an imprint of Taylor & Francis Group, an Informa business

No claim to original U.S. Government works

ISBN-13: 978-1-57808-411-1 (set)
ISBN-13: 978-1-57808-384-8 (Part A)
ISBN-13: 978-1-57808-410-4 (hbk)(Part B)

Visit the Taylor & Francis Web site at
http://www.taylorandfrancis.com

and the CRC Press Web site at
http://www.crcpress.com

FOREWORD

Although ferns were not included in Trimen's original five-volume 'A Handbook to the Flora of Ceylon', they form a significant part of the vascular flora of the island.

A good deal of fern research was carried out contemporaneously to, but independently of, Trimen's work. Publications, though, were widely scattered throughout the literature or constituted major works in their own right, i.e. that of R.H. Beddome 1863-1883. The Overseas Development Agency (ODA) agreed to finance, the preparation of, an additional volume, in the new work 'A Revised Handbook to the Flora of Ceylon', in order to provide an accessible taxonomic base for studies on this element of Sri Lankan vegetation.

This volume has been compiled from contributions by authors of widely differing experience. The previous work of W.A. Sledge 1956-1982, too, is a highly erudite treatment that was a welcome guide to many. Variability in the depth of accounts in the present publication provides, in itself, a valuable insight into the ways such descriptions can be approached. No apology is made for including some longer texts for the reader to benefit from specialist knowledge.

Sri Lankan ferns are of special importance due to the number of endemics to the island. Of the 351 fern species treated in this book 58 are endemic to the island, and an additional 30 species are specific to only Sri Lanka and Southern India. The statistics thus show about one quarter of the fern flora of Sri Lanka to be endemic. Approximately 50% of all fern taxa are part of the floras of South Asia, South East Asia and China.

(References at end of Introduction)

W.D. CLAYTON
MONIKA SHAFFER-FEHRE
May 2006

ACKNOWLEDGEMENTS

It is a pleasure to record our thanks to our sponsors and, in particular, to the Overseas Development Agency (ODA) who financed the additional Volume XV of the 'A Revised Handbook to the Flora of Ceylon'.

It is our sad duty to announce the death of David Philcox who was a contributor to 'A Revised Handbook to the Flora of Ceylon' for a considerable number of years. He gave of his expertise throughout a wide range of families in several volumes. We remember his ebullient character and his generous advice and support when it was required for matters of the work.'

We thank Josephine Camus, pteridologist of The Natural History Museum, London, for reading the text of the Introduction and adding support through valuable comments.

Dr Nicholas Hind, Royal Botanic Gardens, Kew helped with generous advice. His suggestions were gratefully received and incorporated.

M. S-F. deeply appreciates the kindness of Prof. R.J. Johns, Royal Botanic Gardens, Kew, for stimulating her interest in the ferns of Sri Lanka and for providing the chance for her to contribute to 'A Revised Handbook to the Flora of Ceylon'.

We like to record our special thanks to Mr Peter J. Edwards, curator of Ferns at the Royal Botanic Gardens, Kew. His advice, always given readily, amounted to the most invaluable support in completing the work on many taxa treated in this book. His selfless giving of his experience and his advice on additional helpful information to include in the text, were an inspiration of how scientific collaboration should be approached.

Last, but by no means least, we wish to express our thanks to all the valuable assistant collectors in the field as they endured botanists' whims, smoothed language problems and, in turn, exasperated and rescued us.

W.D. CLAYTON
MONIKA SHAFFER-FEHRE
May 2006

CONTENTS

PART A

PART B

INTRODUCTION

(BY MONIKA SHAFFER-FEHRE* & P.J. EDWARDS**)

Ferns and fern-allies have an ancient history in common. Their fossil records show that they achieved early the development of a vascular system for water conduction and structural support, and also evolved a cuticle for water retention and various types of stomata for water and gas regulation. In addition to these evolutionary gains, taxa of this assemblage share, uniquely amongst vascular plants, the mechanism of "Alternation of Generations". Their sexual reproduction proceeds by means of spores (see below), but apart from these factors the connection between ferns and fern-allies is tenuous and, in the main, arbitrary.

The evolutionary ages of fern-allies and ferns, based on the fossil record, differ. Fern-allies bear leaves (microphylls). Ferns bear fronds that, simple or ± dissected, equate to an angiosperm leaf.

Lycopodiaceae, Selaginellaceae, with microphylls and adaxial sporangia, and the Isoetales have affinities with the Lycopsida from the Devonian (410 MYBP - million years before present -) and Upper Carboniferous (330 MYBP).

Equisetaceae date from the Permian (250 MYBP), with nodal leaf sheaths and peltate sporangiophores, have silica in their make-up that supports large plants in the absence of secondary thickening; they recall the Calamites of the Carboniferous (300 MYBP).

Three major groups of ferns deliniate the earth ages: The ancient EU sporangiate ferns can be traced back to the Carboniferous (345 MYBP). These early ferns are characterized by a spore capsule that originates from more than one cell and contains numerous spores (from 500 to several 1000). Curiously, the eusporangiate Ophioglossaceae can only be traced to the Upper Cretaceous and the Tertiary (136 MYBP); this may bear out, however, that fragile plants are less easily fossilized.

The modern LEPTO sporangiate ferns have a sporangium that originates from a single cell and contains various numbers of (but less than 100) spores. The Osmundaceae (Upper Permian 240 MYBP), are close to the origin of leptosporangiate ferns. There i) follow Schizaeaceae, Gleicheniaceae and Polypodiaceae (Jurassic 190-195 MYBP) and Cyatheaceae (Upper Jurassic). Note has to be taken of *Aspidistes* from the Yorkshire Jurassic (48 spores).

---

* & ** Royal Botanic Gardens, Kew, U.K.

Based on evidence from: persistant indusium, sporangial structure and 1-celled abaxial glands, Holttum found affinities connecting *Apsidistes* to *Coryphopteris* in the Thelypteridaceae. Centre of distribution for Thelypteridaceae is at present Sri Lanka and Malesia. The eusporangiate fern *Marattia* (Marattiaceae), with today's centre of distribution in Malesia, also has fossil representatives in the beds of the Yorkshire Jurassic.

AQUATIC ferns are the most recent, highly evolved heterosporous, leptosporangiate ferns with the Salviniaceae (incl. *Azollopsis*) (Upper Cretaceous 136 MYBP) and the Marsileaceae (Tertiary [Oligocene] 38 MYBP). All three groups of ferns can be index fossils at geological boundaries.

Most South Asian species range eastwards to varying extents, often to China and Japan, or through Malesia (South East Asia) and a few reach Northern Queensland (Australia). Sri Lanka and the Indian peninsula form a tectonic unit; this "Deccan Plate" is of Gondwanic origin. During the Cretaceous and Early Eocene the Deccan drifted northwards and joined the Eurasian landmass in the Eocene. During these aeons of time (85 MY) the climate in Sri Lanka is thought to have been comparable to Madagascar, the Seychelles and Malesia (Ashton & Gunatilleke 1987); concerning today's climatic conditions, Sri Lanka is divided into the dry- and the wet zone. The south-west of the island, the wet zone (c. 15,208 km²), contains the majority of the 15 floristic areas that relate to the endemic angiosperms (Gunatilleke and Ashton, 1987) and ferns. Among the plants of the World Heritage Site, Sinharaja, a 'Lowland Wet Rain Forest' and of the Knuckles-Range with 'montane' and 'submontane' flora, some taxa have been referred to as relicts of the Deccan-Gondwana flora. This is true also for the Peak Wilderness and Horton Plains which include a large range of ecotopes: 'Lowland Tropical Evergreen Rain Forest', 'Submontane' and 'Montane Tropical Wet Evergreen Forest' and the wet 'Patana' grasslands. All these contain a high percentage of endemic taxa. All floristic zones of Sri Lanka, "The Resplendent Land", are found among the collecting sites for varying numbers of fern taxa described in this volume.

The criteria used for the various systems of the classification of ferns have, historically, included the stem anatomy, shape of fronds and their lesser or greater dissection, the position and shape of sori and that of the indusia, the venation and many more characteristics. With increasing development of microscopy the spores and the morphology of the perispores (electron microscopy) can be studied. Yet the classification is, even now, by no means finalized (see 3rd column of the appended list of taxa). Of late, ferns have also been studied by the methods of biochemistry and gene sequencing.

It was only in the late 18th century that the basic details of the complicated fern reproductive cycle were recognized. The sexual (generative or gametophytic) stage of the "alternation of generations" has the spore germinating and growing usually into a small, heart-shaped thallus of tissue or a mass of threads, the gametophyte or prothallus. The gametophyte generates antheridia in the vicinity of the rhizoids and archegonia closer to the notch. These organs give haploid (denoted by n) gametes, i.e. the egg cells and the antherozoids. When a film of water is available the flagellate spermatozoids are attracted chemically and swim to the archegonium of a second, genetically different prothallus and fertilize the egg.

On the event of fusion of these two the diploid zygote (denoted by 2n) comes into existence and gives rise to the sporophyte which has the full (diploid =2n) complement of chromosomes. The sporophyte represents the asexual generation. It will generate spores. A few taxa can also reproduce vegetatively by growing a sporeling from the tip of the frond that touches the soil, from axils of pinnae from gemmae or bulbils and from vegetative cells of the gametophyte (one form of apogamy). The fusion of gametes from 2 spores of different genetic make-up gives the chance to progress genetically. Such "outbreeding" has made adaptation to different environments possible. This has enabled ferns and fern-allies to conquer the terrestrial habitat, grow on different substrates such as on rock or on bark as epiphytes (c. 20% of pteridophytes are epiphytes). A few taxa float on and live in the water.

Gametophyte and sporophyte, the alternate generations, may survive independently for long periods of time. Few are known to persist for decades purely in the gametophytic state. The ability to exist in stasis is an added advantage to the survival strategy of ferns.

Peradeniya and the Classification of Fern Genera with reference to Sri Lanka

Professor M.D. Dassanayake contributes the following paragraph:

George Gardner was Superintendent of the Peradeniya Gardens from 1843 to 1849. He did much collecting in Sri Lanka. His specimens are also to be found at The Natural History Museum (BM), London, and at the Royal Botanic Gardens, Kew (K). George Henry Kendrick Thwaites succeeded Gardner as superintendent of the Gardens in 1849. He and his assistants collected assiduously and by 1853 he found that the number of duplicate specimens in the herbarium (PDA) had risen to more than 6000. Using these he compiled several numbered sets of "Ceylon Plants" specimens, the C.P. series. (Some of these were specimens left from Gardners' collection.) Thwaites kept one set at Peradeniya, distributed the others to the principal herbaria abroad, and obtained specimens for Peradeniya (PDA) in exchange.

Thwaites gave the same C.P. (Ceylon Plants) number to all specimens which he considered to be of the same species, regardless of when or by whom they were collected. Specimens, therefore, which bear the same C.P. number may not actually be duplicates. Most of the specimens of the C.P. series at PDA are without labels. The name of the actual collector, or of Thwaites is not given. On some sheets, however, the collectors' name, mostly Gardner's is written in pencil and is now barely legible.

The C.P. series forms the basis of Thwaites' book "Enumeratio Plantarum Zeylaniae" (1864). He seems to have regarded the C.P. set at PDA to be his standard set. Trimen (1893) states that "it must be regarded as the type series".

NB: Herbarium acronyms follow Holmgren *et al.* 1990.

In contrast to the ALPHABETIC arrangement of families throughout this volume, the following synopsis of the taxa treated in this publication lists families and genera TAXONOMICALLY, following Sledge (1982). **Family numbers** (in parentheses behind the name) are given according to Brummitt (1992). The opinion of authors as to the number and / or composition of genera in a family can vary; the 3rd column accounts for this in part, <u>as far as it is relevant to Sri Lanka</u>. The **genera** are given in alphabetical order for quick reference (2nd column); their number within the family in this particular publication is given in '(parentheses)'. *Italicised generic names* (4th column) recall genera in which taxa of the central column were classified in the most recent past. This may also be of help to herbarium curators.

Families	Genera		Alternative (sub) families into which some genera* have been placed by others and commonly used alternative *generic names* employed in the last 50 years.

FERN ALLIES

Families	Genera		Alternative
PSILOTACEAE (254)	Psilotum		
LYCOPODIACEAE (250)	Huperzia	(1)*	*Lycopodium*
	Lycopodiella	(3)*	"
	Lycopodium	(2)	
SELAGINELLACEAE (251)	Selaginella		
ISOETACEAE (252)	Isoetes		
EQUISETACEAE (253)	Equisetum		

EUSPORANGIATE FERNS

Families	Genera		Alternative
OPHIOGLOSSACEAE (300)	Botrychium		
	Helminthostachys		
	Ophioglossum		*Ophioderma*
MARATTIACEAE (301)	Angiopteris	(2) *	Angiopteridaceae
	Marattia	(1)	Angiopteris

LEPTOSPORANGIATE FERNS

OSMUNDACEAE (302)	Osmunda	(1)	
	Todea	(2)	
SCHIZAEACEAE (307)	Anemia	(3)	
	Lygodium	(2)	
	Schizaea	(1)	
PARKERIACEAE (309)	Ceratopteris		
GLEICHENIACEAE (304)	Dicranopteris* one of 5 sections in *Gleichenia*		
HYMENOPHY-LLACEAE (315)	Crepidomanes	(5)*	*Trichomanes*
	Didymoglossum	(4)*	"
	Gonocormus	(9)*	"
	Mecodium	(1)*	*Hymenophyllum*
	Meringium	(2)*	*Hymenophyllum*
	Microgonium	(6)*	*Trichomanes*
	Microtrichomanes	(8)*	"
	Pleuromanes	(7)*	"
	Selenodesmium	(3)*	"
CYATHEACEAE (318)	Cyathea		
PTERIDACEAE (311)	Acrostichum	(6)*	Adiantaceae (313),
	Idiopteris	(8)*	subfam.Pteridoideae
	Pteris	(7)*	"
	Actiniopteris	(13)*	Adiantaceae (313),
	Adiantum	(14)*	subfam.Adiantoideae
	Anogramma	(5)*	"
	Cheilanthes	(11)*	"
	Coniogramme	(3)*	"
	Doryopteris	(9)*	"
	Hemionitis	(1)*	"
	Parahemionitis	(2)*	"
	Pellaea	(12)*	"
	Pityrogramma	(4)*	"
	Taenitis	(10)*	"
VITTARIACEAE (314)	Antrophyum	(2)*	Antrophyaceae
	Monogramme	(3)*	*Vaginularia*
	Vittaria	(1)	
DENNSTAEDTIACEAE (322)	Dennstaedtia	(1)	
	Histiopteris	(4)	
	Hypolepis	(3)	
	Lindsaea	(7)*	Lindsaeaceae

	Microlepia	(2)	
	Pteridium	(5)	
	Sphenomeris	(6)*	Lindsaeaceae
ASPLENIACEAE (327)	Asplenium		
THELYPTERIDACEAE (329)	Amauropelta	(7)*	*Thelypteris*
	Ampelopteris	(9)*	*Cyclosorus*
	Amphineuron	(6)*	"
	Christella	(8)*	"
	Cyclosorus	(4)*	"
	Macrothelypteris	(5)*	*Thelypteris*
	Metathelypteris	(1)*	*Thelypteris*
	Parathelypteris	(2)*	*Thelypteris*
	Pneumatopteris	(3)*	*Cyclosorus*
	Pronephrium	(10)*	"
	Pseudocyclosorus	(11)*	"
	Pseudophegopteris	(12)	—
	Sphaerostephanos	(13)*	*Cyclosorus*
	Stegnogramma	(14)*	"
	Thelypteris	(16)	—
	Trigonospora	(15)*	*Cyclosorus*
WOODSIACEAE (323.01) (Athyriaceae)	Athyrium	(3)	
	Deparia	(2)*	*Lunathyrium*
	Diplazium	(4)	
	Hypodematium	(1)	
DRYOPTERIDACEAE (323.02)	Arachniodes	(3)*	Dryopteris Group
	Diacalpe	(1)*	"
	Dryopsis	(6)*	"
	Dryopteris	(2)*	"
	Polystichum	(4)*	"
	Ctenitis	(5)*	Tectaria Group
	Lastreopsis	(7)*	"
	Pteridrys	(8)*	"
	Tectaria	(9)*	"
LOMARIOPSIDACEAE (325)	Bolbitis	(1)	
	Elaphoglossum	(3)	
	Teratophyllum	(2)	
DAVALLIACEAE (326)	Davallia	(1)*	*Humata* (in part)
	Leucostegia	(2)	
OLEANDRACEAE (324)	Arthropteris	(2)	
	Nephrolepis	(1)*	Nephrolepidaceae
	Oleandra	(3)	

POLYPODIACEAE (331.02)	Belvisia	(8)	
	Drynaria	(1)	
	Lepisorus	(5)	
	Leptochilus	(4)	
	Microsorum	(3)	
	Pleopeltis	(6)	
	Pyrrosia	(2)	
	Selliguea	(7)*	*Crypsinus* *Phymatopteris*

GRAMMITIDACEAE (330)	Calymmodon	(7)
	Chrysogrammitis	(3)
	Ctenopteris	(4)
	Grammitis	(1)
	Prosaptia	(6)
	Scleroglossum	(2)
	Xiphopteris	(5)

LOXOGRAMMACEAE (331.02)	Loxogramme	Polypodiaceae (331.02)

BLECHNACEAE (328)	Blechnum	(2)
	Doodia	(1)
	Stenochlaena	(3)

LEPTOSPORANGIATE FERNS (AQUATIC)

MARSILEACEAE (332)   Marsilea

SALVINIACEAE (333)   Salvinia

AZOLLACEAE (334)   Azolla

REFERENCES and GENERAL LITERATURE

Ashton, P. & Gunatilleke, C.V.S., 1987. New Light on the Plant Geography of Ceylon.1. Historical plant geography of the Lowland Endemic Tree Flora. *J. Biogeogr.* **14**(3): 249-285.

Beddome, R.H., 1863-64. Ferns of Southern India. Madras; Ganz Brothers

Beddome, R.H., 1883 Handbook to the ferns of British India, Ceylon and the Malay Peninsula. Calcutta, Thacker. (The dates for Beddome encompass the time frame of major works.)

Brummitt, R.K., 1992. Vascular Plant Families and Genera. 804 p. Royal Botanic Gardens, Kew, The Board of Trustees.

Camus, J.M., Jermy A.C., & Thomas B.A. 1992. A world of ferns. 1-112 p. London: The Natural History Museum.

Eagar, R.M.C., 1995.The Geological Column. (5th edn.): Manchester. A Manchester Museum Publication. (Double-sided poster.)

Fraser-Jenkins, C.R., 1984. An introduction to fern genera of the Indian subcontinent. *Bull. Brit. Mus. (Nat. Hist.), Bot. Ser.* **12** (2): 37-76.

Gunatilleke, N. & Gunatilleke, C.V.S., 1995. C P D Site IS 11, Knuckles, Sri Lanka. 22: 123-126. *In*: Centres of plant diversity, a guide and strategy for their conservation. Vol. 2, Asia, Australasia and the Pacific. Davis, S.D., Heywood & V.H., Hamilton, (eds). Oxford. World Wide Fund for Nature (WWF) The World Conservation Union (IUCN).

Gunatilleke, N. & Gunatilleke, C.V.S., 1995. C P D Site IS 12, Peak Wilderness and Horton Plains, Sri Lanka. 22: 127-131. loc. cit.

Gunatilleke, N. & Gunatilleke, C.V.S., 1995. C P D Site IS 13, Sinharaja, Sri Lanka. 22: 132-134. loc. cit.

Gunatilleke, C.V.S. & Ashton, P.S. 1987. New Light on the Plant Geography of Ceylon.2. The Ecological Biography of the Lowland Endemic Tree Flora. *J. Biogeogr.* **14**(4): 295-327.

Holmgren, P.K. et al., 1990. Index herbariorum. 693 p. 8th edn. New York Botanic Gardens, International Association for Plant Taxonomy.

Jayasekara, P.W.B., 1997. A systematic Study of the Hymenophyllaceae. xii, 132. M.Phil. thesis. Unpubl., Univ. of Peradeniya (Manuscript at Royal Botanic Gardens, Kew)

Kubitzki, K. (ed.), 1990. The Families and Genera of Vascular Plants:1. Pteridophytes and Gymnosperms. xiii, 404 p. K.U. Kramer & P.S. Green (vol. eds.) Berlin, Springer Verlag,

Lovis, J.D., 1977. Evolutionary patterns and Processes in Ferns. 4: 229-402. *In:* Advances in Botanical Research. Preston, R.D. & Woolhouse, H.W. London, Academic Press

Pichi Sermolli R.E.G., 1996. Authors of scientific names in Pteridophyta. Royal Botanic Gardens, Kew.

Pryer, K.M. & Smith, A.R., 2001. The Tree of Life webpage. (see below)

Sledge, W.A., 1982. An annotated check-list of the Pteridophyta of Ceylon. *Bot. Journ. Linn. Soc.* **84**: 1-30. This check-list follows Sledge's treatments of many families in separate papers.

Stewart, W.N. & Rothwell, G.W., 1993. Palaeobotany and the evolution of plants. x, 521 p. 2nd edn.,Cambridge University Press, Cambridge.

Thwaites, G.H.K., 1864. Enumeratio Plantarum Zeylaniae. viii, 483 p. London, Dulau & Co.

Trimen, G., 1843-1896. A Hand-book to the flora of Ceylon. 5 vols. London, Dulau & Co.

Consult also the Tree of Life web page by Pryer K.M and Smith, A.R. http://tolweb.org/tree?group=leptosporangiate_ferns&outgroup=Filicopsida Tree of Life design and icons copyright© 2001 David Maddison & Wayne Maddison.

KEY to the FERNS and FERN-ALLIES of SRI LANKA
(by Peter J. Edwards*)

Preamble

The following artificial key leads variously to families, genera, or species. The three reasons for this are the wide variety of form in some families, a lack of space and, making a virtue out of necessity, the aim to speed identification by the use of a "shortened key". The key partly recalls that in "Pteridophyte Flora of the Western Ghats—South India" by Manickam & Irudayaraj.

The key assumes that you will be looking at mature plants, juvenile ones may key out in the wrong place or not at all. Be aware of the fact that quite young plants may produce one or more sori and hence **appear to be adult**. Such young plants can, however, have quite a different appearance when compared to an adult. Note also that indusia are best judged <u>before</u> the sporangia are fully mature.

N.B. A 'x 10' lens will be essential at some couplets, particularly at 22, 36, 38, 45, 52, 57, 64; <u>ideally a **x 10** lens should always be at hand</u>. When using the key be aware that there are **exceptions to some statements**, do not give up, go back in the key, try again and keep persevering. Good luck!

Any collections, and that means all sets separated off as duplicates, should be fully representative of the plant: including stipe, rhizome, and at least one complete frond, ideally attached to rhizome, and always fertile material (with sufficient sori). **Familiarize yourself, if at all possible, in a herbarium concerning what material is required in a collection set, and which parts, by their absence from such a set, prevent good research.**

When mounted, both sides of the lamina should be shown. Carefully assess individuals of a population in order to understand the extent of natural variation.

The 254-entry glossary contains more items than indicated by 'TS'=transverse section, 'v' ='very'.

Fronds and leaves: The former term is used for the true ferns (Filices), the latter for the fern allies (*Isoetes, Equisetum, Selaginella* and Lycopodiaceae) (the only taxa in Sri Lanka) excluding *Psilotum*, which has no leaves at all. <u>added to help in emphasizing some groupings</u>
<u>with the stated character in common</u>
Terms used variably:1-pinnate **or** pinnate **or** simply pinnate / 2-pinnate **or** bipinnate.

---

*Royal Botanic Gardens, Kew, U.K.

THE KEY

1   Plant a grass-like tuft growing in or at the edge of fully exposed seasonal pools or in wet seepages; small, white, spherical megaspores in 'pocket' at base of many leaves .... **Isoetes**

1   Not as above... **2**

2   Plants floating and/or rooted in fresh water...................... **3**

2   Plants not growing in (fresh) water and/or erect with whorls of branches **6**

3   Plants floating, leaves 0.2 - 2 cm long ± triangular and finely lobed, or orbicular and/or entire **4**

3   Plants floating, leaves shaped otherwise and/or on mud / wet soil **5**

4   Fronds to 1.5cm long, ± triangular with pinnately arranged 0.5-1mm long lobes, filamentous submerged fronds absent .... **Azolla**

4   Fronds 1-2cm long, orbicular, in opposing pairs, flat or folded along midrib, filamentous, submerged brown fronds present .... **Salvinia**

5   Plant floating or on mud, sterile fronds dimorphous broadly to linear, 2- to 3- pinnatifid part 20-30cm long, fertile fronds with the lobes much more narrow .... **Ceratopteris**

5   Plant on mud/wet soil, rhizome slender, creeping, bearing many erect fronds consisting of a petiole 2-18cm tall bearing 4 clover-shaped leaflets; sporangia (present in dry conditions) in small brown sporocarps at the base of the petiole .... **Marsilea**

6   Plant ± erect, with complete or incomplete whorls of branches (sometimes absent) at regularly spaced intervals, sheathed nodes bearing scale leaves, sporangia in terminal cones .... **Equisetum**

6   Not as above... **7**

7   Small plant without roots or leaves, stems dichotomously branched, sporangia in clusters of 3 in axils of scale leaves .... **Psilotum**

7   Not as above... **8**

8   Much-branched herbaceous plants bearing 2 ranks of very small (<1-5.5mm) leaves, some branches bearing cones (strobili, made up of sporophylls) at their apices .... **Selaginella**

8   Not as above... **9**

9   Terrestrial with long-creeping or trailing stems or pendulous epiphytes ; leaves small (5-35mm long), narrow, arranged spirally or in whorls, cones on long, erect peduncles or at branch tips .... **Lycopodiaceae** also the only homophyllous *Selaginella* .... **Selaginella rupestris**

9   Not as above... **10**

10 Sporangia borne on several pairs of fertile pinnae situated at the centre of an otherwise sterile frond..................................... **11**

10 Sporangia borne on abaxial surface of or near margin of ultimate segment; sometimes on modified fronds.............................. **15**

11 Fertile spikes on 'panicles' originating near the base of the blade segment from the 2-fid or 3-fid frond **12**

11 1-pinnate fronds, some with several pairs of fertile pinnae at centre of an otherwise sterile frond, sterile pinnae linear-lanceolate .... **Osmunda collina**

12 Erect fertile 'panicles' (two); rhizomes scaly at ground surface densely covered in long brown hairs , roots wiry .... **Anemia**

12 Erect fertile portion a spike, rhizome short, fleshy, without scales, subterranean, roots fleshy **13**

13 Blade simple, veins anastomosing .... **Ophioglossum**

13 Blade compound, veins free **14**

14 Fertile segment a simple spike .... **Helminthostachys**

14 Fertile segment a branched 'panicle'-like frond .... **Botrychium**

15 Stipes large (>3cm diam.), fleshy, with scattered white longitudinal short lines (aerophores); a pair of large, succulent brown stipules present at the base of each **16**

15 Stipes, (<1cm diam.), usually with ± continuous pale line (lateral aerophore) on both sides (rarely absent or obscure), stipules absent **17**

16 Sporangia of each group fused together like a cockroach egg mass..... .... **Marattia**

16 Sporangia separate, in a close double row.............. **Angiopteris**

17 Fronds twining, climbing to 10m+, of indefinite growth..... **Lygodium**

17 Fronds not twining, each apparently of limited growth............ **18**

18 Lamina usually 1 cell thick (rarely 2), translucent, indusium of 2 marginal flaps or tubular, sporangia exserted on ± long sorophore (?) (becoming bristle-like), ferns of moist, shady habitats .... **Hymenophyllaceae**

18 Lamina more than 2 cells thick; typical 'fern-plants' of many different habitats **19**

19 Terrestrial ferns with hairy, long-creeping rhizomes, stipes hairy or glabrous **20**

19 Terrestrial, epilithic or epiphytic ferns, rhizomes various & stipes bearing scales, or scales mixed with hairs **22**

20 Fronds branching in a series of elegant pseudo-dichotomous forks, a dormant apex at each fork; stipe glabrous .... **Dicranopteris linearis**

20 Fronds not thus branching **21**

21 Sori continuous along edge of leaflet, protected by a reflexed margin (false indusium), stipe stramineous, hairy .... **Pteridium revolutum**

21 Sori* at end of a single vein, protected by small flap or cup-like true indusium, stipes red-brown & v.glossy or paler & hairy............. .... **Dennstaedtiaceae**

(*v. small in *Hypolepis glandulifera*, best seen when sporangia unripe)

22 Fertile & sterile fronds distinctly dimorphic. (not incl. the 2 species of *Drynaria* with stiff brown fronds at base of 'normal' fronds) **23**

22 Fertile & sterile fronds not (or not strongly) dimorphic; (but including plants with specialized, stiff brown fronds at base of 'normal' fronds in 2 spp. of *Drynaria* only) **30**

23 Lamina simple, entire (rarely laciniate), litho- or epiphytes **24**

23 Lamina simple-pinnatisect to 1-pinnate, ...
terrestrial, epilithic, scandent - to climbing plants **25**

24 Lamina orbicular to ovate, thickly coriaceous, <5cm long, rhizome wiry; epiphytic or epilithic . . . . **Pyrrosia heterophylla**

24 Lamina lanceolate, entire or occasionally laciniate, c. 15cm long, not coriaceous, rhizome thick, epiphytic . . . . **Leptochilus decurrens**

25 Fronds pinnatisect with distinct broad adnate lobes, leaflets linear; sori short to elongated, indusiate; terrestrial plants . . . . **Blechnaceae** (N.B.*Blechnum colensoi* and *Doodia dives* only)

25 Not as above **26**

26 Rhizome creeping, terrestrial, to 80cm tall,1-pinnate, sporangia exindusiate, in a single continuous band, equidistant between midrib and margin . . . .
. . . . **Taenitis blechnoides**
. . . . (when strongly dimorphic)

26 Not entirely as above **27**

27 Lamina pinnatisect, rough to the touch, rhizome erect, sori oblong, in 2 or 4 rows per pinna, indusiate; terrestrial . . . . **Doodia**
. . . . (*D.caudata* and *squarrosa* only)

27 1-2-pinnate, rhizome short- to long-creeping, sori acrostichoid, with or without reflexed margin **28**

28 1-pinnate, terrestrial or epilithic, short-creeping rhizome . . . . . . **Bolbitis**

28 Not entirely as above **29**

29 Terrestrial with thick (5-8mm) long-creeping / scandent, stramineous rhizome, with few roots and scales, sori acrostichoid, in open places or secondary forest . . . . **Stenochlaena palustris**

29 Hemi-epiphyte, wiry (< 2mm) green to brown prickly rhizome with many roots & variable no. of small (<10cm long) 1-pinnate (often one-sided) fronds, when climbing with progressively thicker rhizome (>10mm) and often larger pinnate -pinnatisect fronds both closely appressed to a tree trunk, large (20-90cm) sterile, 1-pinnate sterile and fertile fronds often present above these . . . . **Teratophyllum aculeatum**

30 Small (< 12cm) annual plants with minute rhizome, bearing weakly dimorphic fronds, sterile ones 1-pinnate-pinnatisect, broadly cuneate segments , fertile ones 2-3-pinnatifid, segments more narrow . . . .
. . . . **Anogramma**

30 Not as above **31**

31 Terrestrial, rhizome creeping, plants to 80cm tall, 1-pinnate, sporangia exindusiate; in a single continuous band, equidistant between midrib and margin . . . . **Taenitis blechnoides**
. . . . (when weakly dimorphic)

31 Not entirely as above **32**

32 Terrestrial or epilithic, lamina pinnatisect to 1-pinnate, rough to the touch, pinna stalks v. short, sori elongated, in 2 or 4 parallel rows per pinna, indusiate . . . . **Doodia**

32 Not entirely as above **33**

33 Sori indusiate, linear along veins, or of irregular shape (reniform, 'J'-shaped), sometimes all on one frond **34**

33 Sori not linear, or if so, exindusiate **36**

34 Scales clathrate, often iridescent, 2 vascular strands at base of stipe joining upwards, forming 'X' in cross section . . . . **Asplenium**

34 Scales not clathrate, vascular strands join to 'U'-shape in upper stipe. . .
. . . . **35**

35 Indusia conspicuously reniform, 'J'or 'U'-shaped to linear (can be all on the same frond); fronds 1-pinnate to 3-pinnate, usually 20-200cm long, veins free or anastomosing . . . . **Woodsiaceae**

35 Indusia small, reniform only, thin, fugacious; fronds deeply 3-pinnatifid, 50-150cm long, veins free . . . . **Deparia boryana**

36 Stipe jointed closely to rhizome (look for slight swelling on stipe often with stramineus colour above and brown below) or jointed to conspicuous phyllopodia, both leaving a neat scar when dehisced; older fronds can be broken off cleanly; plants creeping to scrambling entire-leaved terrestrial, **or** 2-3-pinnate terrestrials or epiphytes , **or** small pinnatifid/pinnatisect epiphyte **37**

36 Stipe not jointed to rhizome; **or** if jointed, stipes in two rows & with rounded sori sunken into prominent deep pouch in lamina margin (*Prosaptia* only); plants small linear-leaved epiphytes, **or** small grass-like terrestrials, **or** larger 2-3-pinnate terrestrials with or without white or yellow powder on underside of fronds **41**

37 Sori indusiate, rhizome long-creeping / scrambling **38**

37 Sori exindusiate, (or, if indusiate, indusium linear and closely on either side of costa) rhizome of various forms **40**

38 Fronds simple, sori near midrib, rhizome long-creeping to 4.5m, scrambling or suberect . . . . **Oleandra**

38 Fronds pinnatifid to pinnately divided, sori near margin, at end of veins
. . . . **39**

39 Rhizome radial (TS), short-creeping, bearing hairs and scales, roots all around rhizome, indusia attached at base only, scales attached along a broad basifixed base . . . . **Leucostegia immersa**

39 Rhizome dorsiventral (TS), medium-long creeping, bearing scales only, (peltate or pseudo-peltate), roots usually on lower surface of rhizome only; indusia attached at their sides and/or base . . . . **Davallia**

exindusiate(to 49 excl. *Blechnum*)

40 Veins copiously anastomosing, distinct, with included veinlets, sporangia in discrete circular to elongated sori, **or** pseudo-acrostichoid on narrowed apical portion of simple fronds; small to large epiphytes or lithophytes with thin to coriaceous lamina . . . . **Polypodiaceae**

40 Veins free or at most joining margin (NB veins indistinct), sporangia covering whole undersurface of fertile fronds (acrostichoid condition); small epiphytes with very coriaceous lamina . . . . **Elaphoglossum**

Frond jointed to rhizome (to 38)

Fronds not jointed to rhizome (to 64 (end))
(except for *Prosaptia* (Grammitidaceae))

41 Epiphytic or epilithic; rhizomes compact **or** very short-creeping, sori circular to elongated, or in many oblique, parallel lines ...(***Loxogramme*** only) **or** in a long, marginal groove, **or** submerged in a prominent marginal pouch; scales not clathrate, spores green . . . . **Grammitidaceae**

41 Not as above **42**

42 Epiphytic or epilithic, sori in unequal-length irregular lines, **or** submerged in a deep marginal or sub-marginal groove extending for ± half- of length of lamina, **or** superficial on veins, scales anastomosing, clathrate, iridescent **or** v. small linear fronds < 0.5cm wide, sori in 1 or more long clusters on central vein . . . . **Vittariaceae**

42 Not as above **43**

43 Small (< 30cm tall) erect, terrestrial with linear grass-like fronds, fertile apical crest digitate . . . . **Schizaea digitata**

43 Not as above **44**

44 Small to large pinnatisect to 1-pinnate, linear sori on both sides of costa indusiate . . . . **Blechnum**
. . . . (*B. occidentale* & *B. orientale* only)

44 Not as above **45**

45 Sporangia exindusiate, spread along many veins (lamina still visible between) **or** in a single ± continuous intramarginal line, **or** acrostichoid; stipes & pinnae not jointed to rhachis **46**

45 Sporangia otherwise, indusiate or with a protective, reflexed margin covering sori; stipe or pinnae jointed to rhachis **50**

46 Small (< 30 cm) terrestrial, with simple hastate fronds. . . . .
. . . . **Parahemionitis arifolia**

46 Not as above **47**

47 Terrestrial 20-50cm tall , fronds 1-3 pinnate, pinna / pinnules, ovate-cordate to oblong-lanceolate. In tea plantations and other disturbed habitats . . . . **Hemionitis tomentosa**

47 Not as above **48**

48 Large fern of mangroves (rarely in marshy ground inland), rhizomes large, swollen, erect, fronds 1-pinnate, very coriaceous, sporangia acrostichoid ( ) . . . . **Acrostichum**

48 Not as above **49**

49 Fern of wet forests, with short-creeping rhizome, fronds 1-3-pinnate, sori arranged along many of the veins; stipes stramineous . . . . **Coniogramme serra**

49 Ferns of road banks, tea plantations and other disturbed habitats, white or yellow waxy farina on undersides of most fronds; stipes dark red-brown to black . . . . **Pityrogramma**

exindusiate (to 40) (excl. _Blechnum_)

indusiate or with false indusium (to 64 (end))

50 Small (fronds < 35cm long) terrestrial, scandent, climbing fern stipe jointed to rhizome, fronds remote . . . . **Arthropteris palisotii**

50 Not as above **51**

51 Small to moderate-sized ferns (fronds 40-250cm long); rhizome inconspicuous, small, erect, with many long wiry lateral stolons, pinnae jointed to rhachis (in dry conditions only stipe and rhachis may remain) . . . . **Nephrolepis**

51 Not as above **52**

52 Sori on veins protected by a compact to ± continuous reflexed margin (false indusium) **53**

52 Reflexed margin absent, sori protected by a true indusium of various shapes, or exindusiate (in a few Thelypteridaceae only) **57**

53 Sori on veins which continue into the thick, reflexed margin (false indusium), the shape of segments may be: fan, wedge, trapezoidal or dimidiate, fronds 1-3 pinnate, 30-50 cm long . . . . **Adiantum**

53 Reflexed margin very thin (scarious), lacking veins, not soriferous ultimate segments not as above, fronds 1-3 pinnate or pinnatisect to quadripinnatifid **54**

54 Fronds flabellate, 5-26cm long, consisting of many dichotomously branched linear segments, sori in long, submarginal line, covered by continuous false indusium . . . . **Actiniopteris radiata**

54 Not as above **55**

55 Fronds simple, pseudopalmately lobed (decurrent laminar wings connecting laminar lobes, _Doryopteris_ only); sori interrupted along margin; or 1-pinnate to quadripinnatifid, ± white farina and/or scales on undersurface, fronds < 50 cm long, stipes dark brown to black, glossy . . . **Cheilanthes**

55 Not the above combinations of characters, fronds (1-)2-3 pinnate **56**

56 Lamina glabrous (but some sparse fine hairs in *Pteris argyraea* which is also the only fern with white-patterned pinnae) 1-pinnate, basal pair of pinnae forked into two unequal linear parts .... **Pteris**

56 Lamina hairy (at least on veins abaxially and/or pinnule stalks and/or rhachis) 1-3 pinnate, basal pinnae not consisting of two unequal forked, linear segments .... **Pellaea**

57 Sori linear, at ends of veins close to margin, linear indusium attached at base of sorus **58**

57 Sori not close to margin, usually not at ends of veins **59**

58 Fronds small (20-35cm long) 1- or 2-pinnate; pinnae mostly dimidiate, sori oblong or linear along margin of leaflets; **or** to 60cm long with the basal pair of pinnae forked into two unequal linear segments(= *Idiopteris hookeriana*) .... **Lindsaea & Idiopteris**

58 Fronds larger (40-60cm), 3- to 4-pinnate, ultimate segments cuneate, sori ± orbicular .... **Sphenomeris chinensis**

reflexed margin covering sori, short to long (to 53)

no reflexed margin covering linear sori (to 64 (end))

59 Fronds 1-pinnate, veins anastomosing, basal pinna not reduced, indusia fugacious .... **Athyrium cumingianum**

59 Not as above **60**

60 Two vascular strands at base of stipe (break and pull slowly!), these uniting at top of stipe, forming deeply concave, single strand; unicellular acicular hairs and/or multicellular hairs present **61**

60 Many vascular strands present at base of stipe, unicellular acicular hairs totally absent **63**

61 Hairs on fronds multicellular, not acicular .... **Deparia boryana**

61 Many unicellular acicular hairs present, at least on main axes **62**

62 Base of stipe swollen, covered by very narrow, long, glossy, orange scales; basal pinnae much enlarged on basiscopic side (restricted to rocky habitats in dryer areas) .... **Hypodematium crenatum**

62 Base of stipe rarely a little swollen, scales above base (c. 3-5cm zone) not glossy nor very long. Basal pinnae not much enlarged on basiscopic side. (Disturbed habitats, or near to and sometimes in water.) ....
.... **Thelypteridaceae**

63 'Tree' ferns, trunks 0.3 - 4.0m high, vascular strands in stipe numerous, not in a single arc .... **Cyathea**

63 Not tree ferns, vascular strands in stipe a single arc **Dryopteridaceae** .
.... **64**

64 Midrib of principal axes of ultimate and penultimate segments raised; very short 2-celled hairs (ctenitoid hairs) present in the grooves ....
.... **Dryopteridaceae**
.... (*Tectaria* Group)

64 Midrib of principal axes of ultimate and penultimate segments grooved, grooves connecting, very short, septate hairs absent . . **Dryopteridaceae** . . . . (*Dryopteris* group)

(For further character states of the *Tectaria* and *Dryopteris* groups see key in text.)

POLYPODIACEAE

(by P. Hovenkamp*)

Bercht. & C. Presl, Poir. Rostlin. 1 (1820) 272; Ching, Sunyatsenia 5 (1940) 257; Copel., Gen. Fil. (1947) 174; Abeywickrama, Ceyl. J. Sc. Sect. A 13 (1956) 26, Sledge, Bull. Br. Mus. nat. Hist. (Bot.) 2 (1960) 133; Hennipman et al. in Kramer & Green, Fam. & Genera of Vasc. Pl. (1990). —Type genus: *Polypodium.*
Drynariaceae Ching, Acta Phytotax. Sin. 16 (1978) 1. —Type genus: *Drynaria.*

Rhizome creeping, dorsiventral, with two alternating dorsal rows of phyllopodia and two alternating lateral rows of buds; scales basifix, pseudopeltate to peltate, clathrate or isotoechous, margin entire to ciliate. Anatomy: ground tissue parenchymatous or rarely sclerenchymatous, with or without strands of sclerified cells, stele dictyostelic, composed of 3-many vascular strands, with or without a sclerified circumvascular sheath. Fronds often dimorphic, articulated to the phyllopodia, sessile to stipitate, erect or appressed, simple to pinnate, pedately or dichotomously divided, rarely bipinnatifid, often covered with deciduous or persistent scales or hairs. Veins forked, free (rarely) to copiously branched and anastomosing. Fertile areas often contracted, frequently on separate fronds. Sori exindusiate, rounded, transversally or longitudinally elongated or forming irregular acrostichoid patches, sporangia short- to long-stalked, capsule with vertical, interrupted annulus, soral trichomes similar to the laminar ones or modified, sometimes more persistent, acicular sporangial trichomes sometimes present. Spores mostly 64 (rarely 8, 16 or 32), monolete.

D i s t r. Worldwide, with the greatest diversity in tropical Asia.

TAXONOMY

The treatment here generally follows the one by Hennipman et al. (l.c.) except for the genera *Colysis* (here included in *Leptochilus*), *Phymatosorus* (here included in *Microsorum*). Following Nooteboom., the distinction between *Microsorum* and *Leptochilus* is upheld, although *Leptochilus* in this circumscription is not natural.

---

*Rijksherbarium., Leiden, Netherlands.

KEY TO THE GENERA

1 Rhizome scales not clathrate
  2 Lamina with stellate hairs . . . . **2. Pyrrosia**
  2 Lamina without, or with inconspicuous simple hairs
    3 Separate base fronds present . . . . **1. Drynaria**
    3 Base fronds not present . . . . **7. Selliguea**
1 Rhizome scales clathrate
  4 Sporangia in separate, round or elongated sori, fertile lamina not contracted
    5 Sori at least when young covered with umbrella-like scales
      6 Lamina glabrous or nearly glabrous, rhizome with black sclerenchyma
      strands . . . . **5. Lepisorus**
      6 Lamina with scattered scales at least below, rhizome without scleren-
      chyma . . . . **6. Pleopeltis**
    5 Sori never covered with umbrella-like scales . . . . **3. Microsorum**
  4 Sporangia in lines, fertile lamina strongly contracted
    7 Sori on separate fertile fronds . . . . **4. Leptochilus**
    7 Sori on a contracted apical part of the fronds . . . . **8. Belvisia**

1. DRYNARIA

Drynaria J. Sm., J. Bot. (Hooker) 3 (1841) 397 (nom. cons.); 4 (1842) 60; Bedd., Ferns Southern Ind. (1863) 63; Bedd., Handb. Ferns Brit. Ind. (1883) 338; Abeywickrama, Ceyl. J. Sc. Sect. A 13 (1956) 27; Roos, Drynarioideae (1985) 255. *Polypodium* subgen. *Drynaria* Bory, Ann. Sci. nat. 5 (1825) 464; Baker, Syn. Fil. (1868) 366; *Phymatodes* sect. *Drynaria* C. Presl, Tent. Pterid. (1836) 197; Hook. & F.A. Bauer, Gen. Fil. (1842) T. 21. Type species: *Drynaria quercifolia* (L.) J. Sm.

Drynaria sect. *Poronema* J. Sm., Hist. Fil. (1875) 108. Type species: *Drynaria diversifolium* (R.Br.) J. Sm. (= *Drynaria rigidula*)

Epiphytic, epilithic or terrestrial. Rhizome up to 3 cm thick, sometimes more, short- or long-creeping, phyllopodia not elevated, rhachises often persistent. Anatomy: vascular strands 15 to many (50–100), in cross-section arranged in 1–2 flattened circles with a dorsal protrusion; sclerenchyma strands absent or present. *Rhizome scales* appressed to squarrosely spreading, pseudopeltate or peltate, margin toothed. Fronds dimorphic, with base- and foliage fronds. Base fronds sessile, rounded to ovate-elliptical, lobed up to 2/3. Foliage fronds stipitate, pinnatifid, with conspicuous nectaries at the base of the segments, apex aborted. Pinnae articulated to the rachis and to each other by an abscission vein, deciduous. Venation highly complex, with nume-rous small areoles containing excurrent and recurrent free veinlets, sometimes terminating in a hydathode. Fertile parts similar to sterile or slightly narrowed, sori small, irregularly scattered or in rows along veins. Sporangia glabrous. Spores echinate.

KEY TO THE SPECIES

1 Sori in 2 regular rows along the veins, rhizome scales usually spreading, gradually narrowed from base to apex . . . . **1. D. quercifolia**
1 Sori scattered or in irregular rows, rhizome scales with an appressed base and a squarrose, needle-like tip . . . . **2. D. sparsisora**

1. Drynaria quercifolia (L.) J. Sm. *Drynaria quercifolia* J. Sm., J. Bot. (Hooker) 3 (1841) 398; 4 (1842) 61; Bedd., Ferns South. Ind. (1863) 63, Pl. 187; Suppl. Ferns South. Ind. & Br. Ind. (1876) 24; Handb. Ferns Brit. Ind. (1883) 343, Fig. 191.

Polypodium quercifolium L., Sp. Pl. (1753) 1087; Hook., Sp. Fil. 5 (1864) 96; Baker in Hook. & Baker, Syn. Fil. (1868) 367.
Phymatodes quercifolia C. Presl, Tent. Pterid. (1836) 198; Hook. & F.A. Bauer, Gen. Fil. (1842) T. 21. Type: Herb. *Hermann 382* (BM), Sri Lanka.

Epiphytic, spirally climbing, occasionally epilithic or terrestrial. Rhizome 2–3 cm thick or more, short-creeping, phyllopodia up to 10 cm distant; base fronds contiguous or separate, rhachises persistent. Anatomy: vascular strands many, equally sized, arranged in 1–2 rows, without dark bundle sheaths, sclerenchyma strands absent. Rhizome scales brown to blackish, spreading, pseudopeltate or peltate, 6–20 × 0.5 1 mm., index 10 linear, upwards strongly dentate, tapered to long, narrow apex, acute to nearly acicular, midrib mostly absent. Base fronds sessile, shallowly lobed, (10–)15–40(-50) × 10–30 (–40) cm., index 1–1.4, margin entire. Foliage fronds stalked, stipe up to 15–30(–35) cm long, not or inconspicuously winged, lamina pinnatifid to 0.2–0.5 cm from costa, 40–100(–150) × 15–50 cm., index 2.5–4. Pinnae without basal constriction, all equally long, 1–25(–30) × 2–4.5 cm., index 4–6, margin entire, apex acute, free veinlets simple, infrequent or absent, hydathodes absent. Sori in 2 regular (sometimes irregular) rows parallel and close to the veins, round, 1–2 mm., slightly sunken.

E c o l. Epiphytic, usually exposed, on wayside trees etc., at low elevations to 1200 m.

D i s t r. South East Asia to Australia.

S p e c i m e n s   E x a m i n e d. KANDY DISTRICT: Kandy, *Freeman 318 C* (BM). BADULLA DISTRICT: Badulla, *Freeman 319 D* (BM) ("a low-country fern but I have seen it up to 914 m.). AMPARAI DISTRICT: Lahugula Tank, SE corner, inside forest on dry land, in small colony at edge of forest 25 July 1967, *Muller-Dombois & Comanor, 67072556* (L.). KALUTARA

DISTRICT: Pelawatta, on tree trunk by roadside, Jan 20 1951, *Ballard 1521* (K). Road south of Udugama, epiphytic, *Sledge 904* (K). RATNAPURA DISTRICT: Adam's Peak, up to 750 m., on ground. 9 March 1954, *Walker T 730,731* (BM); hillside above Potupitiya, forest patch, much degraded, common epiphyte, 450 m., 6° 26 N , 80° 30 E, 4 Dec. 1976, *Faden & Faden 76/483* (K). GALLE DISTRICT: Udugama Rd., 19/12 mile post, on tall tree by roadside, c. 4 m.from ground 21 Jan.1950, *Ballard 1544* (K). WITHOUT LOCALITY: Cult in Kew: voucher for Manton, p. 158; *Gardner 1142* (K); *s.n.* (BM). *Walker s.n.*, s.d.

2. Drynaria sparsisora (Desv.) T. Moore Index Filic. (1862) 348; C. Chr., Ind. Fil.(1906) 249.

Polypodium sparsisorum Desv., Mag. Ges. Naturf. Freunde Berlin (Berl. Mag.) 5 (1811) 315. Type: not traced.
Polypodium linnei Bory, Ann. Sci. nat. 5 (1825) 464 p.p., T. 12; Baker in Hook. & Baker, Syn. Fil. (1868) 368.
Drynaria linnei Bedd., Ferns Brit. Ind. (1869) Pl. 315; Handb. Ferns Brit. Ind. (1883) 343. Type: *Gaudich. s.n.*, s.d., Rawak near Java (P, not seen).

Epiphytic, spirally climbing, sometimes terrestrial. Rhizome 1–3 cm thick or more, short-creeping, phyllopodia up to 10 cm distant; base fronds contiguous, mostly imbricate, rhachises not persistent. Anatomy: vascular strands many, equally sized, arranged in 1–2 rows, without dark bundle sheaths, sclerenchyma strands absent. Rhizome scales peltate, with appressed base, squarrosely spreading, often caducous acumen, dark brown-blackish with a light, lacerate margin, elongated, 1.5–11 × 1–2.5 mm., index without the caducous acumen 1–8.5, midrib present. Base fronds sessile, lobed to 1/3, (10–)15–35 × (10–)15–25 cm., index 1–1.5, margin entire, apex rounded. Foliage fronds stalked, stipe (5–)10–18 cm long, conspicuously winged, lamina pinnatifid to 0.2–0.5 cm from costa, 30–80 × 15–30 cm., index 2–4. Pinnae with or without basal constriction, all equally long, to 10–20 × 1.5–3.5 cm., index 3.5–6(–8), margin entire, apex acute to caudate, free veinlets simple or absent, hydathodes absent. Sori 2–7 in each areole, irregularly scattered or in 2 irregular rows between the connecting veins, round, 1–2 mm across, slightly sunken, often distinctly pustulate on upper surface.

E c o l. Epiphytic or epilithic.

D i s t r. South East Asia to Australia.

N o t e. Juvenile foliage fronds often have a somewhat dilated frond base.The narrow, needle-like acumen of the rhizome scales often disappears from old scales, leaving older parts of the rhizome completely covered by characteristic, short, dark, appressed scales. Juvenile specimens tend to have longer, more narrowly subulate acumens, which also appear to be absent from older parts. Although the scales are usually quite different from those of

D. quercifolia, however, occasionally some more intermediate forms occur, and juvenile, sterile specimens cannot always be identified with certainty. Although Sledge doubts the occurrence of this species in Sri Lanka, its presence is confirmed by the presence of several collections which, though juvenile and sterile, appear to be typical *D. sparsisora*. They have been variously identified as *D. sparsisora* or *D. quercifolia* by Jarrett and Roos.

S p e c i m e n s   E x a m i n e d. KANDY DISTRICT: Kadugannawa, 62 mile-post, on Colombo-Kandy Rd., on rock in ravine, Dec. 12 1950, *Ballard 1095* (K, juvenile: sparsisora det Jarrett and Roos, quercifolia det Sledge). RATNAPURA DISTRICT: Belihul Oya, in crevice in granite in stream by rest house, 3 Jan 1951, Jan 3 1951, *Ballard 1370* (K, juvenile: sparsisora det Jarrett, quercifolia det Sledge and Roos); Belihul Oya, on dead tree trunk near Rest House, 3 Jan 1951, *Ballard 1371* (coll. *de Silva*) (K, juvenile: sparsisora det Jarrett, quercifolia det Sledge and Roos). LOCALITY UNKNOWN: *anon s.n.*, s.d., herb. Hooker (K, fertile specimen).

2. PYRROSIA

Pyrrosia Mirb., Hist. Nat. Gen. 4 (1803) 70; Abeywickrama, Ceyl. J. Sc. Sect. A 13 (1956) 26; Hovenkamp, Leiden Bot. Ser. 9 (1986). Type species: *Pyrrosia chinensis* Mirb. (= *Pyrrosia stigmosa*).

Cyclophorus Desv., Mag. Ges. Naturf. Freunde Berlin 5 (1811) 300;. *Niphobolus* Kaulf., Enum. (1824) 124, *nom. superfl.*; Giesenh., Niphobolus (1901). Type species: *Cyclophorus adnascens* (Sw.) Desv. (= *Pyrrosia lanceolata*).

Drymoglossum C. Presl, Tent. Pterid. (1836) 227. Type species: *Drymoglossum piloselloides* (L.) C. Presl (=*Pyrrosia piloselloides*).

Epiphytic, epilithic or terrestrial, in small tufts or forming extensive clones. Rhizome to 3 mm thick, short- to long-creeping, appressed to or just immersed in the substratum., contiguous or up to 9 cm apart, sparsely to profusely branching from lateral buds. Anatomy: vascular strands 3–12, ground tissue parenchymatous or with a distinct sclerenchyma sheath, sclerenchyma strands absent to many. *Rhizome scales* appressed to spreading, pseudopeltate or peltate, to 14 × 3.3 mm., not clathrate, entire, dentate or ciliate, hyaline to brown, central region often darker. Fronds mono- to dimorphic, nearly sessile to stalked, simple, coriaceous, often succulent. Venation: veins mostly distinct, connective veins forming one to several series of rectangular areoles, veinlets simple, forked or more copiously branched and anastomosing, free veinlets excurrent, recurrent or to all directions. Upper surface with or without hydathodes. Indument composed of stellate hairs with straight and on the lower surface sometimes also with curly rays, the straight rays often forming a distinct upper layer, short (0.2 mm) and wide ("boat-shaped") to long (1 mm) and acicular. Sori round to elongated or forming a longitudinal coenosorus, superficial to deeply sunken. Paraphyses similar to the lamina-indument or differentiated. Sporangia glabrous. Spores with a variously sculpted perispore.

KEY TO THE SPECIES

1 Rhizome to 1 mm thick, long-creeping, phyllopodia usually 1-3 cm apart or more
  2 Sporangia in separate round sori
    3 Rhizome scales ciliate . . . . **4. P. lanceolata**
    3 Rhizome scales entire . . . . **1. P. ceylanica**
  2 Sporangia in elongated coenosori . . . . **3. P. heterophylla**
1 Rhizome 1-3 mm thick, short-creeping, phyllopodia up to 1 cm apart
  4 Fronds longly stipitate, ovate-elliptic . . . . **5. P. pannosa**
  4 Fronds sessile or shortly stipitate, narrowly lanceolate
    5 Rhizome with very conspicuous, high phyllopodia, scales appressed, shiny, blackish. . . . . **2. P. gardneri**
    5 Rhizome without conspicuous phyllopodia, scales spreading, dull brown or blackish . . . . **6. P. porosa**

1. Pyrrosia ceylanica (Giesenh.) Sledge, Bull. Br. Mus. nat. Hist. (Bot.) 2 (1960) 133; Abeywickrama, Ceyl. J. Sc. Sect. A 13 (1956) 26 (invalid comb.). —*Niphobolus ceylanicus* Giesenh., Niphobolus (1901) 216, fig. 19.— *Cyclophorus ceylanicus* C. Chr., Ind. Fil. (1906) 198. Type: *G. Wall, s.n.* (1887); herb. Christ (holo P; iso M., fragments), Sri Lanka.

Rhizome long-creeping, with ventral groove, 0.8-1 cm thick, phyllopodia 1.5-4 cm apart, lateral buds situated between the phyllopodia. Anatomy: ground tissue parenchymatous, sclerenchyma sheath distinct, a single sclerenchyma strand present situated centrally in the parenchyma; vascular strands ± 4. Scales peltate, 3.6-7 4 × 0.5-0.9 mm; shining light brown, entire. Fronds moderately dimorphic, distinctly stipitate. Fertile fronds: stipes 0.5-3 cm., - ± × as long as lamina; lamina, index 4-10, widest at or below middle, 3.5-8 × 0.7-1.4 cm., base cuneate, apex acute to obtuse. Sterile fronds: stipes 0.5-1 cm., -1/5 × as long as lamina; lamina, index 3-4; 3.5-5 × 1.2-1.4 cm; otherwise similar to fertile ones. Venation: secondary veins distinct, with tertiary veins forming regular areoles; included veins simple, free, excurrent Hydathodes absent. Anatomy: stipe with + 3 central and 0-2 lateral vascular strands; lamina ± 1 mm thick, upper epidermis with distinctly projecting cells with thin to slightly thickened walls, hypodermis absent but a distinct, moderately thick water-tissue present, palisade and spongy parenchyma distinct, lower epidermis with slightly to strongly thickened cell-walls; stomata slightly to distinctly sunken, pericytic. Indument monomorphic, a dense mat, persistent, whitish brown; hairs 0.3-0.7 mm in diam., with appressed, boat-shaped rays. Sori apical, closely packed, sunken; several in a row in each soriferous areole, not confluent; ± 1 mm in diam.; developing from the apex downwards, when old individually distinct, exserted from the indument. Sporangia on stalks ± 1 × as long as the capsule, capsule + 0.3 mm high, with 17-19 indurated annulus cells. Paraphyses present, in a central bundle in the sorus, with short, relatively wide, straight rays. Spores bisculptate.

E c o l. Epiphytic or epilithic.

D i s t r. Southern India, Sri Lanka.

S p e c i m e n s   E x a m i n e d. RATNAPURA DISTRICT: Balangoda-Rassagala Rd., mile post 6/5, stone wall near bridge, common, 750 m., *Faden & Faden 76/300* (K). WITHOUT LOCALITY: anon. no. 173, p.p.; *Beckett s.n.*, s.d, (K); *Chevalier s.n.*, s.d, near Colombo, common on trees (BM); *Chevalier s.n.*, s.d.; *C.P. 3293* (Giesenhagen, March 1900); *Thwaites C.P. 993* (B, BM., SING), *C.P. 3293* (BM., P); *Wall s.n.* (1887) (L).

2. Pyrrosia gardneri (Mett.) Sledge, Bull. Br. Mus. nat. Hist. (Bot.) 2 (1960) 134; B.K. Nayar & S. Kaur, Comp. Bedd. Handb. (1974) 81.

Polypodium gardneri Mett., Polyp. (1856) 129, Thwaites, Enum. Pl. Zeyl. (1864) 395; Hooker, Sp. Fil. 5 (1863) 51; Baker in Hook. & Baker, Syn. Fil. (1867) 352.
Niphobolus gardneri J. Sm., Cat. Cult. Ferns (1857) 12; Hook., Fil. Ex. (1859) pl. 68; Bedd. Ferns S. India (1864-5) 81, pl. 241; Bedd., Handb. Ferns Brit. Ind. (1883) 331, fig. 181.
Niphobolus gardneri Kunze ex T. Moore, Ind. Fil. (1857) lxxvi, nom. nud; T. Moore, Ind. Fil. (1861) pl. 61 fig. 1-5.
Cyclophorus gardneri C. Chr., Ind. Fil. (1906) 199; Willis, Cat. Pl. Zeyl. (1911) 123. —Type: *Gardner 53* (holo K, iso in P), Sri Lanka.

Rhizome short, without ventral groove, + 2.5 mm thick, phyllopodia 0.5-1 cm apart, lateral buds situated on or between the phyllopodia. Anatomy: ground tissue parenchymatous, sclerenchyma sheath distinct, sclerenchyma strands scattered through inner parenchyma, many; vascular strands + 9. Scales peltate, 1.6-4.2 × 1-1.6 mm; base + entire; acumen shiny black with a conspicuous hyaline or light brown margin, shortly and densely ciliate. Fronds monomorphic, distinctly to indistinctly stipitate; stipes 2-16 cm; lamina, index 10-20; widest at or above middle, 13-51 × 1.2-4.1 cm., base and apex gradually narrowed. Venation: secondary veins very distinct, tertiary veins forming regular areoles; included veins simple, or occasionally forked, free or occasionally anastomosing; free veins many, excurrent. Hydathodes distinct, scattered over lamina, + superficial. Indument dimorphic, a dense mat, persistent, light brown; upper layer composed of hairs 0.6-1.4 mm in diam., with erecto-patent, acicular rays, + mixed with a lower layer composed of hairs with mainly woolly rays. Sori apical or covering lamina entirely, shortly spaced, superficial; several in a row in each soriferous areole, not or occasionally slightly, confluent along veins; 1-2 mm in diam.; developing from apex downwards, when old individually distinct, exserted from indument. Sporangia on stalks 1.5-3 times as long as capsule 0.2 mm high, with 14-16 indurated annulus cells. Paraphyses not differentiated. Spores irregularly verrucate to cristate.

E c o l. Epiphytic or epilithic, in wet forest, low elevations to 1150 m

D i s t r. Sri Lanka

S p e c i m e n s   E x a m i n e d. ANURADHAPURA DISTRICT: 22-24 Mar 1905 summit of Ritigala N.C.P. Ritigala Strict Nature reserve, Weweltenna plain, in humus over rock outcrop,c. 590 m., 28 Sep 1972, *Jayasuriya 864* (BM., U, US); Ritigala Strict Nat Res., on plain below summit, side of old building, 700 m., 4 May 1974, *Jayasuriya & Premadasa 1646* (BM, U, US). KURUNEGALA DISTRICT: Doluwakanda Hill, Forest, frequent, low epiphyte or lithophyte, 600 m., 5 Jan 1977, *Faden & Faden 77/52* (US). KEGALLE DISTRICT: Kelani Ganga, forested hills, epiphytic in large tree, 180 m., 20 Nov. 1974, *Davidse & Sumithraarachchi 8536* (L); Kelani River, Rest House, epiphytic on tree near stream, 11 Nov 1975, *Sohmer & Waas 10565* (GH). KANDY DISTRICT: Macrae 454, on trees and rocks, near Kandy (L, K); Gardens Peradeniya? *Chevalier s.n.*, s.d. Kandy, on trees and rocks (BM); Kandy, *Thwaites 1154* (P); Ambagamwa village, Dec. 1950-Jan. 1951, *Manton 157* (voucher for n=37); P 157 Ambagamwa, voucher for cytological study n=37 (BM); Ambagamuwe; on tree. 19 Jan. 1954, *Walker T 133* (BM); 27 Jan., 1957, Hantane, *Appuhany* (L.); Rangala, Mile post 21/4, very large boulder in shade of trees along edge of tea plantation, 930 m., 9 Nov 1974, *Davidse 8253* (L). Rangala, Mile post 21/4, very large boulder in shade of trees along edge of tea plantation, 930 m., *Davidse 8253* (L); (several specimens *De Silva s.n.* 3 Jan 1927, Kekuna Estate, Mukalane); Kalugammana; 28 Aug. 1963 *Amaratunga 721* (L); KALUTARA DISTRICT:, Meegahatenne (2 mi S of Moragalla), on tree trunk,.c. 4 m. up, by roadside, 20 Jan 1951, *Ballard 1517* (K). RATNAPURA DISTRICT: Sinharaja, North Ensalwatta, in prim. wet evergreen forest, epiphytic on tree, overhanging stream., 8 Nov 1975, *Sohmer & Waas 10469* (L); Sinharaja Forest, Weddagala, occ., terrestrial along logging in tall rain-forest, 2100-2300 m., 27 Oct 1976, *Fosberg & Waas 56527* (L); Sinharaja Forest, Weddagala, occ., on rotten wood, prob. fallen from tree, 2100-2300 m., 27 Oct 1976, *Fosberg & Waas 56461* (L); lower slopes of Beragala, rare, on large rocks in degraded wet montane forest, 2600-3000 m. 7 Dec. 1977, *Fosberg 57274* (L); Morakele-Sinharaja Forest, epiphytic, in sec. wet forest, logged area, 200 m., 26 Feb 1977, *Waas 2068* ( ); Sinharaja, Weddagalla entrance, epiphytic, 21 Sep. 1977, *Meijer & Gunatilleke 1392* (US); Sinharaja Forest, near Weddagalla, wet evergreen forest, 150 m., Nov. 1978, *Kostermans 27114* (PDA, L); Rassagalla, epiphytic on tree; Southern Prov. Sabaragamuwa, Niriella, 225 m., 4 Nov. 1954, *Schmid 1119* (L). NUWARA ELIYA DISTRICT: *C.P. 3104*, Hakgala. GALLE DISTRICT: Hinidumkande, near Hiniduma, wet evergeen forest, 500 m., 2 Sep. 1974, *Kostermans 25512A* (L); Mt. Hinidumakanda, 750 m., on rocks, 25 Oct. 1975, *Bernardi 15449* (Z). MATARA DISTRICT: Deniyaya, 550 m., 5 Feb. 1954, *Schmid 1139* (BM., G); Diyadawa forest, epiphytic, 11 Nov. 1975, *Kubitzki 77-54* (M., Z). LOCALITY UNKNOWN: *Macrae 457, 877* (BM); *Ferguson 176* (US); *Gardner 263* (B); *Hancock 36* (K, US); 3 Feb. 1819, *Moon 98* (BM); *C.P. 988* (B, BM., BO, BR, K, L, P).); *Kittner 17.8.26*; Laalla, 1150 m., *Naylor Beckett 112* (B).

3. Pyrrosia heterophylla (L.) Price, Kalikasan 3 (1973) 177. —*Acrostichum heterophyllum* L. Amoen. Acad. 1 (1745) 268, pl. 12 f. 1, Fl. Zeyl. (1749) 378, Spec. Pl. 2 (1753) 1067 (excl. Afric. specim.); Burman Fl. Indica (1768) 228. —*Drymoglossum heterophyllum* (L.) Trimen, J. Linn. Soc. 24 (1887) 152; Sledge, Bull. Br. Mus. nat. Hist. (Bot.) 2 (1960) 135. Type: P. Hermann s.n. (BM., Herb. Hermann) Sri Lanka.

Pteris elliptica Willd., Sp. Pl. 5 (1810) 356, nom. illeg. (non Poir., 1840).
    Pteris ceilanica Wikstr., K. svenska Vetensk.-Akad. Handl. (1826) 447.
    Pteropsis elliptica (Willd.) Desv., Mem. Soc Linn. Paris 6 (1827) 218.
    Drymoglossum ellipticum (Willd.) T. Moore, Ind. Fil. (1857) 31, 343
    —Type: *Houtt.*, Nat. Hist. II, 14 (1873) pl. 86 f. 1.
Drymoglossum beddomei C.B. Clarke, Trans. Linn. Soc. Lond. II Bot.
    1 (1880) 576, *nom. nud. Drymoglossum piloselloides* var. *beddomei*,
    Beddome, Handb. Ferns Brit. Ind. (1883) 413, *nom. inval.* art. 34 ICBN
    —*Drymoglossum heterophyllum* var. *beddomei* C. Chr., Dansk bot. Ark.
    6 (1929) 85, pl. 12 f. 2a-c. Type: *Beddome 15* (K), S. India, Nilgiri.
Drymoglossum piloselloides auct. non (L.) C. Presl: Thwaites, Enum. Pl.
    Zeyl. (1864) 381.
Niphobolus nummarifolius auct. non (Sw.) J. Sm.: Bedd., Ferns S. Ind.
    (1863) 62, pl. 186, *p.p.*

Rhizome long-creeping, branched, terete, up to 70 by 0.1 cm., phyllopodia 0.8-3.0 cm apart, lateral buds situated between the phyllopodia. Anatomy: with a subepidermal sclerenchyma sheath, without a central sclerenchyma strand, with 3-4 meristeles. Scales peltate, round to triangular, index 1-6, up to 2.0 × 0.7 mm., base rounded, sometimes constricted at place of attachment to stalk, margin with up to 0.4 mm long cilia, apex attenuating into a long, filiform., deciduous tip, central part dull brown to shiny black, marginal part lighter coloured. Fronds dimorphic. Sterile fronds: 1.0-4.5 cm long, sessile to shortly stipitate; stipe up to 0.1(-0.4) cm long with 2 vascular strands; lamina index 1-4 . 5, widest at the middle, (0 . 6-)1-4 .5 × 0. 8-1 .8 cm., base rounded to short attenuate, margin entire, slightly incurved, apex rounded. Venation: rhachis prominent on either side, otherwise veins immersed, veins anastomosing into a regular pattern consisting of a costal areole and smaller distal ones, with included, simple or forked, recurrent, free veins without hydathodes. Fertile fronds: (1.5-) 2.5-10 cm long; stipe up to 0.4 cm long; lamina index 10-30, 2.5-9.5 × 0.3-0.9 cm., slightly recurved. Indument: fronds with scattered, ± sessile, deciduous, stellate hairs with 4-7 sometimes sinuous rays, hyaline apart from the brownish central part. Coenosori linear, all along the lamina, not or rarely interrupted, submarginal between rhachis and the somewhat recurved margin; receptacle situated on the outermost connecting vein. Sporangia maturing irregularly, persistent, with 16-20 indurated cells mixed with paraphyses. Spores variously warty in part with conical spines.

E c o l. Epiphytic, sometimes epilithic, on wayside trees, tea bushes etc.

D i s t r. Southern India, Sri Lanka.

S p e c i m e n s   E x a m i n e d. KANDY DISTRICT: Kandy, semi-wooded hilltop around house, epiphytic, *50657*, 21 Dec 1968, *Fosberg* (K); Kandy, 25 Mar. 1900, *Giesenhagen 68*, (M). Peradeniya, Bot. Garden, 450 m., *Kostermans 272304* (L); Gardens Peradeniya? *Macrae 900* (BM); 10 Sept. 1977, Hantana, epiphyte, 750 m., *Nooteboom 3045* (?). BADULLA DISTRICT: Padiyapelella-Ellamulla road, culvert 28/8, moss-covered rock, local and uncommon, 850 m., *Faden, Faden & Waas 76/326* (L); 12 Sept. 1969, Badulla-Hakgala Rd, rain trees along road, *Read & Desautels 2267* (L). RATNAPURA DISTRICT: Pelmadulla-Hambantota Rd, milemarker 20/4, creeping on boulders, 18 Nov 1973, *Sohmer et al. 8829* (L); Balangoda-Rassagala Rd., mile post 8, epiphyte on tea, common, 610m., *Faden & Faden 76329* (L). MONARAGALA DISTRICT: 18 July 1972, Monaragala, wayside, epiphytic, *Hepper & De Silva 4734* (L). MATALE DISTRICT: Erawalagalla Mt., at base of cliffs in forest near top of mountain, epiphyte, uncommon, 29 Oct 1974, *Davidse 8102* (L). WITHOUT LOCALITY: *Amaratunga 358, CP 3076* (L).

4. Pyrrosia lanceolata (L.) Farw., Amer. Midl. Nat. 12 (1930) 245; Sledge, Bull. Br. Mus. nat. Hist. (Bot.) 2 (1960) 133; B.K. Nayar & S. Kaur, Comp. Bedd. Handb. (1974) 80.—*Acrostichum lanceolatum* L., Sp. Pl. 2 (1753) 1067; ed. 2 (1763) 1523; N.L. Burm., Fl. Ind. (1768) 228; Trimen, J. Linn. Soc. 24 (1888) 130.—*Candollea lanceolata* Mirb. ex Desv., Mem Soc. Linn Paris 6 (1827) 224, *in synonymy;* Farw., Amer. Midl. Nat.12 (1930) 245.—*Cyclophorus lanceolatus* Alston, J Bot. London 69 (1931) 102. —Type: *Herb. Hermann I, fol. 3* (holo BM., iso L), Sri Lanka.

Polypodium adnascens Sw., Syn. Fil. (1806) 25, 222, pl. 2 fig 2; Hook., Sp. Fil 5 (1863) 47; Thwaites, Enum. Pl. Zeyl. (1864) 395; Baker in Hook. & Baker, Syn. Fil. (1867) 349; Trimen, J. Linn. Soc. 24 (1888) 152.—*Cyclophorus adnascens* Desv., Mag. Ges. Naturf. Freunde Berlin (Berl. Mag.)

5 (1811) 300; Willis, Cat. Ceyl. (1911) 123.—*Niphobolus adnascens* Kaulf., Enum. (1824) 124; Bedd, Ferns S. India (1864-5) 62, pl. 184; Handb. Ferns Brit. Ind. (1883)325; fig 176.—*Pyrrosia adnascens* Ching, Bull Chin. Bot. Soc. 1 (1935) 45; Abeywickrama, Ceyl. J. Sc. Sect. A 13 (1956) 26; B.K. Nayar & S. Kaur, Comp. Bedd. Handb. (1974) 80. —Type: *Rottler s.n, s d.,* (*S.* not seen), India, 'Malabar, mixed with *Pteris piloselloides*'.

Rhizome long-creeping, 1.2-2.1 mm thick, narrow ventral groove, phyllopodia 1-2 cm apart, lateral buds situated between internodia. Anatomy: ground tissue parenchymatous, sclerenchyma sheath distinct, single, central sclerenchyma strand usually present; vascular strands 5(-7). Scales peltate 1-7.8 × 0.3-1.3 mm; base entire to ciliate; acumen light brown, often with a distinct hyaline margin, ciliate; short, ± orbicular to ovate scales usually present. Fronds moderately to distinctly dimorphic, distinctly to indistinctly stipitate. Fertile fronds: stipes up to 5(-9) cm; lamina, index ± 5 to over 20; widest below or about middle, 3.5-31 × 0.3 -3.5 cm., base cuneate to narrowly cuneate, apex obtuse to acute. Sterile fronds: stipes to 3(-5) cm; lamina index -20 (occasionally more); widest below, about or above middle, 2-24 × 0.3-.5(-4.3) cm., base attenuate, cuneate or narrowly cuneate, apex rounded, obtuse-acute. Venation: secondary veins distinct to indistinct, tertiary veins forming more or less regular areoles; included veins simple, occasionally forked or more copiously branched, mainly free; free veins excurrent. Hydathodes absent, rarely present, few and indistinct, scattered over lamina. Stomata strongly sunken, pericytic. Indument monomorphic, a sparse, thin or dense mat persistent to fugacious, whitish to brown; hairs 0.2-1.2 mm with erecto-patent to appressed, boat-shaped to acicular rays. Sori apical to covering lamina entirely, closely packed in a more or less sharply defined patch, distinctly sunken; several to ± 10 in a row in each soriferous areole, not confluent; 0.5-1(-2) mm in diam; developing from apex downwards, when old individually distinct, exserted from indument. Sporangia on stalks 1.5 -2 times as long as capsule, capsule 0.2-0.3(-0.4) mm high, with 13-18 indurated annulus cells. Paraphyses in central bundle, with short, straight rays. Spores irregularly verrucate to distinctly bisculptate.

E c o l. Epiphytic or epilithic, on walls, in plantations etc., low elevation to 900 m.

D i s t r. Palaeotropical

S p e c i m e n s   E x a m i n e d: KURUNEGALA DISTRICT: Kurunegala Rock, 7 Jan 1954, *Walker T 1* (BM); Kurunegala Rock, eiphyte on Euphorbiaceae, 7 Jan 1954, *Walker T2* (BM); Doluwakanda Hill, 600 m., lithophyte, forest, locally common, 5 Jan 1977 *Faden & Faden 77/53* (US). MATALE DISTRICT: Matala-Dambulla Rd, 28/4 mi, on boulder, rubber plantation, 13 Jan 1951, *Ballard 1437* (K). COLOMBO DISTRICT: Colombo, on walls and trees, 16 Mar 1900, *Giesenhagen 16* (L). KEGALLE DISTRICT: Kitulgala, Mahaweli Ganga, 60 m., 31 Dec. 1971, *Benl & Benl, s.n* (M). Kegalla, behind rest house, 180 m., 28 Dec 1971, *Benl & Benl* (M). KANDY DISTRICT: Hantana, 660 m., *Gardner 1153* (BM., BT, K); Kandy, on wayside trees, 25 Mar 1900, *Giesenhagen 17* (L); Peradeniya, on trees, 29 Feb 1912, *Matthew s.n* (K); Peradeniya, 7 Feb 1914, *Mrs. Petek* (L); Peradeniya, 6 June 1927, *Alston 1399* (L); Colombo-Kandy Rd, 43 mi, in Hevea-plantation, on boulder, 4 Nov 1950, *Ballard 1036*(L); Kadugannawa, on Colombo-Kandy Rd, 62 mi., on bank in rubber plantation, 12 Dec 1950, *Ballard 1100* (K); Hunnasgiriya, 900 m., 19 Jan 1954, *Schmid 1000* (G, BM); Peradeniya, rocks by the river, Lady Blakes Drive, 450 m., 15 Feb 1954, *Sledge 1141* (U). RATNAPURA DISTRICT: Balangoda-Mahawalatenna Rd., Damabana, in shade under shrubs on rock outcrop, terrestrial or epiphytic, common, 600 m., *Faden & Faden 76/643* (L). WITHOUT LOCALITY: *Macrae 907* (BM); anon *173*, p.p.; 1899, *Bradford 372* (K); *CP 993* (B, L, K, SING); *Ferguson s.n.*, s.d (cat. 173) (US); *Thomson s.n.*, s.d., (U, B); *Wall s.n.*, s.d. (B, P).

5. Pyrrosia pannosa (Mett. ex Kuhn) Ching, Bull. Chin. Bot. Soc. 1 (1935) 58; Sledge, Bull. Br. Mus. nat. Hist. (Bot.) 2 (1960) 135.—*Polypodium pannosum* Mett. ex Kuhn, Linnaea 36 (1869) 141; Baker in Hook. & Baker, Syn. Fil. ed. 2 *(1874) 512.—Niphobolus pannosus* Bedd., Suppl. Ferns S. & Br. Ind. (1876) 22, *p.p.;* Bedd., Handb. ferns Brit. Ind. (1883) 328, fig. 177, *p.p .—Cyclophorus pannosus* C. Chr., Ind. Fil. (1906) 200; Willis, Cat. Fl. Sri Lanka (1911) 123.—Type: *Thwaites CP1294* (holo B. iso BM., BO, K, L, P), Sri Lanka.

Polypodium lingua, auct. non (Thunb.) Sw., *quoad specim. Zeyl..* Hooker, Sp. Fil. 5 (1863) 44; Thwaites, Enum. Pl. Zeyl. (1864) 395; Baker in Hook. & Baker, Syn. Fil. (1867) 350.

Rhizome short-creeping, ventral groove absent, 1-2(-4) mm thick, phyllopodia 0.2-0.8 cm apart, lateral buds situated on or between internodes. Anatomy: ground tissue parenchymatous, sclerenchyma sheath indistinct to distinct, sclerenchyma strands absent; vascular strands 4-6(-9). Scales pseudopeltate or occasionally peltate, 2.4-4.5 × 0.7-1.6 mm; dull brown, or blackish with distinctly lighter margin, dentate. Fronds monomorphic but the fertile fronds sometimes on longer stipes; stipes 6-19.5 cm., to 1 ½× as long as lamina; lamina, index 4-5; widest below or about middle, 7-12.5 × 1.7-3.7 cm., base cuneate to attenuate, apex obtuse to acuminate. Venation: secondary veins distinct, with tertiary veins forming regular areoles; included veins frequently forked or more branched; free veins many, pointing in all directions. Hydathodes distinct, scattered over lamina or, in fertile fronds, in a marginal zone. Stomata superficial, pericytic. Indument dimorphic, dense mat, persistent, grey-brown; upper layer composed of hairs 0.9-1.6 mm in diam., with erecto-patent, acicular rays, ± mixed with a lower layer composed of hairs with mainly woolly rays. Sori all over lamina or in an ill-defined, often apical patch, shortly spaced, superficial; to ± 10 irregularly scattered through each soriferous areole, occasionally confluent along veins; 0.5-1.5 mm in diam.; developing from apex downwards, when old individually distinct, immersed in indument. Sporangia on stalks to as long as capsule, capsule ± 0.2 mm high, with 15-20 indurated annulus cells. Paraphyses not differentiated. Spores finely and sparsely granulate.

E c o l. Terrestrial or epilithic, shaded, 250-900 m

D i s t r. Sri Lanka

S p e c i m e n s   E x a m i n e d. POLONNARUWA DISTRICT: Gunners Quoin, on shady rocks in forest, 250 m., 12 Oct 1977, *Huber 439* (US). MATALE DISTRICT: Erawalagala Mt., steep, rocky forested Eastern slope, terrestrial on rocks in shade, 420 m., 20 Oct 1974, *Davidse & Sumithraarachchi 8080* (L). KANDY DISTRICT: at foot of descent from Madugoda to Weragamtota, in shade, on rocks, 300 m., 9 Jan 1954, *Sledge 947* (K, U); road to Nawanagalla, 1 mile W of Madugoda, c.1100 m., on rock. 8 Jan. 1954, *Walker T 57* (BM); Weragamtota, near Ratna Ella Falls, on rock. 12 Feb. 1954, *Walker T 397* (BM); St. Martins Estate, trail to Nitre Cave, 14 Nov 1975, *Sohmer & Jayasuriya 10691*(K). BADULLA DISTRICT: Lower Badulla Rd., not uncommon up to914 m., *Wall* (L). AMPARAI DISTRICT: Mt. Wadinagala, Amparai distr., in forest, on rocks and in fissures, 300—600 m., 2 Nov 1975, *Bernardi 15609* (Z). MONARAGALA DISTRICT: Between Kataragama and Buttala, near Paddahelakanda, forest, terrestrial, c. 550 m., 28 Oct 1975, *Bernardi 15530* (Z). LOCALITY UNKNOWN: warmer parts of island, *Naylor Beckett 815* (K); *Robinson 206* (K).

6 Pyrrosia porosa (C. Presl) Hovenkamp var. **porosa**

Pyrrosia porosa Hovenkamp, Blumea 30 (1984) 208. —*Polypodium porosum* Wallich, Cat. (1828) no 266, *nom. nud.*—*Niphobolus porosus* C. Presl, Tent. Pter. (1836) 202; Bedd., Ferns S. India (1864-5) 61, pl. 183.— *Cyclophorus porosus* C. Presl, Epim. Bot. (1851) 130.—*Polypodium porosum* Mett., Polyp (1856) 128; Hook., Sp. Fil. 5 (1863). 48; Thwaites, Enum. Pl. Zeyl. (1864) 395.—Type: *Wallich no 266* (holo PRC, *teste* Holtt., iso in B. BM., BR, K, M., P. US), "India Orientalis"
Niphobolus sticticus Kunze, Linnaea 24 (1851) 257.—*Polypodium sticticum* Mett., Polyp. (1856) 128.—*Cyclophorus sticticus* C. Chr., Ind. Fil. (1906) 201—*Pyrrosia stictica* Holtt., Nov. Bot. Inst. Bot. Univ. Car. Prag. (1968) 31, *nom. superfl.*—Type: *Leschenault 149* (G, not seen, iso P), India.
Pyrrosia nayariana Ching & Chandra, Amer. Fern J. 54 (1964) 62, fig. 1-10.—Type: *Chandra 74310* (LWG, not seen), India.
Polypdlium acrostichoides et syn. homot., auct. non G. Forst., *quoad specim.* Zeyl. Willis, Cat. Pl. Ceyl. (1911) 123; Manton & Sledge, Phil. Tr. R. Soc. Lond. B 654 (1954) 169.

Niphobolus mollis et syn. homot., auct. non Kunze: Sledge, Bull. Br. Mus. nat. Hist. (Bot.) 2 (1960) 134; Bir & Shukla, Nova Hedw. 21 (1972) 195; B.K. Nayar & S. Kaur, Comp. Bedd. Handb. (1974) 81.

Niphobolus fissus et syn. homot., auct. non Blume *(p.p.,* partly this refers to *P. mannii* too): Hook., Sp. Fil. 5 (1863) 49; Baker in Hook. & Baker, Syn. Fil. (1867) 351; Bedd., Suppl. Ferns S. & Br. Ind. (1876) 22; Bedd., Handb. Ferns Brit. Ind. (1883) 330; Bedd, Suppl. Handb (1892) 91.

Rhizome short-creeping, ventral groove absent, 1.6-3.1 mm thick, phyllopodia 0.3-0.7 cm apart, lateral buds situated on or between internodia. Anatomy: ground tissue parenchymatous, sclerenchyma sheath distinct or occasionally indistinct, sclerenchyma strands scattered through inner parenchyma, 15-many; vascular strands 5-13. Scales peltate, 1.4-5.7 × 0.5-2.1 mm; base entire to ciliate; acumen light brown to dull blackish with a distinct lighter margin, ciliate to dentate. Fronds monomorphic, not or indistinctly stipitate; stipes to 13 cm; lamina, index 6 to ± over 20; widest above middle, 9-31 × 0.7-3.5 cm., base gradually narrowed, apex acute to acuminate. Venation: secondary veins usually distinct, with tertiary veins forming regular to irregular areoles; included veins simple or rarely forked, free, excurrent. Hydathodes distinct, scattered over lamina, ± superficial. Stomata superficial to slightly sunken, pericytic. Indument dimorphic, occasionally monomorphic, a dense mat, persistent, brown, or lower layer whitish; upper layer composed of hairs 0.5-1.6 mm in diam., with erecto-patent to occasionally appressed, acicular rays, usually ± mixed with lower layer composed of hairs with mainly woolly rays. Sori apical to covering lamina entirely, closely packed, superficial; several in a row in, or scattered through each soriferous areole, rarely confluent; 1-2 mm in diam.; developing from apex downwards, when old individually distinct, immersed in indument to being exserted from it. Sporangia sessile or on stalks 0.5-1 times as long as capsule, capsule 0.3-0.5 mm high, with 14-25 indurated annulus cells. Paraphyses not differentiated. Spores finely granulate.

E c o l. Epilithic or epiphytic, low elevations to 1650 m.

D i s t r. Northern India to China, Japan, Taiwan, Philippines

S p e c i m e n s   E x a m i n e d. KURUNEGALA DISTRICT: Dambetenne (? Dambadeniya), April 1871, *Hutchinson* (LIV). MATALE DISTRICT: Dambulla between 1908-1932: *Freeman 310 & 311*(BM). KANDY DISTRICT: Haragama Pussalamankada (15 miles from Kandy), c.540 m., on rock face, 18 Jan. 1954, *Walker T 127* (BM); *Faden, Faden & Waas 76/235* (L); Padiyapelella-Ellamulla road, culvert 28/8, moss-covered rock, locally common, 850 m (US); 14 Nov 1975, St. Martins Estate, trail to Nitre Cave, *Sohmer & Jayasuriya 10679* (L);. Hunnasgiriya, 900 m., 19 Jan 1954, *Schmid 999* (G, BM); Corbet's Gap, on rock, 22 Jan. 1954, *Walker T 198* (BM); Kandy-Mahiyangana Rd, east of Madugoda, on tree, c. 100 m., 2 Apr 1969, *Robyns 6983* (US); Madugoda, sunny rocks above road, 720 m., 16 Oct 1977,

Huber 465 (US). BADULLA DISTRICT: Mt. Kokegale, east of Mahiyangana, on forested rocks, 730 m., 11 Nov 1975, *Bernardi 15692* (Z); Ambawela rd, Hakgala junction, mi 5, 1560 m., rock crevice, 19 Nov 1976, *Faden & Faden 76/379* (US). RATNAPURA DISTRICT: Rest House Belihul Oya, in crevices in granite cliff, 3 Jan 1951, *Ballard 1369* (K). NUWARA ELIYA DISTRICT: Hakgala, epiphytic, 23 Mar 1900, *Giesenhagen 42* (L); near Hakgala, 1650 m., 27 Dec 1950, *Holttum SING 39199* (SING); 19 Mar 1954, Nuwara Eliya, Kandapola, *Sledge s.n* (U). MONARAGALA DISTRICT: Between Kataragama and Buttala, Paddahelakanda, on rocks in forest, 550 m., 28 Oct 1975, *Bernardi 15519* (Z). LOCALITY UNKNOWN: 1854, *Bradford 371*(P); *CP 3104* (BO, P, B)); *Ferguson 175* (US); 1884, *Wall s.n.* (L); Common on rocks and trees, 914-1524 m., *Wall s.n., s.d* ( ); *Nietzer s.n.,* s.d., s. loc. (P); *Robinson 205* (K).

3. MICROSORUM

Microsorum Link, Hort. Berol. 2 (1833) 110; Abeywickrama, Ceyl. J. Sc. Sect. A 13 (1956) 27; Bosman, Leiden Bot. Ser. 14 (1991) 69.—Type species: *Microsorum irregulare* (= *M. punctatum*).
Phymatodes C. Presl, Tent. Pterid. (1836) 195, nom. illeg. p.p. excl. type.
Phymatosorus Pic.Serm., Webbia 28 (1973) 457. — Type species: *Phymatosorus scolopendria.*

Epiphytic, epilithic or terrestrial. Rhizome terete or dorsoventrally flattened, creeping, white waxy or not. Rhizome scales appressed to spreading, pseudopeltate or peltate, clathrate or subclathrate, sometimes with hyaline margin, often with a central tuft of multiseptate hairs. Fronds monomorphic to dimorphic, sessile to stipitate, simple or pinnatifid, membranaceous to chartaceous. Venation: veins branching close to the costa or running almost to the margin, connecting veins 1- several, ana- or catadromous; forming 1-several rows of areoles, included veins amply anastomosing; free veinlets excurrent and recurrent, recurrent in marginal areoles, often ending in a hydathode. Sori always separate, round or elongate on veinlets, superficial or deeply sunken, scattered over the lamina or in 1-several rows between costa and margin, sometimes in distinct rows along the veins.

KEY TO THE SPECIES

N o t e: Anatomical characters of the rhizome may be studied using a simple hand lens once a clean cut has been made with a sharp knife. In herbarium specimens an oblique cut will show all necessary details.
All pinnate species may form juvenile simple fronds which may be difficult to identify.
1 Sori superficial
  2 Rhizome with distinct sclerified sheath around vascular strands . . . .
  . . . . **1. M. insigne**

2 Rhizome with many (50-100) sclerenchyma strands, venation with several rows of ± equal areoles between midrib and margin, midrib glabrous or sparsely covered with scales below.

  3 Lamina membranaceous, venation distinct . . . . **2. M. membranaceum**

  3 Lamina subcoriaceous, venation usually obscure . . . **5. M. punctatum**

2 Rhizome without or with few sclerenchyma strands (< 10), venation with a single row of elongated, large, areoles, midrib often densely covered with scales below . . . . **4. M. pteropum**

1 Sori deeply sunken

  4 Rhizome not white-waxy, venation conspicuous, forming distinct marginal vein . . . . **3. M. membranifolium**

  4 Rhizome white-waxy, venation indistinct, not forming marginal vein . . .

  . . . . **6. M. scolopendrium**.

1. Microsorum insigne (Blume) Copel.., Univ. Calif. Publ. Bot. 16 (1929) 112. — *Polypodium insigne* Blume, Enum. Pl. Javae (1828) 127. — *Pleopeltis insignis* Bedd., Ferns Brit. India. (1866) t. 214. — *Colysis insignis* J. Sm., Hist. Fil. (1875) 101. — Lectotype: Zippelius s.n. (L) Java. [*Polypodium dilatatum* Wall., Cat. (1829) 295, nom. nud.; Hook., Sp. Fil. 5 (1864) 85, non Hoffm. (1795)]. — *Pleopeltis dilatata* Bedd., Ferns Brit. India. 1 (1866) t. 122. — *Colysis dilatata* J. Sm., Hist. Fil. (1874) 101. — *Polypodium europhyllum* C. Chr., Index Filic. (1906) 525, p.p. — *Microsorum dilatatum* Sledge, Bull. Br. Mus. Nat. Hist. (Bot.) 2 (1960) 143. — *Kaulinia dilatata* B.K. Nayar & S. Kaur, Companion Beddome's Handb. Ferns Brit. India (1974) 89. — Type: Wallich 295 (K, iso US) Nepal.

Rhizome dorso-ventrally flattened to terete, 2–11 mm wide, short-creeping, internodia 2.5–11 mm long, not white waxy, roots densely set; vascular strands 7–16, with sclerified sheaths, rarely with sclerenchyma strands. Rhizome scales pseudopeltate, sparsely set, appressed to distinctly spreading, ovate, narrowly ovate or triangular, (2–)2.5–7.5 × 0.5–2.5(–3) mm., margin entire to denticulate (occasionally with small triangular lobes), clathrate or subclathrate, central region glabrous. Fronds not or slightly dimorphic, sessile to stipitate, simple or pinnatifid, thin-herbaceous, glabrous. Simple fronds: Stipe to 10 cm long, 0.5–1.5 mm thick. Lamina narrowly ovate to obovate, 2.5–65 × 0.5–6.5 cm., index 4–11, base narrowly angustate, decurrent to a long wing, forming two ridges at base of stipe, margin entire (rarely sinuate), apex acute to acuminate. Dissected fronds: Stipe to 4 cm long, 1.4–6.2 mm thick. Lamina 8–110 cm long (the tapering base not included), 3–55 cm wide, index 3–8.5; widest below to about the middle, lobes 1–12(–14) at each side, longest lobes at position 1–3 from base, connected by a wing up to 2.5 cm wide, 2–27 × 0.3–55 cm., index 3–8.5, widest below or about middle. Venation: veins prominent and distinct, 4–13 mm apart, more or less straight or zigzag, dichotomously branched near the margin (immersed at ¾–⅘ of lamina width), connecting veins 1–3, anadromous (rarely catadromous), forming small primary costal areoles or areoles all more or less equally sized, sometimes irregularly shaped; included veins more or less immersed and indistinct, variously anastomosing, free veinlets simple or once forked. Sori separate, round or elongate along veinlets, superficial or sligthly immersed, 0.5–1.5 × 1.5–5 mm., irregularly scattered on the smallest veinlets or with 2 on connective veins, absent from the basal parts for 0.3–0.5 of total length of lamina, present or absent in marginal areoles, generally present in costal areoles. Paraphyses simple uniseriate hairs with glandular top cells.

D i s t r. South East Asia from Nepal to Japan, Philippines, Lesser Sunda Islands. Apparently not common on Sri Lanka, and possibly extinct (no recent collections seen). Altitude: c. 1200 m.

E c o l. Usually epilithic, sometimes epiphytic in primary and secondary forest; in or along streams or falls, in undergrowth of shrubs; twice reported from caves; shady, mossy, muddy, and wet places.

S p e c i m e n s   E x a m i n e d. MATALE DISTRICT: *Hutchinson s.n.*, s.d., Hoolankande Pap. (K, LIV) KANDY DISTRICT: *Thwaites C.P. 3973*, forests above Telgamma, about1219 m.

2. Microsorum membranaceum (D. Don) Ching, Bull. Fan Mem. Inst. Biol. 4 (1933) 309; Sledge, Bull. Br. Mus. nat. Hist. (Bot.) 2 (1960) 142; Bosman, a monograph of the fern genus Microsorum (Polypodiaceae) (1991) 91. — *Polypodium membranaceum* D. Don, Prodr. Fl. Nepal. (1825) 2. —*Pleopeltis membranacea* Bedd., Handb. Ferns Brit. Ind. (1883) 355. —[*Colysis membranacea* J. Sm., Cult. Ferns (1857) 11, nomen. illeg.]. — Type: *Wallich* (K, iso B) Nepal.

Rhizome dorso-ventrally flattened to cylindrical, 3–10 mm wide, not white waxy, with scattered strands of sclerenchyma, vascular strands 18–23, without sclerified sheaths, sclerenchyma strands 50–100, roots densely set. Scales pseudopeltate, densely set, decaying quickly, slightly spreading, ovate or triangular, 1.5–9 mm long, 1–3 mm wide, margin entire, apex acute, clathrate except for a narrow hyaline margin, central region often bearing multiseptate hairs when young. Phyllopodia ± distinct, up to 9 mm apart. Fronds not or slightly dimorphous, membranaceous, glabrous. Stipe winged for a considerable part, margin entire or irregularly sinuose; stipe absent or to c. 15 cm long, 3–5 mm thick. Lamina simple, elliptic to ovate to narrowly elliptic to narrowly ovate to linear, 25 – 77 cm long, 3.5–13 cm wide, index 2–11, widest below the middle, base narrowly angustate. Venation: veins prominent and distinct, 5–15 mm apart, ± straight, branched near margin, connecting veins 4–8, catadromous, often not clearly differentiated, forming a row of ± equally sized, often irregularly shaped areoles, sometimes with a row of small primary costal areoles, included veins prominent and distinct, variously anastomosing, free veinlets simple to twice forked. Sori separate, scattered or in two indistinct, irregular rows between the veins, round, sometimes close together and nearly confluent, superficial or sligthly immersed, over entire surface of lamina or absent from basal parts for 0.5 of total length of lamina, mostly irregularly scattered on smallest veinlets, 1 or 2 on each veinlet, not on connecting veins, 1–2 mm diam., round, absent in marginal, generally present in costal areoles. Paraphyses simple uniseriate hairs with glandular topcells, paraphyses 3 –celled. Annulus 16–21 –celled. Indurated cells 12–14.

D i s t r. India to China, Taiwan

E c o l. Epiphytic, epilithic, or terrestrial in evergreen or deciduous broad–leaved (sub)tropical forest, often in valleys or ravines. Altitude 600–2600 (–4000) m.

N o t e s The fronds appear to be seasonally deciduous.

S p e c i m e n s   E x a m i n e d. MATALE DISTRICT: *Walker T 168*, Hoolankande, c.1350 m Wet rock. 20 Jan. 1954 (BM). KANDY DISTRICT: Labukelle, *Freeman 333, 334* (BM); *Sledge 534*, Dec. 8 1950, near Kandy, jungle at Oodawella, 1219 m (K); Stream gully in forest above and S.E. of "Hunas Falls Hotel" W.side of Hunnasgiriya mountain, S.S.E. of Elkaduwa, N.E. of Kandy, Alt. c.1200-1300 m., 25 Aug 1993, (3 sheets) *Fraser-Jenkins* with *Jayasekara, Samarasinghe & Abeysiri* Field No. *55* (K); P 22, Oodawella jungle, voucher for cytological study n=36 (BM); *Sledge 832*, Jan 7 1951, Corbet's Gap, on boulders in jungle, 1219 m (K, BM). NUWARA ELIYA DISTRICT: Rambodde, Aug 1870, *Hutchinson* (LIV). WITHOUT LOCALITY: *Beddome s.n., s.d.* (K); *Gardner 1145* (K), *1298* (K); *Macrae 906; Skinner s.n., s.d* (K); *Thwaites C.P. 1298* (BM); *Walker 25*, s.d. (K).

3. Microsorum membranifolium (R.Br.) Ching, Bull. Fan Mem. Inst. Biol. 10 (1941) 239. — *Polypodium membranifolium* R.Br., Prodr. (1810) 147. — Type: *Banks* (BM).—*Polypodium nigrescens* Blume, Enum. Pl. Javae (1828) 126. — *Phymatodes nigrescens* J. Sm., Ferns Brit. For. (1866) 94. — *Pleopeltis nigrescens* Bedd., Handb. Ferns Brit. Ind. (1883) 367. — *Microsorum nigrescens* Copel., Occas. Pap. Bernice Pauahi Bishop Mus. 14 (1938) 74. — *Phymatosorus nigrescens* Pic.Serm., Webbia 28 (1973) 459. — Type: *Blume* (L 908.303–605) Java.

Rhizome ± terete, 4–20 mm thick, short- to long-creeping, internodia 1—2 cm long, not white waxy, roots densely set; vascular strands to c. 20, without sclerified sheaths, sclerenchyma strands 100–200. Rhizome scales pseu-dopeltate, sparsely set, appressed, round, ovate, elliptic or triangular, 4–8 mm long, 2.5–3.5 mm wide, margin entire, often eroded, apex rounded, clathrate or subclathrate with very thin cell-walls, cells small, ± isodiametric, central region glabrous or bearing multiseptate hairs. Fronds not or slightly dimorphic, stipitate, membranaceous (often translucent), glabrous. Stipe 19–100 cm long, 2.8–10 mm thick. Lamina pinnatifid (rarely simple), 27–175 × 36–90 cm., elliptic to ovate, index 1–3, widest about middle, base cuneate to narrowly angustate, decurrent to a long wing, lobes 2–20 either side, longest lobes at position 2–10 from base, connected by a wing to 1.5 cm wide, 15–50 × 2.3–7 cm., index 8–10, widest about middle, apex long acuminate; apical lobe longer than upper lateral lobes, 12–48 × 2.5–6.5 cm., widest about or below middle; margin undulate, sometimes wavy. Venation: veins prominent and distinct, 7–13 mm apart, ± straight; connecting veins anadromous, forming small, inconspicuous, primary costal areoles, bordered by one row of conspicuous large areoles; included veins prominent and distinct (small areoles near margin usually surrounded by very prominent veins), marginal vein present, free veinlets simple to twice forked. Sori separate, round or slightly elongated, deeply sunken, visible as protrusions on upper surface, 3.2–6.5 mm diam.; in a single row between midrib and margin, one sorus in, or just outside, each primary costal areole, generally close to costa, at most halfway to margin, on distinct soral veins; on the whole surface of lamina or absent from basal parts. Paraphyses biseriate, non–clathrate.

D i s t r. South East Asia extending to the Pacific.

H a b i t a t & E c o l o g y. Terrestrial, epilithic or (low) epiphytic, often on limestone but also on granite, usually in wet places. Altitude: 0–750 m.

S p e c i m e n s E x a m i n e d. ANURADHAPURA DISTRICT: Ritigala Peak, 732 m., 14 Jan. 1951, *Ballard 1477* (K); Ritigala, near summit in forest, 732 m., 14 Jan 1951, *Sledge 865* (BM); Ritigala strict Nat. Res. Eastern slope, epiphytic, 671 m., 4 May 1974, *Jayasuriya & Premadasa 1638* (BM). KEGALLE DISTRICT: Lower slopes of Alagalla Mt., on tree trunks and rocks in mixture of thicket and sec. growth on steep slope, 762 m., 15 Oct 1978, *Fosberg 57852* (K). KANDY DISTRICT: P 167A, Ambagamwa, voucher for cytological study n=36 (BM); Kandy, Oct. 1870, *Hutchinson* (LIV); Gardens Peradeniya? *Macrae 134, 867* (BM); Kandy, 1868, *Randall s.n.* (*Rawson 322*) (BM); Kandy Catchment, on rock face in shade of sec. jungle, 685 m., 14 Jan 1954, *Sledge 961* (BM); Hunnasgiriya (near Madugoda), c. 900 m. on rock, 16 Jan 1954, *Walker T 105* (BM). RATNAPURA DISTRICT: Adam's Peak, 5 Jan. 1951, *Manton 840* (K); Sinharaja Forest, epiphytic on vertical tree trunks, apparently 2 fronds per rhizome, 28 June 1972, *Hepper, Maxwell & Fernando 4546* (K). WITHOUT LOCALITY: *Moon 250*, Feb 19, 1819 (BM); *Gardner C.P. 1144* (BM); *s.n.* (BM); *Thwaites C.P. 1296* (K, BM)

4. Microsorum pteropum (Blume) Copel., Univ. Calif. Publ. Bot. 16 (1929) 112; Sledge, Bull. Brit. Mus. nat. Hist. (Bot.) 2 (1960) 143; Steenis, Rheophytes of the World (1981) 161. — *Polypodium pteropus* Blume, Enum. Pl. Javae (1828) 125, add. 3. — *Pleopeltis pteropus* T. Moore, Index Fil. (1857) lxxviii; Bedd.. Handb. Ferns Brit. Ind. (1883) 359. — *Kaulinia pteropus* B.K. Nayar,

Taxon 13 (1964) 67. — *Colysis pteropus* Bosman, A monograph of the fern genus Microsorum (Polypodiaceae) (1991) 112. — Lectotype: *Blume* (L) Java, G. Toembal.

Polypodium tridactylum Wall., Cat. (1829) 315, nom. nud.; Hook. & Grev., Icon. Filic. (1831) t. 209. — *Phymatodes tridactyle* C. Presl, Tent. Pterid. (1836) 196. — *Drynaria tridactyla* Fée, Mém. Fam. Foug. 5. Gen. Filic. (1852) 271. — *Pleopeltis tridactyla* T. Moore, Index Fil. (1857) lxxviii. — *Colysis tridactyla* J. Sm., Hist. Fil. (1875) 101. Type: *Wallich 315* (K, iso BM., C, K, UC, US).

Rhizome dorso-ventrally flattened, 0.5-5 mm wide, short-creeping, internodia 1.5-20 mm long, not white waxy, roots densely set. Anatomy: vascular strands 10–14, without sclerified sheaths, sclerenchyma strands few or absent. Scales pseudopeltate, ± densely set, slightly spreading, narrowly ovate or triangular, 1.5-5 × 0.4-1 mm., margin entire, apex acute, clathrate or subclathrate with an opaque central region, which is glabrous or bears multiseptate hairs when young, cells longitudinally rectangular. Fronds not or slightly dimorphic, simple or pinnatifid, sessile to stipitate, thin-herbaceous to membranaceous, lower surface often densely covered with clavate hairs, lower surface of midrib often densely covered with scales. Simple fronds: stipe absent or to 12 cm long, 1-2 mm thick, lamina narrowly elliptic, 3.5-30 × 0.2-5.5 cm., index 5-35, base narrowly angustate, decurrent to a long wing, margin entire, apex acute to acuminate. Dissected fronds: Stipe 1-28 cm long. Lamina 15-45 × 5-25 cm., index 3.5-8, widest about or above middle, lobes 1, rarely 2, at each side, widest about or below middle, basal lobes infrequently with a small basiscopic lobe, 4.5-17 × 0.3-5 cm; apical lobe longer than upper lateral lobes, widest at base to below middle. Venation: veins prominent, distinct, 3-7 mm apart, ± straight or zigzag, dichotomously branched about middle to near margin; connecting veins 1, anadromous, forming a large areole nearly extending to margin, bordered by smaller areoles; costal areole, if present, formed by smaller veins, included venation ± immersed, indistinct or prominent, variously anastomosing, often forming several equally sized areoles within the large areoles, free veinlets 1- or 2-forked. Sori separate, round, sometimes elongate along veins, superficial or sligthly immersed, 1-2.5 × 2-7 mm., mostly irregularly scattered on smallest veinlets, on whole surface of lamina, absent from marginal and costal areoles. Paraphyses simple uniseriate hairs with glandular top-cells, 2-3 celled. Annulus 18-23 celled. Indurated cells 12–17.

D i s t r. Himalayas to Indochina, China, Japan, Taiwan, Malesia to New Guinea.

E c o l. Along or in streams, often under water; low elevations, but sometimes up to 1200 m.

S p e c i m e n s   E x a m i n e d. ANURADHAPURA DISTRICT: Summit of Ritigala N.C.P., anon 22 Mar 1905 (juvenile specimens) (L). RATNAPURA DISTRICT: *Sohmer & Waas 8752*, 16 Nov 1973, Adam's

Peak Wilderness, summit on S side. WITHOUT LOCALITY: 4 Feb 1819, *Moon 108* (BM., under amaurolepida) sterile, possibly *Leptochilus*; *C.P. 1301* (various localities) *Thwaites* (BM); Oct 1882).

5. Microsorum punctatum (L.) Copel., Univ. Calif. Publ. Bot. 16 (1929) 111; Sledge, Bull. Brit. Mus. nat. Hist. (Bot.) 2 (1960) 143. *Acrostichum punctatum* L., Sp. Pl. ed. 2 (1763) 1524. — *Polypodium punctatum* Sw., J. Bot. (Schrad.) 1800 (1801) 21, non Thunb. (1784).

Polypodium lingulatum Sw., Syn. Fil. (1806) 30. — *Phymatodes lingulata* C. Presl, Tent. Pterid. (1836) 198. *Pleopeltis punctata* Bedd., Suppl. Ferns S. Ind. (1876) 22. Type: *Fothergill*, China.

Rhizome terete, 4–8 mm thick, short-creeping, internodia 2–30 mm long, usually white waxy, roots forming a thick mat. Anatomy: vascular strands 11–21, without sclerified sheaths, sclerenchyma strands 50—100. Scales pseudopeltate, sometimes peltate, sparsely to densely set, appressed or slightly spreading, ovate, narrowly ovate or triangular, 1.5—8 × 0.5–3 mm., margin entire to dentate, apex acute, clathrate or subclathrate, rarely with a hyaline margin, cells small, ± isodiametric or longitudinally rectangular, central region glabrous or bearing multiseptate hairs. Fronds not or slightly dimorphic, sessile to stipitate, herbaceous to subcoriaceous (sometimes coriaceous), glabrous. Stipe absent or to 12 cm long, 3–8 mm thick. Lamina simple, narrowly ovate, elliptic or obovate to linear, 10–175 × 1.5–15 cm., index 4–20 (–25), base cordate and auriculate to narrowly angustate, decurrent to a long wing forming two ridges at base of stipe, margin entire or undulate (occasionally irregularly lobed), apex rounded, acute to acuminate, lower surface without acicular hairs. Venation: veins ± immersed and indistinct or prominent and distinct, 6–25 mm apart, ± straight or zigzag, dichotomously branched near margin; connecting veins 3—10, catadromous, forming ± equally-sized (sometimes irregularly shaped) areoles, sometimes also one row of small primary costal areoles, included veins ± immersed and indistinct, variously anastomosing, free veinlets simple to twice-forked. Sori separate, round or elongate along veinlets, superficial or sligthly immersed, 0.5–2.5 mm diam., mostly irregularly scattered on smallest veinlets (occas. in part on tertiary veins), on whole surface of lamina or absent from basal parts for up to 0.9 of total length of lamina. Paraphyses simple uniseriate hairs with glandular top cells, 2–4 –celled, annulus 18–22 –celled (–25). Indurated cells 12–16 (–19).

D i s t r. Palaeotropics and subtropics (see Bosman 1991 f. 20).

E c o l. Common on mossy face trunks and on rock boulders.

S p e c i m e n s   E x a m i n e d. RATNAPURA DISTRICT: Adam's Peak, on fallen tree trunk, S ascent, 9 March 1954, *Sledge 1243* (BM); Adam's Peak, up to 750 m., on fallen tree, 9 March 1954, *Walker T 734* (BM). LOCALITY UNKNOWN: *Robinson s.n., s.d,* (K). SITE NOT LOCATED: *C.P. 3799*, 1869, Unatella Est. ? (K); G. Wall, below Pitarella Estate (L).

6. Microsorum scolopendrium (Burm.f.) Copel.., Univ. Calif. Publ. Bot. 16 (1929) 112; Sledge, Bull. Brit. Mus. nat. Hist. (Bot.) 2 (1960) 143. —*Polypodium scolopendrium* Burm.f., Fl. Indica (1768) 232. [*Polypodium phymatodes* L., Manton Pl. (1771) 306, nom. illeg.]. —*Phymatodes scolopendria* Ching, Contr. Inst. Bot. Nat. Acad. Peiping 2 (1933) 63. —*Phymatosorus scolopendria* Pic. Serm., Webbia 28 (1973) 460.
Type: Herb. Hermann (not seen), Sri Lanka.

Polypodium excavatum Roxb., Hort. Bengal. (1814) 75; Calcutta J. Nat. Hist. 4 (1844) 485; Morton, Contr. Nat. Herb. 38, 7 (1974) 343. Type: Rumphius, Herb. Amboin. 6:80, t. 35, fig.2 (fide Morton, l.c.)

Pteris lobata Roxb., Calcutta J. Nat. Hist. 4 (1844) 504, nom. illeg. non Goldm., 1843; Morton, Contr. Nat. Herb. 38, 7 (1974)370. Type: *Chr. Smith s.n.* (BR, fide Morton, l.c.), Moluccas.

Phymatodes banerjiana S. Pal & N. Pal, Amer. Fern J. 53 (1963) 103. Type: *S. Pal H499158* (Iso in K) cultivated.

Rhizome terete, 8–7 mm thick, internodia 1—6 cm long or more, white-waxy, roots sparsely to densely set; vascular strands 9–30, with sclerified sheaths, sclerenchyma strands 10–100. Rhizome scales peltate, occasionally some pseudopeltate, sparsely to densely set, appressed to distinctly spreading, ovate, narrowly ovate or triangular, 2–7 × 0.6–1.4 mm., margin denticulate, apex acute (often caducous), clathrate or subclathrate, cells longitudinally rectangular in narrow apical part, central region glabrous or bearing multiseptate hairs. Fronds not or slightly dimorphous, stipitate, herbaceous, glabrous. Lamina: simple or pinnatifid. Fronds with simple lamina: Stipe 2–30 cm long, 0.8–2.3 mm thick, lamina elliptic to narrowly elliptic, 8–45 × 2–8 cm., index 5–6, base cuneate or cuneate-angustate, margin entire, apex acute to acuminate. Fronds witth pinnatifid lamina: Stipe 4–55 cm long, 0.8–7.2 mm thick. Lamina ovate, 14–41 × 9–30 cm., index 1–1.5; widest below to about middle, lobes 1–9 at each side, longest lobes at position 1–2 from base, 5—20 × 0.7–4 cm., index 5–10, widest at base to about middle, connected by a wing 0.5-2.3 cm wide; upper lobes at an angle of 35–50° to apical lobe; apical lobe longer than upper lateral lobes, 3–20 × 0.8–5 cm., widest at or just above base. Venation: veins ± immersed and indistinct or prominent and distinct. 4–12 mm apart, ± straight or zigzag, dichotomously branching at or near costa to near margin; connecting veins 1–4; anadromous, forming one row of conspicuous, large areoles or several rows of equally-sized areoles, sometimes also a row of small costal areoles; included venation ± immersed and indistinct, sometimes more conspicuous, variously anastomosing, forming a dense reticulation with smaller areoles, free veinlets simple or once-forked. Sori separate, round or elongate, deeply sunken, visible as protrusions on upper surface, 1–6.5 × 6.5–15 mm., 1–several rows between midrib and margin, one sorus inside, or just outside, each primary costal areole, generally close to costa, at most halfway to margin, on entire surface of lamina or absent from basal parts. Paraphyses simple uniseriate hairs with glandular top cells.

D i s t r. Tropical Africa to Asia, Pacific and Australia.

E c o l. Epiphyic or terrestrial, in plantations, on road sides etc. Sea level to 750 m.

S p e c i m e n s E x a m i n e d: COLOMBO DISTRICT: Colombo, June 1870, *Hutchinson* (LIV); Labugama, roadsides, jungle, 50-60 m., 7-8 Jan 1954, *Schmid 823* (BM). KEGALLE DISTRICT: Ginigathena Avissawela Rd., marker 67/7, on tree base on steep slope above road, 5 Nov 1967, *Comanor 534* (K). KANDY DISTRICT: Hantane, c. 900 m., *Gardner 1143* BM); Kandy, common on the bank of the roadway round the lake, *Freeman 330, 331, 332* (BM); Kandy, Roseneath, bank by roadside, c. 650 m., 10 Dec 1950, *Ballard 1084* (K); Garden Peradeniya?, *Macrae 266, 868* (BM); Kaduganawa, epiphytic on trees by roadside, c. 400 m, 12 Dec 1950, *Sledge 590* (BM). RATNAPURA DISTRICT: 4 II 1954 Prov. Sabaragamuwa, Niriella, 225 m., *Schmid 1115* (BM). GALLE DISTRICT: Bentota, sea level, epiphytic on base of coconut palm., 19 Jan 1951, *Ballard 1509* (K); Near Bentota, 19 Jan 1951, *Sledge 881* (K). MATARA DISTRICT: Deniyaya, 550 m., 5 Feb 1954, *Schmid 1143* (BM). LOCALITY UNKNOWN: *Barkly 6, 9* (BM); *Beddome s.n., s.d*, sea level (K); *CP 1297: anon.* (K); *Thwaites* (BM*); Fraser 154, 1849* (BM); P1 voucher for cytological study n=36 (BM); *Rawson 899* (BM); *Robinson s.n., s.d.; Thwaites s.n.,* s.d. (K).

Microsorum indet *Sohmer & Waas 10561*, 11 Nov 1975, Kegalle Distr., Kelani River, Rest House, epiphytic (juvenile sterile specimen)

4. LEPTOCHILUS

Leptochilus Kaulf., Enum. Filic. (1824) 147; Nooteboom., Blumea 42 (1997): 274.
Type species: *L. axillaris* Kaulf.
Dendroglossa C. Presl, Epim. Bot. (1851) 149; Abeywickrama, Ceyl. J. Sc.
    Sect. A 13 (1956) 27 . Type species: *Dendroglossa normalis.*
Anapausia C. Presl, Epim. Bot. (1851) 185; Abeywickrama, Ceyl. J. Sc. Sect.
    A 13 (1956) 27. *Paraleptochilus* Copel., Gen. Fil. (1947) 198. Type
    species: *Leptochilus decurrens* Blume.

Terrestrial, epilithic or epiphytic. Rhizome not white waxy, creeping. *Rhizome scales* pseudopeltate or peltate, clathrate or subclathrate, sometimes with a central tuft of mulstiseptate hairs. Fronds dimorphic, stipitate, simple or pinnatifid, thin-herbaceous to subcoriaceous. Venation: veins branching close to the costa to near the margin, or seemingly, connecting veins ana- or catadromous; included veins amply anastomosing; free veinlets excurrent and recurrent, recurrent in marginal areoles. Sori in longitudinal or transverse coenosori.

KEY TO THE SPECIES

1 Fertile fronds up to half as wide as the sterile ones, sporangia in transverse elongated coeno-sori . . . . **1. L. macrophyllus**
1 Fertile fronds vary narrow, sporangia in longitudinally elongated coenosori . . . . **2. L. decurrens**

1. Leptochilus macrophyllus var. pedunculatus (Hook. & Grev.) Noot. [*Grammitis hamiltoniana* Wall., Cat. (1828) 9, nomen]. — *Ceterach pedunculatum* Hook. & Grev., Icon. Filic. (1829) t. 5. *Selliguea hamiltoniana* C. Presl, Tent. Pterid. (1836) 216; Bedd., Ferns Brit. India. (1867) t. 239. — *Selliguea pedunculata* C. Presl, Epim. Bot. (1851) 146. *Gymnogramme hamiltoniana* Hook., Sp. Fil. 5 (1864) 161; Hook. & Baker, Syn. Fil. (1868) 389. — *Polypodium pedunculatum* Salomon, Nomencl. Gefässkrypt. (1883) 312. — *Pleopeltis pedunculata* Alderw., Malayan Ferns Suppl. 1 (1917) 405. — *Colysis pedunculata* Ching, Bull. Fan Mem. Inst. Biol. 4 (1933) 321; Holtt., Revis. Fl. Malaya 2 sec. ed. (1966) 160.
Type: *Wallich 9* (K) India, Sylhet.

Rhizome 2–5 mm wide, roots densely or sparsely set. Anatomy: sclerenchyma strands (5–)50–150. Scales 3–4 × 0.5–0.8 mm., margin denticulate, apex acute. Fronds dimorphic, herbaceous. Sterile fronds: Stipe absent or to 12 cm long, 1.6–2 mm thick. Lamina simple, elliptic to narrowly ovate, 14–34 × 2–10 cm., base cuneate-angustate to narrowly angustate, decurrent, forming two ridges at base of stipe. Fertile fronds: Stipe 13–45 cm long. Lamina simple, deltoid, ovate, elliptic or narrowly elliptic, 4–24 × 1–5 cm. Venation: veins prominent and distinct, dichotomously branched near margin. Sori in transverse coenosori, rarely in acrostichoid patches between veins.

D i s t r. Continental Asia: Himalayas to China, Sumatra, Java.

E c o l. Epiphytic or epilithic, on rocks on river bank. 350–600 m.

S p e c i m e n s   E x a m i n e d: ANURADHAPURA DISTRICT: Slopes of Ritigala, about 610 m., 22 March 1905; ?summit of Ritigala N.C.P. MATALE DISTRICT: [June 1848, Dumboola] *Gardner C.P. 1298*. KANDY DISTRICT: Ambagamuwa, 1848, *Gardner C.P. 1296*. RATNAPURA DISTRICT: Sinharaja forest, near Weddagalla, epiphytic on vertical tree trunks, 350 m., 28 June 1972, *Hepper et al. 4546* (K); Adam's peak, on rocks on river bank, 350-400 m., 7 March 1973, *Bernardi 14145* (L). NUWARA ELIYA DISTRICT: [1847, N. Eliya] *Gardner CP 1298* (L). WITHOUT LOCALITY: Dec. 1848, *Gardner C.P. 1296*.

2. Leptochilus decurrens Blume, Enum. Pl. Javae (1828) 206; C. Chr., Contr. U.S. Natl. Herb. 26 (1931) 325; Sledge, Bull. Brit. Mus. nat. Hist. (Bot.) 2 (1960) 140. *Anapausia decurrens* C. Presl, Epim. Bot. (1851) 186. *Acrostichum variabile* Hook., Sp. Fil. 5 (1864) 277. *Gymnopteris variabilis* Bedd. in Hook., Fl. Brit. India (1868) t. 272. *Campium decurrens* Copel., Philipp. J. Sci. 37 (1928) 351. *Paraleptochilus decurrens* Copel., Gen. Fil. (1947) 198 t. 7. Type: *Blume* (L 908.286-396).*Leptochilus lanceolatus* Fée, Mém. Foug. 2. Hist. Acrost. (1845) 87 Pl. 47 f. 1. *Dendroglossa lanceolata* Fée, MémFoug. 5. Gen. Filic. (1852) 81, excl. syn. *Gymnopteris féei* T. Moore, Index Fil. (1857) XXIX. *Acrostichum lanceolatum* Hook., Sp. Fil. 5 (1864) 276, nom. illeg. non L. 1753. *Pleopeltis féei* Alderw., Malayan Ferns Suppl. 1 (1917) 405. *Campium lanceolatum* Copel., Philipp. J. 37 (1928) 348, Pl. 5, 2. Syntypes: *Hügel 1348, Perrotet* (not seen).*Gymnopteris féei* var. *pinnatifida* Bedd., Ferns S. India (1864) 71 t. 211. *Acrostichum variabile* var. *laciniatum* Hook., Sp. Fil. 5 (1864) 277. *Campium laciniatum* Copel., Philipp. J. Sci. 37 (1928) 354, plate 5, 1 & Pl. 7. *Leptochilus laciniatus* Ching, Bull. Fan Mem. Inst. Biol. 4 (1933) 344, excl. syn. *G. féei* v. *triloba.* Type: *Gardner (Thwaites C) 1318* (K) Sri Lanka. *Leptochilus thwaitesianus* Fée, Mém. Foug. 10 (1865) 7; Sledge, Bull. Brit. Mus. nat. Hist. (Bot.) 2 (1960) 141. Type: *Gardner (Thwaites C.P.) 1316* (Holo RB, iso BM., K, L, P) Sri Lanka. *Leptochilus zeylanicus* Fée, Mém. Foug. 10 (1865) 8. *Dendroglossa zeylanica* Copel., Gen. Fil. (1947) 199. Type: *Gardner (Thwaites CP.) 1317* (BM., K, L, P) Sri Lanka. *Acrostichum wallii* Baker, J. Bot. 10 (1872) 146. *Leptochilus wallii* C. Chr., Ind. Fil. (1906) 366; Sledge, Ann. & Mag. Nat. Hist 12, 9 (1956) 876; Sledge, Bull. Brit. Mus. nat. Hist. (Bot.) 2 (1960) 141. *Campium wallii* Copel., Philipp. J. Sci. 37 (1928) 348. *Dendroglossa wallii* Copel., Gen. Fil. (1947) 200. Type: *Wall* (K, iso BM., BO) Sri Lanka. *Gymnopteris metallicum* Bedd., Ferns Brit. India. Suppl. (1876) 26, t. 390. *Leptochilus metallicus* C. Chr., Ind. Fil. (1906) 386; Sledge, Ann. & Mag. Nat. Hist 12, 9 (1956) 876. *Campium metallicum* Copel., Philipp. J. Sci. 37 (1928) 347. — Type: *Beddome* (K) Sri Lanka, Haycock Hill.

Rhizome dorso-ventrally flattened, 1–3 mm wide, short, internodia 1–7 mm long, not white waxy, roots densely set. Vascular strands 4–10, with or without sclerified sheaths, sclerenchyma strands absent or to 100, scattered. Scales pseudopeltate, sometimes peltate, sparsely to densely set, slightly spreading, narrowly ovate or triangular, 1.2–5 × 0.2–1 mm., clathrate or subclathrate, cells longitudinally rectangular towards apex, margin denticulate, central region glabrous or bearing multiseptate hairs, apex acute. Fronds strongly dimorphic, simple (sometimes irregularly pinnatifid to bipinnatifid), sessile to stipitate, thin-herbaceous to herbaceous. Sterile fronds: Stipe absent or to 18 cm long, to 1.7 mm thick. Lamina simple, narrowly elliptic, ovate to obovate, 1.7–50 × 0.6–11 cm., lower surface without acicular hairs, base more or less abruptly cuneate or acuminate, or long-decurrent, often forming two ridges at base of stipe. Fertile fronds: Stipe to 50 cm long. Lamina simple, to 40 × 0.1–1 cm wide, ovate to linear. Venation: veins prominent and distinct, 2–12 mm apart, ± straight or zigzag, dichotomously branched below middle to near margin, or each costal areole giving rise to two lateral veins, thus lateral veins seemingly branching at or near costa; connecting veins 1—8, ana- or catadromous, forming ± equally sized areoles or one row of large areoles bordered by several smaller areoles; included veins ± immersed and indistinct, variously anastomosing, free veinlets simple or once-forked. Sori acrostichoid, covering lamina.

D i s t r. India to S.China and N Indochina, Philippines, New Guinea, Indian Ocean: Christmas I.

E c o l. Terrestrial, low climbing and epilithic, montane rainforest, hill evergreen forest, moss forest, often on rocks in stream. Altitude: from 100 to 2500 m.

N o t e s: This is a very variable species, especially in frond size, shape of lamina base, and presence of a distinct stipe. Several ± distinguishable forms have been separated, and Nooteboom (in press) distinguishes *L. minor* as a separate species, but especially on Sri Lanka all possible intermediates can be found. The taxonomic confusion may be partly due to the occurrence of a polyploid complex in this species (Manton & Sledge, 1954).

L. wallii is a very narrow form., with strap-shaped fronds hardly more than 5 mm wide. The fertile fronds are extremely narrow and often only sporadically fertile.

L. minor and *L. thwaitesianus* refer to small forms, often without a distinct stipe and with the lamina base not suddenly contracted into a wing.

L. metallicus is similar to the last form., but has a metallic blue tint when alive. In dry state these plants cannot be distinguished from normal, green, forms.

L. laciniatus is an irregularly furcate, laciniate or pinnatifid form., it might be the result of hybridisation with a pinnate species of *Microsorum*. It is not uncommon on Sri Lanka, but similar forms also occur in India and Burma. Laciniate forms may occur mixed with normal forms, sometimes even on the same rhizome.

Typical form

S p e c i m e n s   E x a m i n e d: MATALE DISTRICT: Jungle, Hoolankande, 1372 m., 20 Jan 1954, *Sledge 1015* (BM); Forested stream- gulley beside and below road 1km W. of top of pass between Rattota and Laggala, N.E. of Matale, Alt. c. 1200m., 15 Sept 1993, (2 sheets) *Fraser-Jenkins* with *Jayasekara & Bandara* Field No. *142* (K); KANDY DISTRICT: Hantane,

Randall (Rawson 3278) (BM); In forest above Le Vallon estate, 9 Feb 1951, *Sledge 1115* (BM., K); Le Vallon Estate, c.1650 m., 9 Feb. 1954, *Walker T 340* (BM); Le Vallon Estate, on rock in jungle, c.1650 m., 9 Feb. 1954, ? *Walker 346, 347* (BM); P 25 voucher for cytological study n=72, Oodawella path up to jungle, same plant as wild fix (BM). BADULLA DISTRICT: Glenanore Estate, Blackwood forest, mile post 15 on Welimada-Haputale Road, forest, near stream., common, 1525 m., c. 6.46 30" N, 80.56 E,., 18 Nov 1976, *Faden & Faden 76/354* (K). RATNAPURA DISTRICT: Ravine below Hunawalkande east of Ratnapura, 457 m., 10 March 1954, *Sledge 1254* (BM) (large form with nearly sessile fronds: 23 × 5.5 cm., narrowed part 6 cm., wingless stipe 1 cm); Pinnawela, in forest, shade, on tree trunk, 30 June 1971, *Balakrishnan 566* (L). NUWARA ELIYA DISTRICT: Forest Slope above Hakgala Botanic Gardens on N.E. Side of Hakgala mountain. E. of Nuwara Eliya, Alt. c. 1700 m., 23 Sept 1993, (2 sheets) *Fraser-Jenkins* with *Jayasekara, Samarasinghe & Abeysiri* Field No. 201 (K); Nuwara Eliya, *Freeman 380* (BM); Nuwara Eliya, Jan 1871, *Hutchinson* (LIV); Hakgala, near pond, on rocks, 2000 m., 19 May 1971, *Jayasuriya 168* (L). WITHOUT LOCALITY: Fl. Zeylanica, *Gardner 14* (BM); Z16 voucher for cytological study n=36, Rocks by stream side near Lonach (BM); *Amaratunga 234* (L); CP *1317* (BM., isotype zeylanicus): clearly the large form, decurrens.

Minor form

S p e c i m e n s   E x a m i n e d: KANDY DISTRICT: Hantane, sec. jungle on shady banks, 1158 m., 8 Dec 1950, *Ballard 1032 B* (K); On rocks in stream through jungle, Hunnasgiriya, 884 m., 16 Jan 1954, *Sledge 967* (BM., US, *"Leptochilus thwaitesianus* Fée)" ("minor" det Nooteboom). BADULLA DISTRICT: Bibile to Monaragala Road, on rock in stream, 22 Feb. 1954, *Walker T 536* (BM). KALUTARA DISTRICT: Botala Kande mt., Pelawatta, on rocks near mountain stream, 300 m., 5 Dec 1975, *Bernardi 15733* (L); (?) *C.P. 1316* (BM., coll. Beddome). RATNAPURA DISTRICT: Hedigala, Sinharaja Forest, on tree trunks in wet forest, 5 Jan 1951, *Ballard 1391* (K, id by Sledge as *Pl. amaurolepida*, which it most certainly is not). Pahala Hewessa, in jungle fringing stream, 20 Jan 1951, *Ballard 1524* (US, *"L. metallicus"*) ("minor" det Nooteboom); Adam's Peak, above Moray Estate, epiphytic on mossy tree trunks, 15 June 1971, *Jayasuriya et al. 205* (L); Sinharaja, Weddegalla, on rock in creek, 21 Sept. 1977, *Meijer & Gunatilleke 136* (L); P 445, Hewissa, voucher for cytological study n=36 (BM); Pahala Hewessa, 90 m., 20 Jan 1951, *Sledge 887* (BM., as metallicus); Pahala Hewessa, Jungle, 90 m., 20 Jan 1951, *Sledge 888* (K, as L. metallicus). NUWARA ELIYA DISTRICT: Nuwara Eliya, *Freeman 381, 383* (BM); Peacock, April 1871, *Hutchinson* (LIV). GALLE DISTRICT: On rocks and trees in deep shade, Haycock Mt., 457 m., 22 Jan 1951, *Sledge 909* Type locality (US, *"Leptochilus metallicus*") ("minor" det Nooteboom); Kanneliya For. Res., lowland ever-green forest, on rocks, mostly in streams. 100–170 m., 7 Dec 1976, *Faden & Faden 76/509* (K, US) ("minor" det Nooteboom); P. 446, Haycock Mt, voucher for cytological study n=36 (BM). WITHOUT LOCALITY: anon (Herb. Smith*), L 908, 289, 320* ("minor" det. Nooteboom*); C.P. 1316* (BM., coll. *Beddome*); *Thwaites C.P. 1316* (BM., type of *thwaitesianus*); "*Acrostichum wallii* Baker", *Wall 10559* (BO) ("minor" det Nooteboom).

Intermediates minor-decurrens:

KEGALLE DISTRICT: Kitulgala, 260 m., 28 Jan 1954, *Schmid 1081* (same as *Schmid 1288*) (BM). NUWARA ELIYA DISTRICT: Nuwara Eliya, 1950 m., 24-27 Feb 1954, *Schmid 1288* (BM) (small plants but with distinctly contracted lamina base and winged stipe). WITHOUT LOCALITY: *Walker s.n.* (K); *Gardner 1157* (K: type of *A. variabile* Hook.); *Hance 95* (BM).

wallii

Here too: MATARA DISTRICT: Morawaka, July 1871, *Hutchinson* (LIV); July 1871, *Wall* (photograph in K); Ad Mooroowa, on rocks, *Wall s.n., s.d.* (BM)

Furcate form

BADULLA DISTRICT: Badulla, *Freeman 382 C* (BM). LOCALITY UNKNOWN: *Wall*?

Laciniate forms

MATALE DISTRICT: Mouragalla Matele, *Rawson 3278, 1858* (BM); East Matale, Aug 1871, *Hutchinson* (LIV); Matale, in forest, 914 m., *Wall s.n., s.d.* (BM). KANDY DISTRICT: Corbet's Gap, in shade, 1219 m., 7 Jan 1951, *Sledge 842* (BM., K). WITHOUT LOCALITY: coll. *Beddome* (K); *Gardner C.P. 1318* (K); *Naylor Beckett 215* (K); *Wall, s.n., s.d.* (K). SITE NOT LOCATED: Dooroonadella (?), irregularly pinnatifid/pinnae, (one specimen of this collection has bipinnate sterile and fertile fronds), *Wall*? (K)

Other forms

KANDY DISTRICT: Hunnasgiriya, on stream bank in shade of jungle, 884 m., "mixed with entire leaved state", 16 Jan 1954, *Sledge 968 B* (K); Knuck-les, Madulkelle area, wet evergreen forest in rivulet, 900 m., 14 June 1973, sterile fronds with a few lobes, fertile fronds simple and with several pinnae on the same rhizome, *Kostermans 25041 A* (K).

5. LEPISORUS

Lepisorus Ching, Bull. Fan Mem. Inst. 4 (1933) 47. *Drynaria* sect. *Lepisorus*
J. Sm. Bot. Mag. Comp. 13 (1846) 7. *Polypodium* subgen. *Pleopeltis* sect.
Lepisorus (C. Chr., Ind. Fil. (1906) L. *Polypodium* subg. *Lepisorus* C. Chr.,
Ind. Fil. Suppl. 3 (1934) 12.
Type species: *Lepisorus nudus* (Hook.) Ching.

Pleopeltis auctt. non Humb. & Bonpl. ex Willd.: Abeyw., Ceyl. J. Sc. Sect. A
    13 (1956) 26.

Rhizome terete, short- to long-creeping, usually unbranched. Anatomy:
vascular strands without sclerified sheaths, scattered sclerenchyma strands
present. Rhizome scales peltate, clathrate or with a thickened midrib. Fronds
stipitate. Lamina simple, coriaceous, sparsely covered with deciduous scales
on the lower surface. Venation usually not distinct, veins and connecting
veins poorly differentiated; veins and included veins forming a mesh of numerous areoles with recurrent and excurrent free veinlets. Sori round or slightly
elongated, in a single row between midrib and margin, covered with peltate
clathrate indusium., at least when young.

KEY TO THE SPECIES

1 Rhizome long-creeping, phyllopodia to 3.5 cm apart, rhizome scales comple-
   tely translucent . . . . **2. L. nudus**
1 Rhizome short, phyllopodia nearly contiguous, rhizome scales with opaque
   dark centre . . . . **1. L. amaurolepidus**

1. Lepisorus amaurolepidus (Sledge) Bir & Trikha J. Bombay Nat. Hist.
Soc. 68 (1971) 68 ("amaurolepida"), Amer. Fern J. 64 (1974) 60; B.K. Nayar
& S. Kaur, Comp. Beddome's Handbook Ferns Brit, India (1974) 84. *Pleopeltis
amaurolepida* Sledge, Bull. Brit. Mus. nat. Hist. (Bot.) 2 (1960) 136, fig. 2.
Type: *Sledge 999*, Ambagamuwa, 575 m (BM). *Polypodium wightianum* auct.
non [Wallich]: Thwaites, Enum. Pl. Zeyl. (1864) 394.

Rhizome to 2 mm thick, short-creeping, phyllopodia contiguous or at most
2 mm apart, in cross-section with scattered strands of sclerenchyma; rhizome
scales appressed to slightly spreading, peltate, triangular, c. 3.3 × 1.5 mm.,
widest at base, dentate, centrally with cell-walls very strongly thickened,
forming an opaque dark pseudocosta, margin clathrate, dentate with protruding
cell-walls; apex acute. Fronds monomorphic, stipe 2.5-4.5 cm long, sometimes
indistinct. Lamina 4-24 cm long, 0.5 - 2 cm wide, widest about or above
middle, base very narrowly attenuate, decurrent on upper part of stipe as two
narrow wings, apex narrowly acute to acuminate, often with scattered elongated

scales on or near costa on lower surface. Sori round, 3-4 mm in diam., at c. 1/2 between costa and margin. Indusium c. 0.5 mm diam., yellowish-brown, clathrate, central cells with strongly thickened cell-walls, sometimes completely opaque, margin dentate. Spores yellowish.

D i s t r.: India and Sri Lanka.

E c o l.: A lithophyte on wet rocks on forest floor, or an epiphyte in humid ambience above 800 m.

S p e c i m e n s   E x a m i n e d: KANDY DISTRICT: P 159 a, Ambagamwa, voucher for cytological study n=70 (BM); Ambagamuwa, epiphytic on tea bushes, 579 m., *Sledge 999* (L); Hewahette, 1890, Mrs. *Jefferies s.n.* (L); Hunnasgiriya (near Madugoda), c. 900 m., on rock. 16 Jan. 1954, *Walker T 104* (BM); Hunnasgiriya (near Madugoda), c. 900 m., 16 Jan. 1954, *Walker T 107* (BM); Corbet's Gap, 1067 m., 22 Jan 1954, *Sledge 1026* (K); Jungle above Le Vallon Estate, 1524 m., 9 Feb 1954, *Sledge 1125* (K); Stream gully in forest above and S.E. of "Hunas Falls Hotel" W.side of Hunnasgiriya mountain, S.S.E. of Elkaduwa, N.E. of Kandy, Alt. c.1200-1300 m., 25 Aug 1993, (2 sheets) *Fraser-Jenkins* with *Jayasekara, Samarasinghe & Abeysiri* Field No. *54* (K). BADULLA DISTRICT: Attampettia Estate, near Mirahalgama, on stumps of tea bushes, 1219 m., *Piggott 2673* (K); Tonacombe Estate, virgin(?) jungle on rock, 23 Feb 1954, *Walker T 566, 567* (BM). RATNAPURA DISTRICT: Adam's Peak, 1524 m., 14 Dec 1950, *Sledge 604 a* (K) "tetraploid"; Belihul-Oya, 650 m., 20 Feb 1954, *Schmid 1220* (BM); Gongala Hill, on rock in jungle stream., 1000 m., 11 March 1954, *Sledge 1270* (K); Suriyakanda, Aberfoyle Estate, telephone transmitting satation, ridgetop, rooted on semi-exposed rock, 1200 m., 16 March 1985, *Jayasuriya & Balasubramaniam 3261* (L); Adam's peak, 1200 m., on rocks and between roots of tea-bushes, *Van Beusekom 1537 A* (L). NUWARA ELIYA DISTRICT: Nuwara Eliya, *Freeman 323* (BM); Castlereagh Est., in tea bushes, 3 July 1927, *Alston 1854* (K); Blackwood Forest, Welimada-Haputale Rd., mile post 15, on rocks along stream., common, 1525 m., 18 Nov 1976, *Faden & Faden 76/347* (K, p.p. p.p.= nudus); Nuwara Eliya, Parawella Falls, epiphytic in forest, 1524 m., 19 March 1954, *Sledge s.n.* (K); Nuwara Eliya, 1850 m., surrounding hills, common in low hill forest, creeping up to halfway on tree trunk, 16 Oct 1969, *Van Beusekom 1393* p.p. (L). WITHOUT LOCALITY: *Thwaites C.P. 1295 p.p.* (BM); cryptogamia *Wight 129* (K).

2. Lepisorus nudus (Hook.) Ching, Bull. Fan Mém. Inst. Biol. 4 (1933) 83. *Pleopeltis nuda* Hook., Exot. Fl. 1 (823) Pl. 63; Sledge, Bull. Brit. Mus. nat. Hist. (Bot.) 2 (1960) 135. *Polypodium nudum* Kunze, Linnaea 23 (1850) 281, nom. illeg. non G. Forst. (1786); Takeda, Not., R. Bot. Gard. Edin. 7: 277(1915). Type: Nepal, *Wallich s.n.* (K).

Rhizome to 2 mm thick, long-creeping, phyllopodia to 3.5 cm apart, in cross-section with scattered strands of sclerenchyma; rhizome scales appressed to slightly spreading, peltate, triangular, 3-5 × 1.4-2.6 mm., widest at base, centrally clathrate with thick cell walls leaving a clear lumen in cells, margin hyaline, entire or irregularly dentate at base and some way up along acumen, rarely entirely to apex, dentate with protruding clathrate walls in upper part of acumen, apex narrowly obtuse or rounded. Fronds monomorphic, stipe 2-7 cm long. Lamina 10-34 cm long or longer, 0.7 - 2.7 cm wide, widest below or about middle, base very narrowly attenuate, decurrent on upper part of stipe as two narrow wings, apex narrowly acute to acuminate, glabrous or nearly glabrous on both sides. Sori round, 2.5-3 mm in diam., at -½ between costa and margin. Indusium light brown, clathrate, translucent, entire, 0.5-1.0 mm diam.. Spores whitish.

D i s t r.: N. and S. India, Sri Lanka, China and Malay Peninsula.

E c o l.: Montane, on mossy tree trunks and wet boulders; also as an epiphyte on tea bushes or on shola trees.

S p e c i m e n s   E x a m i n e d: MATALE DISTRICT: Hoolankande, 1372 m., on mossy rock in shade of jungle, 20 Jan1954, *Sledge 1018* (K). KANDY DISTRICT: Gardens Peradeniya (?) *Macrae 865* (BM). Hantane, on boulder fringing sec. Jungle, 1158 m., 8 Dec 1950, *Ballard 1035* (K); Corbet's Gap, on trunk of jungle tree, 1219 m., 9 Dec 1950, *Ballard 1075* (K); Corbet's Gap, 750 m., along stream by waterfall in forest, creeping on large boulder, *Davidse 8352* (L); Corbet's Gap, on rock. 22 Jan 1954, *Walker T 196* (BM); Stream gully in forest above and S.E. of "Hunas Falls Hotel" W.side of Hunnasgiriya mountain, S.S.E. of Elkaduwa, N.E. of Kandy, Alt. c.1200-1300 m., 25 Aug 1993, (2 sheets) *Fraser-Jenkins* with *Jayasekara, Samarasinghe & Abeysiri* Field No. *53* (K). BADULLA DISTRICT: Nuwara Eliya & Hakgala, common in forests, on rocks & trees, Dec 1889, *Wall s.n.* (L); between Haputale & Bandarawela, Sept. 1890, *anon.* (L); jungle between Haputale and Ohiya, 24 May 1906, *anon.* (L); Haputale, 1300 m., evergreen forest on rock, *Kostermans 23249* (L); Namunukula Mt, 2000 m., 13 March 1971, *Jayasuriya 136* (L); Blackwood Forest, Welimada-Haputale Rd., mile post 15, on rocks along stream, common, 1525 m., 18 Nov 1976, *Faden & Faden 76/347* (K, p.p. p.p.= amaurolepida); R.B. Haputale, 1300 m., evergreen forest on rock, *Kostermans 23249* (L). RATNAPURA DISTRICT: Adam's Peak 1524 m., 14 Dec 1950, *Sledge 604* (K); Pinnawala, S. escarpment, rock along road side, damp shaded portion, 1000 m., 19 Mar 1968, *Comanor 1093* (K); Adam's Peak Wilderness, summit on S side, 16 Nov 1973, *Sohmer & Waas 8740* (L). NUWARA ELIYA DISTRICT: Peacock hill, *Rawson 3227* (leg. *Randall 1870*) (BM); forest slope above Hakgala Botanic Gardens on N.E. Side of Hakgala mountain. E. of Nuwara Eliya, Alt. c. 1700 m., 23 Sept 1993, (2 sheets) *Fraser-Jenkins* with *Jayasekara, Samarasinghe & Abeysiri*

Field No. *200* (K); Temmetiya-kele, trail from Top pass to Kabaragala Estate, forest, occasional on rocks, 1720 m., 6 Nov 1984, *Jayasuriya et al. 3029* (L); Nuwara Eliya, Jan? 1871, *Hutchinson* (LIV); Nuwara Eliya & Hakgala, common in forests, on rocks & trees, Dec 1889, *Wall s.n.* (L); Nuwara Eliya, *Freeman 320* (BM); Nuwara Eliya, *Freeman 321* (BM) irregularly furcate forms; Nuwara Eliya, *Freeman 322* (BM) possibly not nude, a broad form, 3 cm wide, with somewhat elongated sori, but rhizome too fragmentary to assign to a species (L); Hakgala in jungle on bank of stream., 28 Feb. 1906, anon. (L); Hakgala Bot. Garden, epiphtic on wet jungle trees with abundant mosses, 1829 m., 16 Dec 1950, *Ballard 1106* (K); ("larger and thinner than 1106"); Hakgala, epiphytic in montane forest, 1800 m., *Bernardi 15851* (L); Nuwara Eliya. On rock, exposed. 20 Feb. 1954, *Walker T 500* (BM); Nuwara Eliya, on rock, 20 Feb. 1954, *Walker T 502* (BM); Hakgala, in shade of jungle 1676 m., 26 Feb 1954, *Sledge 1220* (K); Nuwara Eliya, surrounding hills, common in low hill forest, creeping up to halfway on tree trunk, 1850 m., 16 Oct. 1969, *Van Beusekom 1393 p.p.* (L); Hakgala 1824 m., jungle epiphyte, *Sledge 637* Diploid (L); Hakgala, montane forest, 1835 m., 30 June 1972, *Hepper 4584* (K, origin of cult. plant); Nuwara Eliya, Horton Plains, epiphytic on tree trunks in dense mossy forest along rather recently cut path, 3 Dec 1970, *Fosberg 53281* (L); Pidurutalagala For. Res., epiphytic, 31 Oct 1973, *Sohmer et al. 839, 840 8424* (L); Diyagama Tea Estate to Horton Plains, *Sohmer & Sumithraarachchi 10,016* (NY); Hakgala Mt., behind Hakgala Bot Garden, forest, common, low epiphyte, 1830-2100 m., 1 Jan 1977, *Faden & Faden 77/7* (K); near Hakgala, km 84 on route A5, on granite wall, c. 700 m., 24 Dec 1980, *Piggott 2676* (K). LOCALITY UNKNOWN: *Gardner 1139* (K); *anon. 242* (Amaratunga collection) (L); no collector, *C.P. 1295* p.p..

6. PLEOPELTIS

Humb. et Bonpl. ex Willd., Sp. Pl. 5: 211 (1810)

Polypodium L. "Pleopeltis group", Hennipman et al. in Kubitzki Fam. Gen. Vasc. Pl. 1:225 (1990)

Rhizome creeping, with closely or widely spaced fronds and clathrate scales. Fronds simple, entire, membraneous to coriaceous, with peltate scales and with freely and irregularly anastomosing veins with included veinlets. Sori round, covered when young with peltate paraphyses.

A warm-temperate and tropical genus of about 40 species.

Pleopeltis lanceolata Kaulf., Enum. Fil. (1824) 245. . Type from South America
Polypodium lanceolatum L, Sp. Pl. 2 (1753) 1082. *Pleopeltis lanceolata*
    Polypodium macrocarpum Willd., Sp. Pl. 5 (1810) 147. *Pleopeltis*

macrocarpa Kaulf., Enum. Fil. (1824) 245; Sledge, Bull. Brit. Mus. nat. Hist. (Bot.) 2 (1960) 139. Type: Herb. Willdenow N 19629, Réunion.

Polypodium lepidotum Willd. ex Schltdl., Adumbr. Pl. (1825) 17 (nom. illeg?); Hook., Sp. Fil 5 (1863) 56. *Pleopeltis lepidota* Bedd., Ferns S. Ind. (1864) 60, pl. 181 (nom. illeg.?). Type: Herb. Willdenow N 19612, South Africa.

Rhizome to 1-2 mm thick, long-creeping, phyllopodia 1.5-2 cm apart, in cross-section without scattered strands of sclerenchyma; rhizome scales spreading, peltate, triangular-ovate, 2.3-3.3 × 1.4 mm., widest distinctly above point of attachment, margin irregularly dentate, central cells with slightly more strongly thickened walls than marginal cells, apex obtuse to rounded. Fronds monomorphic, stipe 1-2 cm long. Lamina 9-10 × 0.6-1 cm., widest about or above middle, base narrowly attenuate, decurrent on upper part of stipe as two narrow wings, apex narrowly obtuse to acute, lamina on both sides with elongated scales scattered over surface. Costa distinct, veins immersed, forming a single row of main areoles with an included network of mostly excurrent veinlets, outer areoles much smaller, in 2-3 rows between main areoles and margin, with very few included recurrent veinlets. Sori 2–3 mm diam., c. half way between costa and margin. Indusium c. 1 mm diam., light-brown, lightly clathrate, margin irregularly dentate. Spores yellowish.

D i s t r.: S. India, Sri Lanka, trop. America, trop. and southern Africa including St. Helena, Madagascar and the Mascarenes.

E c o l.: It occurs as an epiphyte on moss-covered trees at higher elevations.

S p e c i m e n s   E x a m i n e d: NUWARA ELIYA DISTRICT: Ambawelle, May 1871, *Hutchinson* (LIV), Sita Eliya, Hakgala "*Pleopeltis lanceolata*", March 1885, *anon.* (L); Ambawala Estate, on rocks and stumps among the coffee trees "*Pleopeltis lanceolata*" anon. (L); WITHOUT LOCATION: Rock? 1882 (L). SITE NOT LOCATED: Ambalenka, *Thwaites C.P. 3988*, "*Pleopeltis macrocapa*".

7. SELLIGUEA

Selliguea Bory, Dict. Class. d'Hist. Nat. 17 (1825) Pl. 41; Type species: *Selliguea féei*.

Crypsinus C. Presl, Epim. Bot. (1851) 123; Copel., Gen. Fil. (1947) 205; Abeywickrama, Ceyl. J. Sc. Sect. A 13 (1956) 27 . Type species: *Crypsinus nummularius* (nom. superfl. for *Polypodium pyrolaefolium* Goldm. =*Selliguea pyrolifolia*).

Epiphytic, epilithic or terrestrial, small to medium-sized ferns. Rhizome densely set with ± persistent, very variable scales. Fronds monomorphic to dimorphic, ± distinctly differentiated into stipe and lamina. Lamina simple or pinnate to pinnatifid. Texture coriaceous; hydathodes restricted to adaxial side. Venation: at least one row of closed areoles present, usually with included veinlets. Sori 1-5 mm diam., round or elongated, superficial or in deeply sunken pits forming pustules on adaxial surface, in one to many rows between midrib and margin, occasionally forming coenosori. Sporangia stalked, glabrous, intermixed with highly variable numbers of uniseriate paraphyses. Spores brown, monolete, smooth or with spines or globules.

D i s t r. India to Japan, throughout Malesia, extending to Fiji and Australia (Qld.).

Selliguea montana (Sledge) Hovenkamp, comb. nov.

Crypsinus montanus Sledge, Bull. Brit. Mus. nat. Hist. (Bot.) 2 (1960) 145.
   Type: *Sledge 624*, Adam's Peak (BM).
Crypsinus hastatus auct. non Thunb.: Abeywickrama, Ceyl. J. Sc. Sect. A 13 (1956) 27.

Rhizome 4-4.5 mm thick, short-creeping, internodia 0.5-1.5 cm long; in cross-section with many scattered sclerenchyma strands; rhizome scales pseudopeltate, triangular, appressed to spreading, 3-4 × 1–2 mm., acumen suddenly narrowed above the wide base, brown with a lighter margin, irregularly dentate, apex acute. Fronds monomorphic; stipe 6-18 cm long. Lamina 7-25 cm long, pinnatifid, segments in 2-5 pairs, basal ones slightly ascending, uppermost ones more strongly so, width of connecting strip 0.2-0.5 cm; base cuneate, largest segment is 1st-2nd from base, 2.5-12(-18) × 0.8-1.7 (–3.5) cm., gradually narrowing in the upper -½ to a narrowly acute segment apex; apical segment usually larger than largest segment. Main veins on upper surface raised, connecting veins distinct, sometimes veinlets also distinct; free and anastomosing, free veinlets excurrent and recurrent. Hydathodes frequent, calcareous scales not persistent; margin cartilaginous, slightly thickened; often somewhat sinuose, notches regularly present. Sori round, singly between adjacent veins, in one row at -½ between costa and margin, 1–3 mm across, superficial.

N o t es. One of the commonest ferns of high elevations, up to 2438 m.

S p e c i m e n s   E x a m i n e d. KEGALLE DISTRICT: Kelani River, Rest House, epiphytic, 11 Nov 1975, *Sohmer & Waas 10561* (L). KANDY DISTRICT: Wattekelle Hill, Aug 1871, *Hutchinson* (LIV); Knuckles, c.1650 m., 30 Jan 1954, Walker T 253 (BM); Le Vallon Tea Estate, jungle, 9 Feb 1954, *Sledge 1108* (BM., K); mountain ridge between Knuckles and Ritigala no. 2 peak, E. of Bambrella, 9 Nov 1974, *Davidse 8276* (L); Maskeliya, above Moray Estate. 1350-1700 m., 14 Sept 1977, *Nooteboom & Huber 3143* (L); Ambagamuwa?, var. localities, Nov. 1854, *CP 3291*(SING).

RATNAPURA DISTRICT: Central Prov., on trees, 1219 m., *Naylor Beckett 2* (BM); Adam's Peak 1981 m̊., on trees, 14 Feb 1908, *Matthew s.n* (K); Adam's Peak, rocks by stream Ramboda Pass, voucher for cytological study n=36, P 188 (BM); one specimen with 7 pairs of pinne, some of which with a basiscopic lobe; Adam's Peak, voucher for cytological study n=36, *Z 15* (BM). NUWARA ELIYA DISTRICT: Nuwara Eliya, *Freeman 326, 327, 328, 329* (BM); Nuwara Eliya, 1900 m., on trees, April 1899, *Gamble 27589* (K); Nuwara Eliya, Mt. Pedro, epiphytic on jungle trees and moss etc, 26 Dec 1950, c. 2250 m., *Ballard 1240* (K); Kuda Oya on Ramboda Road, 1676-1740 m., on tree trunk in gully, 28 Dec 1950, *Ballard 1301* (K); Nuwara Eliya, surrounding hills, low hill forest, epiphytic on tree trunk a few m. high, 1900 m., 17 Sept 1969, *Van Beusekom 1408* (L); Horton Plains, epiphytic on tree trunks in dense mossy forest along rather recently cut path, 3 Dec 1970, *Fosberg 53284* (K); Horton Plains, forest back of Farr Inn, epiphytic on tree trunks in dense mossy forest, 3 Dec 1970, *Fosberg 53284* (L); Pidurutalagala For. Res., epiphytic. 31 Oct 1973, *Sohmer et al. 8410* (L); Hakgala Mt., behind Hakgala Bot. Garden, on rocks and stunted trees near summit, locally common, 1830-2100 m., 1 Jan 1977, *Faden & Faden 77/13* (K). LOCALITY UNKNOWN: Aug '89'? Sri Lanka, *Bishop House s.n.* (SING); *Thwaites* (BM); Sri Lanka *Ferguson s.n., s.d.* (GH); *Gardner 1296, 1297* (K, both on one sheet); Herb. *Wight 3160* (L); 12 March 1819, *Moon 488* (BM); 1871, *Randall (Rawson 3225)* (BM); *Walker 44* (K); *Walker 1297* (K); *Wall s.n., s.d.* (K). SITE NOT LOCATED: top of Nawinakuli? [Numunukula, Badulla Distr.], 12 March 1907, *Silva?* (L).

8. BELVISIA

Belvisia Mirb., Hist. Nat. Gen. 4 (1803) 65; Hovenkamp & Franken, Blumea 37 (1993) 517. Type species: *Belvisia spicata* (L.f.) Copel.

Hymenolepis Kaulf., Enum. (1824) 146, pl.i, fig. 9, nom. illeg. non Cassini (1817); Beddome, Ferns S. Ind. (1863) 15, pl. 26. Type species: *Hymenolepis ophioglossoides* Kaulf. (=*Belvisia spicata*).
Hyalolepis Kunze, Linnaea 23 (1850) 258, nom. illeg. non D.C. (1837). Type species: *Hyalolepis ophioglossoides* (Kaulf.) Kunze (=*Belvisia spicata*).
Macroplethus C. Presl, Epim. Bot. (1851) 141; Tagawa, Acta Phytotax. Geobot. 11 (1942) 232. Type species: *Macroplethus platyrhynchus* (Kunze) C. Presl (=*Belvisia platyrhynchos*).

Epiphytic, rhizome short- to long-creeping, approx. terete, ventral and lateral roots dense, contain scattered sclerenchyma strands. *Rhizome scales* basifix, pseudo-peltate, ovate to linear-lanceolate, fully clathrate or with a membranaceous margin, reddish to brown, entire to dentate. Fronds simple, entire. Lamina linear-lanceolate to linear, base cuneate, apex gradually tapering to narrow fertile part, olivaceous to brown when dry, dull, pergamentaceous, sparcely scattered scales are soon shed. Venation: midrib distinct throughout sterile and fertile lamina, veins and connecting veins distinct or immersed, poorly differentiated, included veins forming areoles with many scattered free veinlets; free veinlets in costal and marginal areoles all directed towards costa, in other areoles predominantly directed towards costa; included veins and free veinlets immersed. Fertile part linear-lanceolate to linear; one elongated, sometimes interrupted. Sorus at either side of midrib, sorus usually fully covers lamina between midrib and margin, sometimes leaves a narrow zone free along midrib. Sporangia stalked, with 12-16 indurated annulus cells; sporangia interspersed with peltate or basally attached paraphyses. Spores monolete, rugulate.

KEY TO THE SPECIES

1 Rhizome scales completely clathrate, dentate .... **1. B. mucronata**
1 Rhizome scales with an entire, hyaline margin .... **2. B. spicata**

1. Belvisia mucronata (Fée) Copel. var. mucronata

Belvisia mucronata Copel., Gen. Fil. (1947) 192; Sledge, Bull. Brit. Mus. nat. Hist. (Bot.) 2 (1960) 140. *Hymenolepis mucronata* Fée, Mém. Fam. Foug. 5. Gen. Fil. (1852) 81, pl. 6, fig. 1. *Macroplethus mucronatus* Tagawa, Acta Phytotax. Geobot. 11 (1942) 234. *Belvisia mucronata* var. *mucronata* Hovenkamp & Franken, Blumea 37 (1993) 521, fig. 1 c, d; 3c.
Type: *Cuming 92*, Philippines, Luzon (P; iso BM., G, UC, US, W).

Rhizome short-creeping, 2–5 mm thick, internodes not elongated. Rhizome scales ovate-oblong, ovate- to linear-lanceolate, 2.5–8.5 by 0.5–2 mm., index 3–8, margin minutely to distinctly dentate, acumen contracted, apex acute, reddish-brown to black, cell-walls generally thickened. Stipes up to 6 cm long, 1–2 mm thick. Lamina linear to linear-lanceolate, 10–50 by 1–5 cm., index 5–22, narrowed towards base and apex, spikes linear, 3–25 by 0.3–0.7 cm., index 3–85. Sori close to midrib, completely covering lower surface when ripe. Paraphyses with laterally affixed or peltate, round blades, 0.1–0.65 mm diam., brownish to black, margin entire to toothed, thick –walled cells. Spores rugulate, 40–90 by 60 μ.

D i s t r. Throughout Malesia; outside Malesia: Sri Lanka, Indochina, Taiwan, Australia. Pacific.

E c o l. Epiphytic on all kinds of trees or on rocks. In primary and secondary forest. 0–1500m.

S p e c i m e n s E x a m i n e d: LOCATION UNKNOWN: *Gardner C.P. 1303* (K). *Levinge* (G).

2. Belvisia spicata (L.f) Mirb. ex Copel.

Belvisia spicata Mirb. ex Copel, Gen. Fil. (1947) 192; Hovenkamp & Franken,
   Blumea 37 (1993), 524; fig. 1a, b; 2e; 3f. *Acrostichum spicatum* L.f.
   Suppl. Pl. (1781) 444; Thwaites, Enum. Pl. Zeyl. (1864) 381; Baker,
   Syn. Fil. 2nd Ed. (1874) 424. *Schizaea spicata* Smith, Mém. Acad. Sci.
   Turin 5 (1793) 43. *Onoclea spicata* Sw., J. Bot. (Schrad.) 1800 2 (1801)
   299. *Lomaria spicata* Willd., Sp. Pl. 5 (1810) 289. *Hymenolepis spicata*
   C. Presl, Epim. Bot. (1851) 159; Hooker, Fil. Exoc. (1859) pl. 78;
   Bedd. Ferns S. Ind. (1863) 15, pl. 46; Hooker, Sp. Fil. 5 (1864) 280;
   Bedd. Suppl. Ferns S. Ind. (1876) 27. *Taenitis spicata* Mett. in Miquel,
   Ann. Mus. Bot. Lugd. Bat. 4 (1868) 173. *Gymnopteris spicata* C. Presl,
   Tent. Pterid. (1836) 244; Bedd., Handb. Ferns Brit. Ind. (1883) 431;
   fig. 261. *Macroplethus spicata* Tagawa, Acta Phytotax. Geobot. 11
   (1942) 235. Type: *Commerson s.n.*, Mauritius (P).
Hymenolepis revoluta Blume Enum. Pl. Jav. (1828) 201. *Hyalolepis revoluta*
   Kunze, Linnaea 23 (1850) 258. *Taenitis revoluta* Mett., Fil. Hort. Lips.
   (1856) 28. *Taenitis spicata* (L.f.) Mett. f. *angustata* Mett. in Miquel,
   Ann. Mus. Bot. Lugd. Bat. 4 (1868) 173. *Macroplethus revoluta* Tagawa,
   Acta Phytotax. Geobot. 11 (1942) 234. *Belvisia revoluta* Copel., Gen.
   Fil. (1947) 192; Sledge, Bull. Brit. Mus. nat. Hist. (Bot.) 2 (1960) 139.
   Type: *Blume s.n.*, Java (BO, iso L).

Rhizome short-creeping, 2–4 mm thick, internodes not elongated. Rhizome
scales ovate, ovate-lanceolate or narrowly triangular, 1.7–4.1 by 1–1.5 mm.,
index 1.7–3, apex acute to rounded, often recurved; margin usually entire,
rarely dentate; central cells with thickened walls, marginal cells in a 0.2–0.3
mm wide membraneous zone with thin walls. Stipes 0.5–4 cm long, 1–2 mm
thick. Lamina linear-lanceolate to linear, 8–30 by 0.3–2 cm., index 5–70;
narrowed towards base and apex; spikes linear, 2–25 by 0.2–0.4 cm., index
5–100. Sori covering lamina when ripe, situated close to midrib. Paraphyses
with irregularly branched and lobed blades, thick-walled cells. Spores rugulate,
40–90 by 25–60 μ, index 1.5–1.7.

D i s t r. Throughout Malesia; outside Malesia: Tropical Africa, Sri Lanka,
Indochina, Australia: Queensland; New Caledonia, Fiji, Tahiti.

E c o l. Epiphytic or epilithic in primary or secondary forest. Common in
mountainous areas. Up to 3000 m.

N o t e. Superficially, this species is very similar to *B. mucronata*. The
main difference is found in the rhizome scales, which usually have a narrow,
light, entire margin.

S p e c i m e n s   E x a m i n e d. KEGALLE DISTRICT: Kelani River,
Kitulgala Rest House, epiphytic, 11 Nov 1975, *Sohmer & Waas 1056* (L).

KANDY DISTRICT: Hakgala, on tree in gully near Hakgala Garden, 1680 m., 28 Dec 1950, *Ballard 1310* (K); Hantane Hill, above Peradeniya, epiphytic on mossy tree trunks, 1 Feb 1973, *Burtt & Townsend 42a* (K). RATNAPURA DISTRICT: Gongala Hill, above Rakwana-Deniyaya road, 1128 m., 11 March 1954, *Sledge 1257* (BM). NUWARA ELIYA DISTRICT: Epiphyte in jungle at Hakgala, 1676 m., 27 Dec 1950, *Sledge 739* (K); Hakgala, upstream near gardens, epiphyte on tree, 20 Feb 1954, *Walker T 507* (BM); Sita-Eliya, just past mile post 53/13 on Nuwara Eliya-Hakgala Road, forest, epiphyte, two plants E c o l.d, 1720 m., *Faden & Faden 77/24* (K); Hakgala Bot Garden, forest, epiphytic, one plant seen, 1830-2100 m., *Faden & Faden 77/8* (L); Nuwara Eliya "epiphytic", not common,1524-1829 m", *Freeman 384, 385* (BM). LOCATION UNKNOWN: *Thwaites, C.P. 1303* (BM); 1872-3, *Thwaites 1010 (C.P. 1302)*, (W); *Gardner 1303; Ferguson 212* (US); on trees, 19 Feb 1819, *Moon 247* (BM); *Robinson 30* (K); *Walker s.n., s.d.* (G); *Wall 60-127* (BO).

INTRODUCED SPECIES AND DOUBTFUL OCCURRENCES

Polypodium (Phlebodium) aureum L.

This is a tropical American species that is widespread in cultivation. The specimen collected on Sri Lanka is almost certainly a garden escape, but it may be settling.

Adam's Peak Jungle, above Moray estate, 2000 m., *Kostermans 24206* (L).

Leptochilus ellipticus (Thunb.) Noot.

Polypodium ellipticum Thunb., Fl. Jap. (1784) 335. *Selliguea elliptica* Bedd., Ferns Brit. India. (1870) Index. *Colysis elliptica* Ching, Bull. Fan Mem. Inst. Biol. 4 (1933) 333. Type: *Thunberg* (not seen).

In Herb. coll. Beddome in BM one sheet represents *L. ellipticus*, but without any other data than "Sri Lanka" on the label. *L. ellipticus* is widespread from the Himalayas to Korea, Japan and the Philippines. Its presence on Sri Lanka would not be highly surprising, but is not accepted on basis of these few data.

PSILOTACEAE

(by B. Wadhwa*)

Kanitz, Novenyrends Attek. 43. 1887.

Small to moderate-sized, often epiphytic plants, without roots. Rhizome sub-terranean, branched, wiry, short. Aerial stem often dichotomously branched, erect or pendulous, glabrous, slender, angular to sulcate, sometimes flattened, the basal part often with rudimentary scale-like projections, but in upper part scale-like structures developed. Sporangia fused in twos or threes to form synangia, borne on much reduced lateral branches or on ridges of the branches bearing forked scaly projections at base.

A family of a single genus with 2 species, widely distributed in the tropics and subtropics.

PSILOTUM

Sw., J. Bot. (Schrad.) 1800(2): 8, 109. 1801; Sw., Syn. Fil. 117. 1806; Abeywickrama, Ceyl. J. Sci. 13(1): 9. 1956; Thoth. et al., Bull. Bot. Surv. India 12: 280. 1970 (1972); Tagawa & K. Iwats., Fl. Thailand 3(1): 5. 1979; Sledge, Bot. Journ. Linn. Soc. 84: 8. 1982; Dixit, Cens. Ind. Pterid. 20. 1984; K.U. Kramer & P.S. Green, Fam. Gen. Vasc. Pl. 1: 25. 1990.
Lectotype species: *P. nudum* (L.) P. Beauv.

Small to moderate-sized plants; terrestrial or epiphytic. Roots absent. Rhizomes wiry, much branched with gemmae. Aerial stem slender, well-developed, 10-65 cm long, copiously dichotomously branched; branches trique-trous, bearing scaly projections or awl-shaped to squamiform scale-like leaves, without veins. Synangia consisting of 3 sporangia, borne on ridges of the branches, bearing forked scaly projection at base. Spores isosporous, bilateral, oblong or ellipsoid, coarsely rugose.

A genus of 2 species, widely distributed in the tropics and subtropics except in dry areas, extending to S.E. United States, S.W. Europe, Central Japan, S. Korea and Pacific Islands. The delimitation of the species varies with different authors. Represented by 1 species in Sri Lanka.

---

* Royal Botanic Gardens, Kew, U.K.

Psilotum nudum (L.) P Beauv., Prodr. Aetheog. 112. 1805; Trimen, Syst. Cat. Fl. Pl. & Ferns. 118. n. 1066. 1885; Abeywickrama , Ceyl. J. Sci. 13(1): 9. 1956; Thoth. et al., Bull. Bot. Surv. India 12: 281. 1970 (1972); Tagawa & K. Iwats., Southeast As. St. 5: 26. 1976; Fl. Thailand 3(1): 5. 1979; Sledge, Bot. J. Linn. Soc. 84: 8. 1982; Dixit, Cens. Ind. Pterid. 20. 1984; Manicham et Irudayaraj., Pterid. Fl. W. Ghats, S. India 45. 1992.

Lycopodium nudum L., Sp. Pl. 1100. 1753. Type from India.

Psilotum triquetrum Sw., J. Bot. (Schrad.) 1800(2): 109. 1801; Syn. Fil. 117. 1806; Thwaites, Enum. Pl. Zeyl. 378. 1864; Hosseus, Beih. Bot. Centr. 28(2): 367. 1911; Willis, Rev. Cat. Fl. Pl. & Ferns Ceyl. 114. n. 1030. 1911; Tardieu et C. Chr., Fl. Gen. I-C. 7(2): 596. f. 64, 4-5. 1951; Larsen, Dansk Bot. Ark. 23: 59. 1963. Type: same as for *P. nudum* (L.) P. Beauv.

Rhizome creeping, dichotomously branching at irregular intervals, up to 20 mm in diam., densely beset with brown to dark-brown slender, unicellular rhizoids. Aerial stems fasciculate, erect o-patent or pendulous, 10-65 cm in height, green to dark-green, glabrous, iso- or aniso-dichotomously branching several times in upper portion, with several distinct ridges and grooves, ultimate unbranched stem 0.5-1.5 mm in diam., the branches triangular in cross section. True leaves absent; scale leaves irregularly arranged throughout the branches, pale green, narrowly triangular to lanceolate with subulate apex. Synangia borne adaxially to the projections or at the axils of scale leaves, glabrous, c. 2 mm in diam., green at first, yellow when mature. Spores numerous, reniform, oblong or ellipsoid, coarsely rugose, lemon-yellow.

Chromosomes n = 104 (Abraham & Ninan in Caryologia 18: 537-539. 1965; Ghatak in Nucleus 20: 105-108. 1977); 2n = 156 + 3 (Ninan in Cellule 57: 307-318. 1956; and 2n = 208 (Abraham & Ninan in Caryologia 18: 537-539. 1965).

D i s t r. Throughout India in hilly regions, Borneo, Malaysia, Burma (Myanmar) and Sri Lanka.

E c o l. Usually epiphytic on mossy tree-trunks or in shady places in dense forest or as lithophytes forming small clumps along stream banks, between 1,200 m and 2,000m altitudes.

S p e c i m e n s E x a m i n e d. POLONNARUWA DISTRICT: Adam's Peak and Habarana, March 1846, *Gardner s.n.* (PDA). KANDY DISTRICT: Kobonilla Hill, Rangala, Sept. 1888, *s.coll. s.n.* (PDA). BADULLA DISTRICT: Ohiya, 1,650m., 2 Sept 1978, *Huber 886* (PDA); Top of Namunukula, 12 March 1907, *J.M. Silva s.n.* (PDA); Uma Oya, from kumbak tree, 10 Dec 1927, *Silva 288* (PDA). RATNAPURA DISTRICT: Sinharaja, North Ensalwatta along Mada Ella stream., 8 Nov 1975, *Sohmer & Waas* (K, PDA).

NUWARA ELIYA DISTRICT: Hakgala Mt., behind Hakgala Botanic Gardens, 1,830-2,000 m., 1 Jan 1977, *Faden & Faden 77/9*(PDA); Kabaregala Estate, Cardamon Plantation, Field No. 4, 7° 04'N 80° 45'E, 1,350m., 30 Oct 1984, *Jayasuriya & Gunatilleke 2990* (PDA); Hakgala Forest area, on way to Hakgala Peak, 1 Nov 1973, *Sohmer, Jayasuriya & Eleizer 8509* (PDA); Hakgala, 23 Aug 1926, *A.M.S. 138* (PDA); Forest of Barmuella Oya, above Parawella Falls, Kandapola near Nuwara Eliya, 1,445m., 9 March 1954, *Sledge 1333* (K); Hakgala, 27 Dec 1930, *Senaratne s.n.* (PDA). COLOMBO DISTRICT: Mutwal, on large kumbuk tree, *s.coll. s.n.* (PDA); near Ja-ela on coconut tree, *s.coll. s.n. 239* (PDA). Locality unknown: April 1954, *CP 1420, 1272* (PDA); *Walker s.n.* (PDA); *s.coll. CP 1420* (PDA*); Col._W. Robinson s.n.* (K).

PTERIDACEAE

(by C.R. Fraser-Jenkins*, T.G.Walker**, B.Verdcourt***)

Pteridaceae ["Pterioideae"] Reichenb. *fil.*, Handb. Nat. Pflanzensystem: 138. 1937, nom. cons.
Type *Pteris* L.

Acrostichaceae Mett. ex Frank in Leunis, Syn. Pflanzenk., ed. 2, 3: 1458. 1877.
Sinopteridaceae Koidz., Act. Phytotax. Geobot. 3: 50. 1934, nom. inval.
Gymnogrammaceae Herter, Rev. Südamer. Bot. 6: 130. 1940, nom. nud. vel comb. inval.
Negripteridaceae Pich.Serm., Nuov. Giorn. Bot. Ital., ser. 2, 53: 160. 1946.
Trismeriaceae Kunkel, Ber. Schweiz. Bot. Ges: 72: 36. 1962, nom. nud.
Actiniopteridaceae Pich.Serm., Webbia 17: 5 1962.
Cryptogrammaceae Pich.Serm., Webbia 17: 299. 1963.
Hemionitidaceae Pich.Serm., Webbia 21: 487. 1966.
Cheilanthaceae B.K. Nayar, Taxon 19: 233. 1970.

Misapplied names. Adiantaceae (C. Presl) Ching, *Parkeriaceae* Hook., *Taenitidaceae* (C. Presl) Pich.Serm.

Plants terrestrial, with small or large, generally fern-like fronds. Rhizome usually ± small, erect, though becoming massive and decumbent in some species, surrounded and encased by old stipe-bases which are occasionally thickened as storage organs (in some *Pteris* species), with lanceolate scales at its apex and usually at the very base of stipes. Stipe terete or ± sulcate adaxially, with one or two internal vascular strands near its base, bearing scales and/or multicellular hairs at least at stipe-base, often dark; lamina simple or up to four times pinnate, arrangement of the pinna-lobes or pinnules catadromous or anadromous, veins free or anastomosing, surfaces of axes and lamina naked, or bearing scales, fibrils and/or hairs (sometimes very densely so), often bearing ± densely packed white, yellow or orange-coloured ceraceous

*c/o A.M. Paul, Botany Department, The Natural History Museum, Cromwell Road, London SW7 5BD, UK. (*Cheilanthes & Doryopteris*)
**25,Lyndhurst Road,Newcastle upon Tyne,UK. (*Idiopteris, Pteris*)
***Royal Botanic Gardens, Kew, U.K. (*Actiniopteris, Acrostichum., Adiantum., Anogramma, Coniogramme, Hemionitis, Parahemionitis, Pellaea, Pityrogramma & Taenitis*)

glands, mainly below, which result in a ± intensely coloured abaxial farina containing numerous different flavonoid compounds. Sori either marginal, with or without a folded-under, marginal, thin, protective flap (pseudo-indusium) that may be continuous or interrupted into a number of ± separate lobes, or without a pseudo-indusium and ± spreading down veins towards the segment-midrib, sometimes away from the margins, sometimes becoming confluent, sometimes well-protected by scales. Spores trilete, tetrahedral, often ± globose, with an outer layer (a sporoderm) which is often separable by squashing the spores and is usually minutely sculpted in a way characteristic to each species. Chromosome base number x = 29, 30; some aberrant numbers occur occasionally in some genera.

A large family of about 31 genera, though often much overestimated (Pichi Sermolli (1977) listed 51 genera) due to to over-splitting of *Cheilanthes* and other polymorphic genera; containing about 700 species and occurring world-wide, especially in old- and new-world subtropical montane areas, the *Cheilanthoideae* being particularly abundant in semi-arid regions. Ten genera (one introduced) occur in Sri-Lanka.

The family has frequently been split into a number of ± closely related smaller families. In the present treatment these have been submerged into the Pteridaceae. Three families which are more clearly distinct, but are allied to the Pteridaceae within the order Pteridales form an exception. These are the Parkeriaceae Hook. and Taenitidaceae (C. Presl) Pich.Serm., which are both morphologically and cytologically distinct, and the Adiantaceae (C. Presl) Ching., which though cytologically similar are too distinct morphologically to be included in the Pteridaceae. When, as has sometimes been done, they are all included together, the correct name for the family would be the Parkeriaceae, which is the legitimate name (see Pichi Sermolli 1970 and Tryon, Taxon 29(1): 161. 1980) despite *Parkeria* Hook. being a taxonomic (but not nomen-clatural) synonym of *Ceratopteris* Brongn. Indeed the name Pteridaceae is itself a nomen superfluum as it contained the type-genera of quite a large number of unrelated pre-existing families, as circumscribed by its author, and should more correctly be applied to the earliest of those families as a syn-onym. Thus in the present circumscription the name Acrostichaceae would have been the correct name. However recent approaches to taxonomy have led to the name Pteridaceae being conserved with the formerly incorrect type of *Pteris*, in keeping with the need to use this obvious name.

The less significant families which have been combined into the Pteridaceae include the Sinopteridaceae containing the large group of cheilanthoid ferns. All these families are basically similar in their main morphological features even though they contain a wide and sometimes rather discontinuous range of form even within one genus. The two chromosome base-numbers of x = 29 and x = 30 can also occur even within one section of the same genus. It is hardly practical or meaningful to separate all the families. There is also a series of genera, such as *Cryptogramma* and *Onychium.,* intermediate between the extremes of, for example, *Hemionitis* and *Pteris.*

The genera with marginal sori and a pseudo-indusium were formerly treated as quite distinct from those without a pseudo-indusium and with sori spread out over the lamina, but Pichi Sermolli (1977) and, more recently, Ranker (Syst. Bot. 15: 442-453. 1990) have concluded that the acrostichoid condition is only a development of relatively minor taxonomic significance from the genera with marginal sori and that the former Hemionitidaceae are closely related to the erstwhile Sinopteridaceae (as is *Acrostichum* to *Pteris*). An intermediate condition exists in the exindusiate American species of *Cheilanthes* erroneously placed by Tryon (Contrib. Gray Herb. 179: 1-106. 1956 and Taxon 29(1) 160-161. 1980) within *Notholaena* due to mistypification, Tryon's proposals and typification twice being rejected (see Pichi Sermolli, Webbia 37(1):112-117. 1983). Pichi Sermolli (Nouv. Giorn. Bot. Ital., ser. 2, 53: 129-169, 1946 and 1977) also placed two African and Arabian species, which probably belong in the same group, in a separate genus *Negripteris* Pich.Serm. and even created a new family and order for them., not accepted here.

KEY TO THE GENERA

1. HEMIONITIS

L., Sp. Pl.: 1077. 1753 & Gen. Pl. ed. 5: 485. 1754; R.M. Tryon et al. in Kubitzki, Fam.& Gen. Vasc. Pl. 1: 245. 1990.
Type: *H. palmata* L.

Gymnopteris Bernh. in Schrad. J. Bot. 1799 (1): 297. 1799.

Rhizome erect or short-creeping; scales light brown, thin, unicolorous or with dark hardened centres, gradually grading into ± brown hairs. Fronds all similar or dimorphic, the fertile ones erect and the sterile ones forming a rosette. Stipes with 1 vascular strand near the base, terete to grooved. Lamina entire to deeply pedately 7-lobed or 1-2(-±3)-pinnate; veins free or partly to completely anastomosing, without included veinlets, the ends nearly reaching the margin, sometimes clearly enlarged. Sporangia borne along the veins in long usually branched lines; indusium absent. Spores trilete, tetrahedral-globose, crested, echinate or variously warted.

Seven species in tropical America one of which has occurred as an escape in Sri Lanka. The final resting place for the Old World species still mostly known as *H. arifolia* (Burm.) T. Moore appears undecided. The genus *Gymnopteris* is still widely considered separate.

Hemionitis tomentosa (Lam.) Raddi, Fil. Br.: 8, t. 19. 1825. Type: Brazil, *Commerson* or *Dombey* (P-LAM., syntype; BM., photo., Morton negative 2760 & Tryon collection)*.

Asplenium tomentosum Lam., Encycl. Meth. 2: 308. 1786.

Gymnopteris tomentosa (Lam.) Underw. in Bull. Torr. Bot. Cl. 29: 627. 1902; Sledge in J. Linn. Soc. Bot. 84: 11. 1982.

Gymnopteris rufa sensu Manton in J. Linn. Soc. Bot. 56: 89. 1958 non (L.) Bernh. ex Underw.

Rhizome ascending, rather thick, ± 1 cm wide, covered with clear brown soft scales, long drawn-out at apex. Fronds tufted, (30-)45-50 cm tall, pinnate, mostly bipinnate at base. Stipes 5-25 cm long, terete, smooth, chestnut, densely shortly brown spreading pubescent (hairs colourless in life fide Faden); rhachises similar; lamina ovate-lanceolate, 15-30 cm long, 5-10 cm wide. Pinnae in 5-12 pairs, petiolulate, membranous to coriaceous, ovate-cordate to oblong-lanceolate, up to 4 cm long, 1.2 mm wide, some pinnatifid at base; terminal pinna lanceolate to deeply 3-lobed; secondary rhachises 1-2.5 cm long; petiolules of upper pinnae 2-5 mm long; nerves free, repeatedly forked. Sporangia closely arranged obliquely along the nerves over entire lower surface of pinnae. Spores almost black, rounded tetrahedral, shortly spinulose.

D i s t r. Widely distributed in Brazil; also in Peru, Paraguay and Argentina and as an escape in Sri Lanka.

E c o l. As a weed in tea plantations and shady disturbed areas near streamlets; 560-690 m.

N o t e. Matthew's specimen was collected in 1906. *H. rufa* (L.) Sw. and *H. tomentosa* (Lam.) Raddi are closely related and in parts of South America there appears to me difficulty in deciding which species some populations belong to, but in general they are easily distinguished. This does not concern the Sri Lanka material.

2. PARAHEMIONITIS

Panigrahi in Pteridophytic Flora of Orissa. Abstr. & Souv. Booklet, Nat. Symp. Current Trends in Pteridology: 13. 1991 (nom. invalid.) & in Indian Fern J. 9: 244. Jan. 1993 & in Amer. Fern J. 83: 90. July-Sept. 1993. Type: *P. arifolia* (Burm.) Panigrahi

Rhizome erect, shortly creeping or ascending; scales blackish brown with pale margins but pale concolorous when young, linear. Fronds tufted on the rhizome, simple, dimorphic, discolorous; stipes rather thick, blackish, shining,

---

*) Lamarck states 'où elle a été observée par MM Commerson & Dombey'; it is not known who collected the specimen in P-LAM but since there is some information in the description not shown by this specimen I have assumed Lamarck also saw another.

with sparse scales at base but often densely hairy; hairs septate; vascular bundle solitary, U- to V-shaped; lamina rather thick with slightly revolute margins; midrib prominent, black; basal primary veins sometimes visible on lower surface but venation obscure but visible by transmitted light; areolae numerous, elongate, without included veinlets. Lamina of sterile fronds ovate-cordate or rarely hastate, the lower lobes rounded. Fertile fronds more erect, longer and less numerous, 1-few; lamina hastate or ± sagittate, the basal lobes pointed; sporangia borne along the veins, exindusiate.

It has been recognised for some years that the single Asian species did not fit well into *Hemionitis* otherwise known only from the New World nor into any Old World genus, although resembling *Doryopteris*. As Panigrahi suggests a separate genus seems to be the best way to deal with the problem.

Parahemionitis arifolia (Burm.) Panigrahi in Indian Fern J. 9: 244. 1993 & in Amer. Fern J. 83: 90. 1993.
Type: Philippines, Luzon, Petiver, Gazophylacium: t.50, f.12. (1764) ((11) t.50, f.12 (1767)).

Asplenium arifolium Burm., Fl. Ind.: 231. 1768.
Hemionitis cordata Roxb. ex Hook. & Grev., Ic. Fil., t.64. 1828; Thwaites, Enum. Pl. Zeyl.: 382 (1864); Hook., Sp. Fil. 5: 192. 1864. Lectotype: India, near Calcutta, *Wallich* 6132 (cat. no. 44) (K, general herb.) (There is Roxburgh material at the BM).
Hemionitis cordifolia Roxb.; Wallich List. n44. 1828, nomen nudum.

Hemionitis arifolia (Burm.) T. Moore, Ind. Filic.: 114. 1859; Beddome, Handb. Ferns Brit. India: 413, f.245. 1883; Holtt., Fl. Malaya 2 (ed. 2): 596. 1968; Nicolson et al., Interpret. Van Rheede's Hort. Malab.: 26. 1988; W.C. Shieh in Fl. Taiwan ed. 1, 1: 314, t.110. 1975 & in ed. 2, 1: 250, t.105. 1994.

Stipes of sterile fronds (1-)4–10(-20) cm. long, often shaggy hairy, ovate- or elliptic-cordate. Laminae 3–15 cm. long, 1.5–12 cm. wide, sparsely to densely hairy on lower surface, often bearing buds at base. Stipes of fertile fronds (8-)10–30 cm. long. Laminae 3.5–16.5 cm. long, 1.8–13.5 cm. wide, usually hastate, with widely diverging acute basal lobes and mostly wide sinus, sometimes sagittate with narrow sinus or occasionally resembling sterile fronds, more densely hairy beneath and densely areolate. Sori on all the reticulate areolar veinlets beneath, often densely covering entire surface or appearing to be in densely placed lines.

D i s t r. S. and N.E. India, Sri Lanka, Burma, Indo-China to China (Hainan) and Taiwan, Malaysia to the Philippines. Increasingly cultivated as a house plant in Europe etc.

E c o l. Shady places by roads and paths on dry banks, often between stones, rock outcrops with *Euphorbia antiquorum*; 595–1650 m.

N o t e. A photograph of a specimen in the Herbier Delessert, Collection Burman (Geneva) (BM., US-Morton Neg. No. 3863) marked by Morton as the 'holotype' of *Asplenium arifolium* Burm. shows that this sheet has been annotated by Alston as *Acrostichum aureum*.
The type is, however, as indicated above.

S p e c i m e n s E x a m i n e d. VAVUNIYA DISTRICT: About 2 mi. SW. of Nedukeni on Puliyankulam road, mi. post 21/3, *Davidse & Sumithraarachchi 9068* (PDA). KURUNEGALA DISTRICT: Magazine Hill, *Amaratunga 181* (PDA). MATALE DISTRICT: Rattota, Aug.1821, *C.P.1309* (LIV). COLOMBO DISTRICT: Cotta Road, *Chevalier s.n.* (BM). KEGALLE DISTRICT: 0.5 mi. W. of Kadugannawa, between railway tunnels 10 and 11, *Grupe 204* (PDA). KANDY DISTRICT: Roseneath, *Ballard 1019* (K); *1089* (K, PDA); Corbett's Gap, *Ballard 1067* (K); Under trees by stream just above"Gem Inn" above Hewaheta road at Talwatta c. 1½ km E.of Kandy, Alt c.600m., 21Aug 1993, (3 sheets) *Fraser-Jenkins* Field No.*8* (K); Rangalla to Corbett's Gap, 24/12 mile post, *Ballard 1414* (K); Trincomalee - Kandy Road, 82 mile post, *Ballard 1499* (K); Kandy, *Matthew s.n.* (K); Peradeniya, *Petch s.n.* (PDA); Hunnasgirya, *Schmid 982* (BM); Galaha, *Schmid 1029* (BM). BADULLA DISTRICT: Badulla Road, mile 71/6, *Ballard 1331* (K); Badulla, *Freeman 366A, 367B* (BM); Haputale, *A.W.S. (? Schimper) s.n.* (PDA); Hangiliella, between Welimada and Badulla, *Sledge 779* (K). KALUTARA DISTRICT: Kalutara (Cultura), *Moon s.n.* (BM). RATNAPURA DISTRICT: Belihul-Oya, near Rest House, *Ballard 1365* (K); Balangoda-Manawalatenna road, Damachana, *Faden & Faden 76/642* (K). NUWARA ELIYA DISTRICT: Maturata, *A.W.S. s.n.* (PDA); Hakgala, *Sledge 724* (BM). Moneragala District: 7 mi. N. of Siyambalanduwa, E. of Kandakabella Hill, *Fosberg & Sachet 53081* (K, PDA, US); Ruhuna National Park, Block 3, N. of Kataragama, *Wirawan 637, 649* (K, PDA, US). GALLE DISTRICT: Pointe de Galle, *Winterbottom s.n.* (K). WITHOUT LOCALITY: *Bradford in Herb. Hance 110* (BM); *Gardner in C.P. 1309* (BM., K, PDA); *Gardner s.n., Moon s.n.* (BM); *Rae s.n. in Lindley Herb.* (K); *Robinson C 40* (K); *no collector '27'* (K); '202' (PDA); *Wall s.n.* (BM); *Macrae 899* (BM).

3. CONIOGRAMME

Fée, Mém. Fam. Foug. 5 (Gen. Fil.): 167, t. 14, fig. 1-2. 1852; Tryon et al. in Kubitzki, Fam. Gen. Vasc. Pl.: 246. 1990.
Type: *C. javanicum* (Blume) Fée, nom. cons.

Rhizome short- to ± long-creeping with spaced monomorphic fronds; scales brown, ± lanceolate, rather stiff. Stipes with one large vascular bundle near base, grooved on adaxial side; lamina 1–3-pinnate; ultimate pinnules mostly large, mostly opposite or subopposite but with a terminal pinnule, entire to sharply serrulate, mostly glabrous; veins mostly free or if anastomosing near costa then without included veinlets. Sori arranged linearly along the veins, the lines sometimes branched, exindusiate. Spores trilete.

About 20 mostly poorly differentiated Old World species; one in Africa, the rest in India (mainly north), Sri Lanka, Java, China and Japan and Pacific to Hawaii and Samoa. A species described from Central America is actually a *Hemionitis*.

Coniogramme serra Fée, Mém. Fam. Foug. 5 (Gen. Fil.): 167. 1852; Sledge in Bot. J. Linn. Soc. 84: 11. 1982.
Type: Sri Lanka, *Gardner* 21 in 1847 (ubi?, holotype, BM., isotypes)

Grammitis serrulata sensu Thwaites, Enum., Pl. Zeyl.: 386. 1864 non
     Gymnogramme serrulata Blume
Syngramme fraxinea sensu Beddome, Ferns Br. India & Ceylon: 386, fig.
     222. 1883 non *Diplazium fraxineum* D. Don = *C. fraxinea* (D. Don) Diels

Rhizome thick, creeping for ± 25 cm. Fronds up to 80 cm. tall. Stipes to ± 60 cm., glabrous save for scales at extreme base. Lamina triangular in outline, bipinnate; pinnules lanceolate, 6-16 cm. long, 1.6-4 cm. wide, acuminate at apex, cuneate or truncate at base, closely sharply serrate at margin; glabrous or with some pubescence at base of midrib beneath; hairs 2-3-celled; petiolules 0.3-1.5 cm long. Nerves simple or forked at or just before midrib and sometimes forked again further towards margin, slightly dilated and paler above for a very short distance near margin and some running into the teeth. There are sometimes small elliptic abortive pinnules 1.4 cm long, rounded and ± 6-serrate at apex.

D i s t r. Apparently restricted to Sri Lanka.

E c o l. Wet forest floors including dense mossy forest with *Hedyotis* shrubs; (1200-)1720-2270 m. (Thwaites gives 5000-4000 ft./1524-1219 m.).

S p e c i m e n s   E x a m i n e d. BADULLA DISTRICT: Badulla, *Freeman 353C* (BM). RATNAPURA DISTRICT: Galagama, *Gardner in C.P. 3264* (PDA). NUWARA ELIYA DISTRICT: Forest Slope above Hakgala Botanic Gardens on N.E. Side of Hakgala mountain. E. of Nuwara Eliya, Alt. c. 1700 m., 23 Sept 1993, (3 sheets) *Fraser-Jenkins* with *Jayasekara, Samarasinghe & Abeysiri* Field No. *203*(K); Hakgala Peak, *Ballard 1122* (K, PDA); Hakgala Botanic Garden and slopes of Hakgala Mt., *Faden & Faden 76/273* (K, PDA, US); Horton Plains North Entrance, *Fosberg & Mueller-Dombois 50016* (K, US); Horton Plains, forest at back of Farr Inn, *van Beusekom & van Beusekom 1489* (PDA); *Fosberg & Sachet 53283* (K, US); Nuwara Eliya, *Freeman 351A, 352B, 354D* (BM); NE. Adams Peak (label bears name *Gymnogramme fairholmei* Gardner MS named after his companion) *Gardner 1225 pro parte* (K); Horton Plains *Schmid 1381* (BM); Hakgala, *Sledge 643* (BM); Ramboda Pass - Maturata track, *Sledge 1347* (K); "Horton Plains and Maturatte District" fide *Thwaites*. A sheet '189', no collector given has this data. WITHOUT LOCALITY: *Gardner 1225 pro parte* (K); *Thwaites in C.P.3264* (BM., K), *T. Moore Herb.* (K); *Mrs. Walker s.n.* (K).

N o t e. Old sheets have the names *Gymnogramma javanica* Blume var. and *G. caudata* C. Presl

4. PITYROGRAMMA

Link, Handb. Gewächse 3: 19. 1833; R.M. Tryon in Contrib. Gray Herb. Harv. Univ. 189: 52-76. 1961 & et al. in Kubitzki, Fam. Gen. Vasc. Pl. 1: 237, fig. 123 H-K. 1990.
Type species: *P. chrysophylla* (Sw.) Link.

Rhizome erect or creeping, with tufted or spaced fronds and with linear attenuate brown scales. Stipe chestnut, with 2 almost parallel vascular strands, glabrous except for a few scales towards base. Fronds 2–4-pinnatifid, membranous to thinly coriaceous with white, pink, yellow or orange waxy powder on the lower surface but glabrous above; veins free. Sori borne along the length of the veins, without indusia or paraphyses. Spores tetrahedral.

A genus of about 16 species, mostly tropical American but with 4–5 in Africa; one introduced from America is now very widely naturalised in the Old World. This and two other introduced species occur in Sri Lanka.

Apart from the species dealt with below, two sheets, *Schmid 933* (BM) from Peradeniya Botanic Garden and *Freeman 346T* (BM) from Badulla, both from cultivated material, appear to be *Pityrogramma pulchella* (Moore) Domin, a Venezuelan plant. Tryon (Contrib. Gray Herb. Harv. Univ. 189: 73 (1962)) treats this as a dubious taxon. Domin has 6 varieties, some with much divided ultimate segments; *Freeman 346* has rather finely divided segments. Another cultivated plant was sent to Kew by T.G. Weather of Kandy as *P. chrysophylla* (Sw.) Link cv. *heyderi* Domin. There are endless cultivars of *Pityrogramma* and Domin recognised a great many species and varieties nearly all dismissed by Tryon. They are beyond the scope of this Flora and doubtless the two mentioned are not the only ones in cultivation in Sri Lanka.

KEY TO SPECIES

1 Lamina broadest near middle, ie middle pinnae the longest; stipes mostly very short . . . . **3. P. sulphurea**
1 Lamina broadest near the base; stipes usually ± equalling the lamina save in very small plants
  2 Ultimate segments and pinnae acute to acuminate at apex, ± lanceolate, with white or golden yellow waxy powder beneath . . . . **1. P. calomelanos**
  2 Ultimate segments and pinnae ± rounded at apex, the segments obovate-cuneate, with white waxy powder beneath . . . . **2. P. dealbata**

1. Pityrogramma calomelanos (L.) Link, Handb. Gewächse 3: 20. 1833, (as *calomelas* ? sphalm.); R.M. Tryon in Contrib. Gray Herb. Harv. Univ. 189:

60. 1961; Holtt., Fl. Malaya Ferns ed. 2: 593, fig. 348. 1968; Manickam & Irudayaraj, Pterid. Fl. W. Ghats: 94. 1992. Lectotype: S. America, Linnean Herbarium 1245. 19 (lectotype, LINN)*

Acrostichum calomelanos L., Sp. Pl.: 1072. 1753.
Gymnogramma calomelanos (L.) Kaulf, Enum.: 76. 1824.
Ceropteris calomelanos (L.) Underw. in Bull. Torr. Bot. Club 29: 632. 1902.

Rhizome short, erect or procumbent, 8 mm in diam., with tufted fronds and with concolorous light-brown entire linear rhizome-scales up to 4 mm long. Fronds erect to arching, firmly herbaceous to thinly coriaceous. Stipe up to 33 cm long, blackish chestnut or very dark purplish, shining at maturity, with a few scales at the base. Lamina up to 37(-60) × 14(-27) cm., oblong-lanceolate, 2-pinnate to 3-pinnatifid with lowest pinnae not reduced. Pinnae up to 9(-17) × 2.2(-5) cm., lanceolate, acute-acuminate, with pinna-segments oblong-trapeziform to lanceolate, the larger up to 0.3-3 × 0.3-0.8 cm., slightly auriculate, serrate (apparently entire if margin involuted), acute or acuminate, set at an acute angle to costa and with white or yellow powder on dorsal surface. Rhachis blackish chestnut, shining at maturity, strongly grooved above. Sori up to 3 mm long, set along veins, exindusiate, often covering lower surface completely in mature plants.

KEY TO VARIETIES

1 Lamina with white, pale lemon or pinkish waxy powder beneath, rarely glabrous . . . . var. **calomelanos**
1 Lamina with bright yellow to orange-yellow waxy powder beneath . . . . . .
. . . . var. **aureoflava**

1. var. **calomelanos**; Manickam & Irudayaraj, Pterid. Fl. W. Ghats: 94, t. 69. 1992. Lamina with white, pale lemon or pinkish waxy powder beneath.

D i s t r. Originally W. Indies and S. America now pantropical and beyond, China, India, Malesia, Australia, Bougainville, New Hebrides, Tahiti, Hawaii etc., Africa, Comoro Is., Seychelles, Madagascar, Mascarene Is.

E c o l. Roadside banks, path-sides, secondary jungle, also weed in tea plantations; 540–1200 m.

V e r n. Silver fern (E)

S p e c i m e n s   E x a m i n e d. KANDY DISTRICT: Roseneath, *Ballard 1022,1088* (K); Hunnasgiryia, *Schmid* 983 (BM); near Kandy.

---

*Schelpe gives this as holotype but there are several syntypes and I take this to be a lectotypification.

Oodawella tea plantations, *Sledge 538* (K). BADULLA DISTRICT: Haputale, ?*A. de Alwis 12* (K); Badulla, *Freeman 345E* (BM), *344D* (BM); Ella, *Hepper & de Silva 4787* (K); near Mirahalgama, Attampettia Estate, *Piggott 2667, 2670* (K). RATNAPURA DISTRICT: S. of Ratnapura, Deepdean, *Fosberg 56620* (PDA); Sudagalla, *Hancock 32* (K). MONEREGALA DISTRICT: Near Muppane, Kumaradola Estate, *Miss Alston 2* (BM).

2. var. aureoflava (Hook.) Weath. ex Bailey, Man. Cult. Pl.: 64. 1924; R.M. Tryon in Contrib. Gray Herb. Harv. Univ. 189: 61. 1961; Manickam & Irudayaraj, Pterid. Fl. W. Ghats: 95, t. 70. 1992. Lectotype: 'Hot valleys of Ecuador', *Seemann* 948 (suggested as lectotype by Tryon but not seen, neither could I find it at Kew; in Kew copy of Gard. Ferns 948 had been altered to 945)

Pityrogramma chrysophylla auctt. et adnot. in Herb. non (Sw.) Link.
Gymnogramma calomelanos var. *aureoflava* Hook., Gard. Ferns: t. 50 & text. 1862.
Pityrogramma austroamericana Domin, Publ. Fac. Sci. Univ. Charles 88: 7. 1928 & in Kew Bulletin 1929: 221. 1929. Lectotype: Bolivia, Larecaja, Challapampa, *Mandon* 1549 bis (lectotype K, isolectotype GH).

Lamina with bright yellow to orange yellow waxy powder beneath.

D i s t r. Costa Rica, Venezuela, Columbia, Ecuador, Galapagos Is., to Bolivia, NW. Argentina and Brazil; now introduced into India, Java, Tahiti, Zaire, Zambia and S. Africa.

E c o l. Steep roadside banks, under boulders, rock crevices, shady roadsides and tea plantations sometimes occurring together with var. *calomelanos*; 940–1950 m.

V e r n. Golden fern (E)

S p e c i m e n s   E x a m i n e d. KANDY DISTRICT: Uda Pusselawa, *Freeman 342B, 343C* (BM); Galaha, *Schmid 1037* (BM., G); Oodawella tea plantation, *Sledge 520* (BM., K). BADULLA DISTRICT: Mile post 16 on Welimada to Haputale road, Blackwood Forest, Glenanore Estate, *Faden 76/665* (K, PDA); Ella, *Hepper & de Silva 4786* (K, PDA); near Mirahalgama, Attampettia, *Piggott 2668* (K). NUWARA ELIYA DISTRICT: Near Ramboda, *Ballard 1135* (K, US); Ramboda Pass, *Ballard 1155* (K, P, US); Ramboda road, Kuda Oya, *Ballard 1288* (K, US); Ramboda *Ballard s.n.* (K); Maskeliya to Balangoda, between Boguwantalawa and Maratenna, Mt. Kotiyagala, *Bernardi 15975* (G, PDA); Newara Eliya, *Schmid 1344* (BM., G). MONERAGALA DISTRICT: Uva Province, Rendapola, *Ballard 1234* (K). DISTRICT NOT KNOWN: Badulla road, mile 71/6 *Ballard 1323* (K, P, US). WITHOUT LOCALITY: *Amaratunga 240* (PDA).

2. Pityrogramma dealbata (C. Presl) R.M. Tryon in Contrib. Gray Herb. Harv. Univ. 189: 62. 1961; Sledge in Bot. J. Linn. Soc. 84: 11. 1982; Mickel & Beitel in Mem. N.Y. Bot. Gard. 46: 281, figs. 37 E, F. 1988. Lectotype: Panama, *Haenke* sheet 24358a (PR, lectotype; GH, photo.)

Gymnogramma dealbata C. Presl, Rel. Haenk. 1: 18, t. 3, f. 1. 1825.
Pityrogramma chrysophylla (Sw.) Link var. *praestans* Domin in Rozpravy II.
   Tridy Ceske Akad. 38 (4): 33 t. IV (1929). Type: Cult. Hort. Parker, in
   Coll. T. Moore (K, holo., not found).
Pityrogramma peruviana auctt. non Desv.

Rhizome erect with scales light brown, 2–3 mm long, 0.1–0.3 mm wide, entire margin, apically prolonged into a hair-like often divided tip. Fronds tufted, up to 35(–80) cm long, 12 cm wide; stipe dark reddish brown, up to half the length of frond, glabrous save for a few scales at extreme base. Lamina lanceolate, bipinnate-pinnatifid to 3-pinnate or 3-pinnate-pinnatifid in large specimens, up to 35 × 16 cm; ultimate segments obovate- or ovate-cuneate, 0.3-1.3 × 0.2-0.6 cm., blunt, coriaceous, toothed at margin, white-floury beneath; veins free, depressed above. Sori arranged along veins, appearing as black specks scattered among white powder. Spores reddish brown.

D i s t r. Mexico to Panama; introduced into Sri Lanka.

E c o l. Shady roadside banks, rock crevices, rubber plantations; 450-900 m.

S p e c i m e n s   E x a m i n e d. KANDY DISTRICT: Colombo to Kandy road, 62 mile post, Kadugannawa, *Ballard 1093* (K); Hunnasgyria, *Schmid 984* (BM., G); Kadugannawa, *Sledge 586* (BM., K). BADULLA DISTRICT: Badulla road, mile 77/3, *Ballard 1321* (K); Badulla, *Freeman 347G, 348H, 349I, 350J* (BM).

3. Pityrogramma sulphurea (Sw.) Maxon in Contrib. U.S. Nat. Herb. 17: 173. 1913; R.M. Tryon in Contrib. Gray Herb. Harv. Univ. 189: 71. 1961; Sledge in Bot. J. Linn. Soc. 84: 11. 1982; Proctor, Ferns of Jamaica: 205. 1985. Type: Jamaica, *Sw.* (S - PA, holotype, US, fragment).

Acrostichum sulphureum Sw., Prodr. Veg. Ind. Occ.: 129. 1788.
Gymnogramma sulphurea (Sw.) Desr. in Ges. Naturf. Freunde Berl. Mag. 5:
   305. 1811.

Rhizome short-lived for two to three years only, creeping or suberect, mostly ± 1 cm long; scales few, brown, linear- to triangular-lanceolate, 2–8 cm long. Fronds tufted, ascending, 15–40 cm long; stipes shining purple-brown, 2–10 cm long; lamina linear-lanceolate to ovate-lanceolate, 12–30 cm long, 3–12 cm wide, acuminate-attenuate at apex, abruptly narrowed at base, 2-pinnate-pinnatifid to 3-pinnate, thinly to densely covered with lemon-yellow powder beneath; rhachis narrowly winged towards apex; pinnae narrowly triangular, 1.5-6 cm long, 1-2.6 cm wide, blunt to acuminate at apex, short-stalked; pinnule trapeziform to broadly ovate-oblong, unequal at base, the longer ones obliquely pinnatifid or pinnate at base; ultimate pinnules or segments flabellate-cuneate, ± 5 × 4 mm., with lobules or teeth sharply retuse. Sporangia confined to ultimate branches of veins.

D i s t r. Cuba, Jamaica, Hispaniola, Porto Rico; introduced into Sri Lanka.

E c o l. Roadside banks, tea plantations; 600-900 m.

S p e c i m e n s   E x a m i n e d. KANDY DISTRICT: Hunnasgiryia, *Schmid 948* (BM); Kadugannawa, Allagalla Estate, *Sledge 1154* (K); Kandy, Oodawalla, *Sledge 520A* (K). BADULLA DISTRICT: Badulla, *Freeman 341A* (BM). RATNAPURA DISTRICT: Sudagalla, *Hancock 31* (K). WITHOUT LOCALITY: (labelled Hainan in error) *Hancock Kew No 101* 1828 (? date) (K).

5. ANOGRAMMA

Link, Fil. Sp.: 137 .1841; R.M. Tryon et al. in Kubitzki, Fam. Gen. Vasc. Pl. 1: 237. 1990.

Rhizome erect, very short, usually with very few thin brownish or whitish hairs or scales. Fronds delicate, not dimorphic. Stipe dark with 1–2 vascular bundles near the base, rarely with few to many hairs. Lamina 1–4-pinnate, anadromous, glabrous or rarely somewhat pubescent; veins free, ending near or behind the margin, the ends not enlarged. Sporangia borne along the veins in short, sometimes branched lines; exindusiate.

A very widely distributed genus of about 5 species, extending from Mexico to Peru in the New World and SW. and S. Europe to N. India, Sri Lanka and to Taiwan, Australia and New Zealand also Azores to Cape Verde Is. throughout Africa and Madagascar. Scarcely to be distinguished from *Pityrogramma* by technical characters, but the annual rhizome, which does not bear leaf-bases of previous years and the unusual gametophyte that produces a dormant tubercle from which new growth occurs each season, show it is biologically very different. Pichi Sermolli refers it to his family *Hemionitidaceae*.

Anogramma leptophylla (L.) Link, Fil. Sp.: 137. 1841. Sledge in Bot. J. Linn. Soc. 84: 11. 1982. Lectotype:* Spain, Portugal or Provence, Linnean Herbarium 1251. 56 (lectotype LINN.)

Polypodium leptophyllum L., Sp. Pl.: 1092. 1753.

---

*Pichi Sermolli (followed by Schelpe etc.) gives the lectotype as S. Europe, *Tournefort* no. *5337* (P-JUSS) after a lengthy and erudite discussion. but Linnaeus could not possibly have seen this

Asplenium leptophyllum (L.) Sw., Obs. Bot.: 403. 1791.

Osmunda leptophylla (L.) Sav. in Encycl. Méth. 4: 657. 1797.

Acrostichum leptophyllum (L.) Lam. & DC., Fl. Franç. 2: 565. 1805.

Grammitis leptophylla (L.) Sw., Syn. Fil.: 23, 218, t. 1, fig. 6. 1806; Thwaites, Enum. Pl. Zeyl., Addenda

Gymnogramma leptophylla (L.) Desv. in Ges. Naturf. Berl. Mag. 5: 305. 1811; Beddome, Ferns Br. Ind. & Sri Lanka: 382, t. 220. 1883.

Hemionitis leptophylla (L.) Lag., Gen. & Sp.: 33, 1816.

Discranodium leptophyllum (L.) Newman, Hist. Brit. Ferns ed. 3: 13. 1854.

Pityrogramma leptophylla (L.) Domin in Publ Fac. Sci. Univ. Charles 88: 9. 1928.

Annual 2.5–10(–23) cm. tall, with few to several (1–6 usually but up to 20–30 in Africa) erect membranous or herbaceous tufted fronds from the minute rhizome; scales pale, linear, ± 1 mm long, entire. Stipes chestnut, ± shiny, glabrous or with few small scales at base. Lamina oblong-ovate to narrowly triangular, up to 7 × 2.8(–5) cm., 2–3-pinnatifid or rarely pinnate in very small plants; pinnae ovate-triangular in outline; ultimate segments broadly or narrowly cuneate, emarginate or shallowly lobed, glabrous on both surfaces; rhachis chestnut or straw-coloured, narrowly winged above, glabrous.

D i s t r. Europe (including Britain (Guernsey & Jersey)), Mediterranean, Asia Minor, Iran, India, Sri Lanka, Java, Nepal, Australia, New Zealand, Africa from Cape Verde Is., W. Africa to Ethiopia and South Africa, Madagascar, also Mexico to Ecuador and Peru and it appears best to follow Tryon's earlier choice of the Linnaean Herbarium specimen.

E c o l. Hillsides; 1500 m. In India usually at 1800-2400 m.

S p e c i m e n s   E x a m i n e d. BADULLA DISTRICT: Near Badulla, Hingumgama, *Freeman 337A, 338B, 339C, 340D* (BM). NEWARA ELIYA DISTRICT: Above Hakgala Gardens, *sine coll. C.P. 3934* (PDA); Hakgala, *Nock* (fide Freeman). WITHOUT LOCALITY: *Robinson 46a* (K); *Skinner s.n.* (K).

6. ACROSTICHUM

L., Sp. Pl.: 1067. 1753 & Gen. Pl., ed 5: 484. 1754; R.M. Tryon et al. in Kubitzki, Fam. Gen. Vasc. Pl. 1: 252. 1990.

Type species: *A. aureum* L.

Terrestrial or swamp-loving plants, with erect or procumbent massive rhizomes bearing large tufted scales, thick fleshy roots and large tufted fronds. Fronds sometimes slightly dimorphic, with fertile narrower and shorter than sterile, simply pinnate with large entire coriaceous sessile (uppermost fertile) to petiolate pinnae, glabrous or minutely pubescent; veins closely anastomising, without included veinlets. Sporangia borne over the undersurface on and between the veins, interspersed with numerous short capitate paraphyses, exindusiate, the margin coriaceous; spores tetrahedral-globose, trilete.

Pantropical genus of 3–5 species along tropical and subtropical coasts; one pantropical and variable, one Old World and one New World; two occur in Sri Lanka.

KEY TO SPECIES

1 Sterile pinnae rounded truncate or emarginate at the apex, mucronulate; simple frond of very young plant with ligulate lamina about 22 × 1.7 cm . . . . **1. A aureum**
1 Sterile pinnae narrowly acuminate at apex; simple frond of very young plant with ± broadly lanceolate lamina about 8 × 2.5 cm. . . . . **2. A. speciosum**

1. Acrostichum aureum L., Sp. Pl.: 1069. 1753; Thwaites, Enum. Pl. Zeyl.: 300.1864; Beddome, Ferns Br. Ind. & Ceylon: 440, t. 268. 1883; Diels in Engl. & Prantl, Pflanzenfam. 1 (4): 336, fig. 174. 1899; Holtt. Fl. Malaya, 2, ed. 2: 409, fig. 239. 1968; Sledge in Bot. J. Linn. Soc. 84: 12. 1982. Lectotype (chosen by Proctor in Howard, Fl. Lesser Antilles 2: 152. 1977): *Linnean Herbarium* 1245.5 (LINN. lecto., microfiche!)

Acrostichum inaequale Willd., Sp. Pl. ed. 4, 5: 115. 1810. Lectotype: India, *Klein* (B-WILLD. *19542*) (BM., photo.).
Chrysodium inaequale (Willd.) Fée, Mém. Fam. Foug. 2: 100. 1845.
Chrysodium aureum (L.) Mett., Fil. Hort. Bot. Lips.: 21. 1856.

Rhizome forming tussocks in swamps, scales hard, subulate, ± 1–4 cm. long, up to 1.8 cm. wide, median area thick brown to black, borders narrow, pale, clathrate. Fronds tufted, leathery, 1.5–3 m. long, up to 40 cm. wide, erect, with fertile pinnae borne towards apex of frond which is occasionally forked; sterile pinnae narrowly oblong-lanceolate, 8–50 cm. long, 1–7.5(–10) cm. wide, rounded to truncate or emarginate at apex, mucronate, unequally cuneate at base, entire or irregularly undulate, coriaceous, glabrous; petioles to 2 cm. long; costa raised and prominent beneath, reticulate venation apparent; fertile pinnae similar but smaller and often sessile; stipe brown, up to 50 m. long, covered with peg- or spine-like bladeless pinna-bases, shallowly channelled above; fertile pinnae covered beneath with golden-brown (also said to be purple-brown) sporangia save for the costa.

D i s t r. Pantropical.

E c o l. Usually on landward side of salt-water mangrove swamps, also coconut groves, grassland near swamps, canal edges; recorded inland from marshy ground at foot of tea estate; sea-level.

Specimens Examined. PUTTALAM DISTRICT: Kalpitiya, *Amaratunga 106* (PDA); Puttalam—Palavi road, *Cooray 69100604 R* (K, PDA, US). TRINCOMALEE DISTRICT: Mutur, *Worthington 1660* (K). COLOMBO DISTRICT: Muthuraja Wela, just N. of Uswetakeiyawa, *Amaratunga 157* (PDA); *Faden & Faden 76/417* (K, PDA). KALUTARA DISTRICT: Kalutara (Caltura), *C.P. 1315* (LIV). ?Moneragala DISTRICT: Kumbukkan Oya, *Comanor 1049* (PDA). GALLE DISTRICT: Bentota, *Ballard 1506* (K); Pointe de Galle, *Beddome s.n.* (K);? *Contut-Lacour s.n.* (K). NOT TRACED: Sanasgama (64-11), *Ballard 1506* (K). WITHOUT LOCALITY: *C.P. 1315* (PDA); *Frazer 187* (K); *Gardner 1162, 1315* (K); *Walker s.n.* (K); *Wall s.n.* (BM). *Sir & Lady Barkley s.n.* (BM); *König s.n.* (BM); *Macrae 220* (BM); Bandarawela *Newman in Sledge P398* (BM).

N o t e. *Comanor 1049* has more crisply prominent venation than any other specimen I have seen.

2. Acrostichum speciosum Willd., Sp. Pl., ed. 4, 5: 117. 1810; Watson, Mal. Forest Records 6: 151-157. 1928; Holtt., Fl. Malaya, 2, ed. 2: 410, fig. 240. 1968. Lectotype: India, *Klein* (B-WILLD 19541).

Rhizome scales ± 8 mm. long. Fronds up to c.1.5 m. tall. Sterile pinnae up to c. 30 cm. long, 4 cm. wide, narrowly acuminate at apex, cuneate at base; petiolules up to 1 cm. long. Fertile pinnae up to c. 18 cm. long, 2–3 cm. wide, mucronate-caudate at apex, projection ± 5 mm. long. Young plants with simple lanceolate lamina c. 8 × 2.5 cm. Otherwise similar to *A. aureum*.

D i s t r. Confined to tropical Asia and Australia. Only one unlocalised specimen has been seen from Sri Lanka.

E c o l. According to Holttum in those parts of mangrove swamps more frequently inundated by the tide.

Specimens Examined. WITHOUT LOCALITY. *Walker s.n.* (K).

7. PTERIS

L., Sp. Pl. 2: 1073. 1753; J. Agardh, Rec. Sp. Gen. Pteridis:1.1839; Copel., Gen.Fil.: 60. 1947; Holtt., Fl. Mal. 2: 398.1954; W.C. Shieh, Bot. Mag. (Tokyo) 79: 283.1966; R.M. Tryon & A.F. Tryon, Ferns and Allied Plants: 332 1982; Proctor, Ferns of Jamaica: 267. 1985; Kramer in Kubitzki, Fam. & Gen. Vasc. Plants 1, Pteridos. & Gymnos.: 250.1990;
Type species: *Pteris longifolia* L.

Campteria C. Presl, Tent. Pterid.: 146. 1836; Beddome, Ferns Br. India: 116.1883. Type species *C. rottleriana* C. Presl
Litobrochia C. Presl, Tent. Pterid.:148. 1836; Beddome, Ferns Br. India: 120.1883. Type species *L. denticulata* (Sw.) C. Presl

Terrestrial ferns; rhizome erect or creeping, with scales; fronds clustered, occasionally dimorphic, simply pinnate to several times pinnate but not very finely dissected, sometimes tripartite; basal pair of pinnae (sometimes others additionally) frequently bearing a branch (or several) of similar morphology to pinnae, the terminal pinna similar to lateral pinnae; rhachis deeply grooved on upper surface and connecting with similar grooves on the upper surface of costae; spinules may be present at the junction of costae and costules and on the midribs of pinna lobes; veins free or anastomosing to form a single series of areoles parallel to the costae with other veins free (campteroid venation) or forming a complete reticulum, the areoles lacking free included veins (litobrochioid venation); sori linear, marginal, overlying an inframarginal vein connecting the vein endings; indusium continuous, linear, formed from re-flexed lamina margin; sporangia interspersed with paraphyses normally con-sisting of a uniseriate row of cells; spores tetrahedral or globose, very light brown to almost black in colour, the ornamentation ranging from reticulate to papillose, to ridged to rugose; basic chromosome number $x = 29$.

A pantropical to warm-temperate genus of about 300 species, concentrated predominantly in the tropical and subtropical zones. The taxonomy of the genus is bedevilled by the presence of numerous complexes consisting of very similar species which frequently intergrade morphologically with one another. This is of particular significance in Sri Lanka and the Indian subcontinent. Hybridization between species is common and many of the hybrid forms may be perpetuated by the occurrence of apomixis.

KEY TO THE SPECIES

1 Fronds pinnate
  2 Pinnae numerous, unbranched, very reduced at the base of the frond . . .
  . . . . **1. P. vittata**
  2 Pinnae few (14 or less), at least lowest pair with one or more branches
    3 Fronds dimorphic
      4 Fertile pinnae linear, 0.5-0.6 cm wide, sterile ones with rounded serrate lobes . . . . **2. P. ensiformis**
      4 Fertile pinnae linear, 0.7-1 cm wide, sterile ones linear,1.4-2 cm wide with serrate spinulose margins . . . . **3. P. cretica**
    3 Fronds not or only slightly, dimorphic
      5 Pinnae decurrent on rachis . . . . **4. P.multifida**
      5 Pinnae not decurrent **5.P.multiaurita**
1 Fronds pinnate-pinnatisect
  6 Lamina ternate
    7 Venation free, very prominent long spinules on costae and midveins of segments . . . . **6. P. longipes**
    7 Venation areolate, spinules not long and prominent . . . **7. P.tripartita**

6 Lamina not ternate
  8 Sinus extending to within 2-3 mm of costa
    9 Lowest veins of adjoining segments united into a costal arch . . . . . .
    . . . . **8. P.biaurita**
    9 Veins free
      10 Pinnae having a broad white central band on upper surface . . . .
      . . . . **9. P.argyraea**
      10 Pinnae concolorous . . . . **10. P.confusa**
  8 Sinus extending to within less than 2 mm of costa
    11 Pinnae regularly pinnatisect
      12 Pinnae segments with serrate apices, texture herbaceous. . . . .
      . . . . **11. P. quadriaurita**
      12 Pinnae segments with entire apices, texture coriaceous
        13 Pinnae segments up to 4.5 cm long, widely separated from one
        another. . . . . **12. P. mertensioides**
        13 Pinnae segments less than 3 cm long, closely spaced
          14 Pinnae ending in prominent caudate apex, rhizome erect
            15 Spinules on segments conspicuous, perispore wing of
            spore finely punctate . . . . **13. P. praetermissa**
            15 Spinules on segments rare or absent, perispore wing not
            punctate . . . . **14. P. gongalensis**
          14 Pinnae tapering without a conspicuous caudate apex,
          rhizome thick creeping. . . . . **15. P. reptans**
    11 Pinnae incompletely pinnatisect with some segments absent or
    reduced to lobes . . . . **16. P. × otaria**

1. Pteris vittata L., Sp. Pl. 1074.1753. Type: China, *Osbeck.*

Pteris longifolia auctt. quoad pl. asiat. Beddome, Ferns Southern India:
t.33.1863; Hook. Sp.Fil.2: 157. 1858; Thwaites, Enum. Pl. Zeyl.:
386.1864; Manickam & Irudayaraj, Pter. Fl. W. Ghats-S.India:68.1991.

Rhizome short, creeping, much branched, furnished with abundant light
brown or golden scales. Stipe 14-23 cm long and scaly. Fronds simply pinnate,
46-90 cm long. Pinnae unbranched, up to 40 pairs, 11-17 cm long, 0.3- 1.3
cm wide, reduced in length near base of frond, the lowest often appearing as
auricles; texture herbaceous or subcoriaceous. Veins free, usually once-forked.
Sori continuous from base of a pinna to near its apex. Spores light brown,
reticulate with a papilla within each of the areoles, mean diam. c. 53 mm;
chromosome number n = 58, 2n = 116, sexual tetraploid.

D i s t r. Widespread throughout the Old World tropics and subtropics,
occurring as an adventive in the New World.

E c o l. Often in dry exposed rocky situations and in man-made habitats
such as roadside banks and cuttings.

N o t e. This is often confused with the New World *Pteris longifolia* L. which differs in having pinnae which are clearly articulate to the rachis whereas those of *P.vittata* have pinnae stalks decurrent on the rachis (see Fig.14 in Walker, Bull. Br. Mus. nat. Hist.(Bot.) 13: 177. 1985.

S p e c i m e n s   E x a m i n e d. KANDY DISTRICT: Gonagampitya, c. 15 miles from Kandy, *Walker T448, T449* (BM); Kandy, Jardin Botanique, *Schmid 914* (BM). BADULLA DISTRICT: Milepost 76/6 on Badulla - Kandy road, *Ballard 1328* (K); Badulla, *Freeman 85A,86B* (BM); Road between Wellimada and Badulla, *Manton P315* (BM);Milepost 67/10 on Badulla - Kandy road, *Ballard 1312 (K)*. LOCALITY UNKNOWN: *Thwaites CP3144* (BM., K); *Moon 1819* (BM); *Walker 28* (K). DISTRICT UNKNOWN: Gongalapalata Korale, on bank of Maha Oya, *Sledge 1149*(K).

2. Pteris ensiformis Burm.f., Fl. Ind.: 230.1768; Bedd., Handb. Ferns Br. India: 107. 1883; Holtt., Fl. Mal.2: 399.1954; Piggott, Ferns of Malaysia:224.1988.

Pteris crenata Sw., Schrad. Journ. 1800: 65 1801; Hook., Sp. Fil. 2: 163. 1858; Bedd., Ferns S. India: t.39 1863; Thwaites, Enum. Pl. Zeyl.:386. 1864. Type: Burm. Thes. Zeyl., t.187, based on Sri Lanka material.

Rhizome slender, 0.3 - 0.4 cm wide, creeping, shortly branched; scales lanceolate, entire,dark brown, concolorous. Fronds dimorphic, clustered, lanceolate; veins once-forked, free. Sterile fronds shorter than fertile ones, stipes 6-22 cm long, lamina 8-15 cm long. 4-9 cm wide, pinnate, with 2-5 pairs of pinnae 3.5-7 cm long, 0.5-1.5 cm wide, bearing 1-3 pairs of coarsely toothed pinnules or rounded lobes. Fertile fronds with stipes 12-66 cm long, lamina 21-33 cm long, 6-20 cm wide; pinnate, with 2-6 pairs of pinnae 9-13 cm long, 0.5-0.6 cm wide, bearing 1-3 elongated lobes near base. Sori elongate along margin except to within 1-2 cm of apex; Spores melleous with conspicuous raised tubercles in rows, mean diam. c. 45 µm; n = 58, 2n = 116, a sexual tetraploid.

D i s t r. Widespread in tropical Asia, through Polynesia and Australia.

N o t es. Variegated forms also occur which are widely cultivated. This species may be distinguished from *P.multiaurita* by the dimorphism of the fronds and the spore characters.

S p e c i m e n s   E x a m i n e d. KURUNEGALA DISTRICT: Kurnegalle, *Gardner 1847* (PDA). MATALE DISTRICT: Road up to dam from Nalanda, *Ballard 1440* (PDA). COLOMBO DISTRICT: Abundant at Mount Lavinia, *Mrs Chevalier 1887* (BM). KANDY DISTRICT: Kandy, *Manton P96* (BM); Kandy Catchment Area, *WalkerT419,T420,T421,T424,T426, T427* (BM); Utuwankande, nr. Kandy, *Walker T883* (BM); Giragama, *Amaratunga 355* (PDA); Kumaradola Estate, near Muppane, 2,000 ft alt., *Alston,s.n.,*

25/3/1928 (BM); Stream gully in forest above and S.E. of "Hunas Falls Hotel" W.side of Hunnasgiriya mountain, S.S.E. of Elkaduwa, N.E. of Kandy, Alt. c.1200-1300 m., 25 Aug 1993, *Fraser-Jenkins* with *Jayasekara, Samarasinghe & Abeysiri* Field No. *30* (K); Corbet's Gap, *Walker T189* (BM); Hunnasgiryia, *Walker T118* (BM); Near Peradeniya - Gadawela road at marker 5/2, 453 m alt., *Comanor 506* (K); Ganorwa Forest, *Faden 76/390* (K,PDA); Roseneath, Kandy, *Ballard 1015* (PDA); Galaha, *Schmid 1038* (BM); Above Hewaheta road at Talwatta, 2 1/2 km E. of Kandy, *Fraser-Jenkins 93/5* (BM); Woods c.1km S.E. of "Gem inn ii" above Hewaheta road at Talwatta, c, 1 ½ km E. of Kandy, Alt. C. 600 m., 15 Aug 1993, *Fraser-Jenkins* Field-No. *5* (K); Woods c.7km S.E. of "Gem inn ii" above Hewaheta road at Talwatta, on old road from Talwatta to Kandy, c. 3 km E. of Kandy. (via new road) Alt. C. 600 m., 22 Aug 1993, (2 sheets) *Fraser-Jenkins* Field-No. *18* (K); Old road from Talwatta to Kandy, c.3 km from Kandy, *Fraser-Jenkins 93/18* (BM). BADULLA DISTRICT: Bandarawella, *Manton P397* (BM); Ravine near Lunugala, S. of Bibile, *Walker T557* (BM); Badulla, *Freeman 92A, 92B,93B* (BM); Ella, *Hepper & de Silva 4790* (K). MONARAGALA DISTRICT: 4 miles E. of Bibile, 230m alt., *Fosberg & Sachet 53144* (K,PDA). LOCALITY UNKNOWN: *Gardner 1243* (BM); Sri Lanka, *Wallich,s.n.* (CGE); *s.coll. C.P.1328* (BM., PDA*); Mrs Walker 386* (K); *Gardner 1245* (K).

3. Pteris cretica L., Mantissa Pl.:130.1767; Hook. Sp. Fil.2: 159.1858; Beddome, Ferns S. India: t. 39.1863;. Thwaites, Enum.Pl.Zeyl.: 386.1864; Beddome, Handb. Ferns Br. India:106.1883; Walker, Kew Bull.12: 429.1959; Manickam & Irudayaraj, Pter. Fl. W. Ghats- S.India: 70.1991.
Type: from Crete.

Rhizome short creeping, much branched; scales brown,lanceolate, with entire margins. Fronds clustered, dimorphic; stipes 16 - 90 cm long (about half of this in sterile fronds). Lamina ovate, 32-46 cm long, 10-32 cm wide, pinnate; pinnae, 4-7 pairs, the upper ones shortly adnate to rachis, the lower ones usually branched. Sterile pinnae c. double the width of the fertile and with conspicuously spinulose-serrate margins. Fertile pinnae 13-25 cm long, 0.7-1 cm wide with entire margins. Indusium occupies the margins of the fertile pinnae except for the serrate tips. Spores medium brown, shallowly rugose, mean diam. c.53 µm.

D i s t r. Widely distributed throughout the tropics and subtropics and in Southern Europe. Popular in cultivation and appearing as an adventive in many areas.

N o t es. In Sri Lanka the species with which *P. cretica* is most likely to be confused are *Idiopteris hookeriana (q.v.)* and *P. multiaurita*. The dimorphic character of *P. cretica* and the adnation of the upper pinnae to the rachis readily distinguish it from the latter species.

S p e c i m e n s E x a m i n e d. NUWARA ELIYA DISTRICT: Nuwara Eliya, *Freeman 87A, 88B* (BM); S.boundary of Hakgala, *Clarke s.n.*, 28/2/06 (PDA); Badulla road below Hakgala, *Robinson s.n.* (K); 2-3 km above Ohiya on road to Horton Plains, *Fraser-Jenkins 93/483* (BM). LOCALITY UNKNOWN: *Thwaites C.P.3502* (K, PDA); *Walker, 417* (K); *S.Coll. H127/95* (PDA).

4. Pteris multifida Poir., in Lam. Encycl. Meth. Bot. 5:714.1804; Edie, Ferns of Hong Kong: 232.1978; Khullar, Fern Flora of W. Himalaya 1:270.1994. Type: from a cultivated plant of unknown wild origin in Jardin du Museum d'Histoire Naturelle, Paris.

Pteris serrulata L.f. Suppl.: 445.1781; Hook. Sp. Fil.2:167.1858.

Rhizome short, erect or decumbent. Stipe c. 20 cm long, stramineous to brown. Lamina light green, herbaceous, ovate, 33-36 cm long, 26-28 cm wide, slightly dimorphic, sterile pinnae being somewhat shorter and wider than fertile ones. Pinnae 5-6 pairs with a similar terminal one, several of lower pinnae bearing 1-4 branches, all pinnae and branches ending in a sterile serrate tip up to 3 cm long, pinnae 11-18 cm long, 0.4-0.5 cm wide, all decurrent on rachis forming conspicuous green wing. Veins simple or once-forked. Sori occupying entire margin of pinnae except for serrated tips. Spores light coloured, mean diam. c. 45 μm; n=58, 2n=116, sexual tetraploid.

D i s t r. China to Japan but introduced as an ornamental in many countries, often becoming naturalized (as in Sri Lanka).

N o t es. Readily distinguished from *Pteris cretica*, *P.multiaurita* and *P.ensiformis* by the prominent wing extending along the rachis.

S p e c i m e n s E x a m i n e d. LOCALITY UNKNOWN: *Manton s.n., 1951-2(BM); s.coll.,s.n.* v.v. Hort.Kew 1848 Sri Lanka(BM).

5. Pteris multiaurita J. Agardh, Rec. Sp. Pteridis:12.1839; Walker, Evolution 12:82 1958; Walker, Kew Bull.14: 323.1960; Manickam & Irudayaraj, Pter. Fl. W. Ghats - S.India: 69.1991.

Types: Sri Lanka, *Macrae s.n.*,Herb.Greville (E); Sri Lanka, *Emerson s.n.*, Herb.Greville (E).

Rhizome shortly ascending; scales lanceolate with dark central area, outer cells light brown and with ciliate margin. Stipes 18-50 cm long, stramineous. Lamina herbaceous, broadly ovate, pinnate, 16-40 cm long, 12-23 cm wide. Pinnae 5-14 opposite or subopposite pairs; pinnae simple except for irregular branching varying from one branch in the lowest pair only to one or several branches in up to 5 of the lowest pairs, 9-17 cm long, 0.6-1 cm wide, ending in a serrated acuminate apex. Indusium along entire length of pinnae except for serrated apex. Spores light brown, verrucose, mean diam. c. 35μm; chromosome number n=29, 2n=58, sexual diploid.

D i s t r. Sri Lanka and Southern India.

E c o l. Widespread in Sri Lanka in small populations in light shade. It hybridizes freely with *P. quadriaurita* to give fully fertile genetically segregating populations (see *P. x otaria*). It can also hybridize with *P. confusa* to give rise to stable apomictic offspring.

S p e c i m e n s   E x a m i n e d.KURUNEGALA DISTRICT: Wariapola – Matale(?) *C.P.1328, Brodie s.n.*, 1874 (E). COLOMBO DISTRICT: near Hiyare Reservoir near Kottawa, *Sledge 1379*(K); Labugama, *Schmid 1084* as *P. cretica* (BM). KANDY DISTRICT: Karawita Kande, *Lewis & JMS s.n.*, 22/3/1919 (PDA).Lady Horton's Drive, Kandy, *Manton P45* (BM), *Walker T314, T315,T436,T440, T445,T897,T902* (BM); Lady Horton' Walk, Udawattakelle Forest, Kandy, *Fraser-Jenkins & Jayasekera 93/112*(BM); Roadside Madugoda to Weregamtota, *Walker T28* (BM); Hunnasgirya, near Madugoda, *Walker T102, T112* (BM); Allagalla, near Kandy, *Sledge 1153* (K); Peradeniya, *Petch s.n.*,1913 & 1915 (PDA); Kandy, *Schmid 851*, as *P.cretica* (BM), *Bradford s.n.*,1854 (BM), *Freeman 95B* (BM); Hantane, *Gardner 1132* (BM); Roseneath Road, above Kandy, *Walker T90* (BM); Ganoruwa Forest, *Faden & Faden 76/383* (K): RATNAPURA DISTRICT: near Ratnapura, *Ballard 1382*(K); near Rest House, Belihul Oya, *Ballard 1360* (K); Near Kitulgala,upper Kelani valley, *Sledge 1409* (K). MONERAGALA DISTRICT: Bibile to Moneragala, *Walker T540* (BM). LOCALITY UNKNOWN: *Macrae,s.n.*,Type, Herb. Greville (E); *Macrae 360* (BM); *Emerson, s.n.*,1827, Type, Herb.Greville (E); *Wall s.n.*,Herb.Fraser (E); *Wall C.P.1330, C.P.1351, C.P. 3060* (E); *Wight 1926* (E); *Thwaites 1328* (E); *Gardner 1132*, left h. specimen only (E), *s.coll., s.n.*,1926(E); *Gardner 1128* (K); *Robinson 93a* (K); *Walker 7* (K).

6. Pteris longipes D. Don, Prod. Fl. Nepal:15.1825; Beddome, Handb. Ferns Br. India: 115.1883;Manickam & Irudayaraj, Pter. Fl. W. Ghats- S. India: 72.1991.
Type: Nepal, *Wallich 1819*.

Pteris pellucens J. Agardh, Rec. Sp. Gen. Pteridis:43.1839; Beddome, Ferns S. India: t 32. 1863; Beddome, Handb. Ferns Br. India:115.1883.

Rhizome erect; scales ovate-lanceolate, dark brown with entire margin. Stipe up to 80 cm long, both stipe and rachis furnished with uniseriate hairs. Lamina thin-herbaceous, light green, broadly ovate, c. 40 cm long, 30 cm wide, ternately divided, median branch longer than the two laterals, each branch bipinnatisect. Pinnae up to 15 pairs on median branch, 7-12 cm long, 2-3 cm wide tapering to a serrate apex; pinnae segments up to 20 pairs, 1.2-2 cm long, 0.5-0.8 cm wide, sinus extending almost to costa, sterile segments serrate on margins, fertile segments serrate only at apex. Spinules at junction of pinnae and rachis and on costa very prominent, inconspicuous on the mid-veins of the segments. Veins 6-7 pairs once-forked plus 2-3 pairs of simple veins at apex. Sorus not reaching sinus nor apex. Spores medium brown, coarsely rugose or spinous with the projections arranged in rows, mean diam. c. 38 μm.

D i s t r. Sri Lanka, S. India and the Himalayas.

N o t es. this species is readily recognized by the ternately divided frond, combined with the free venation and very conspicuous spinules on the rachis and costae.

S p e c i m e n s   E x a m i n e d. MATALE DISTRICT: In silvis ad Matale, *Wall C.P.3945* (BM); Glen Forest, Matele(?), *Wall s.n.*(PDA). KANDY DISTRICT: Hantana, *Thwaites C.P.3945* (K). LOCALITY UNKNOWN: *Thwaites C.P.3945* (BM., CGE). DISTRICT UNKNOWN: Bolagalla Garden, *Robinson 1808*(K).

7. Pteris tripartita Sw., Schrad. Journ. 1800/2: 67. 1801. Type: Java, *Thunberg 24968* (Uppsala).

Litobrochia tripartita (Sw.) C. Presl, Tent. Pter.:150. 1836.

Rhizome forming a short stout erect rootstock; scales broad,concolorous. Stipes 75-150 cm long. Lamina deltoid-ovate, tripartite, texture thin when dry, fleshy when living; lamina branches of equal length up to 100 cm long, 25 cm wide, the two lateral branches each bearing a smaller similar branch, all being bipinnatisect and bearing up to 30 pairs of pinnae. Pinnae up to 25 cm long, 2-3 cm wide; segments oblong, up to 1.2 cm long, 0.4 cm wide, apex rounded, crenulate, separated by rounded sinuses 1.5-3 mm above costa. Veins with narrow areoles along costae and costules. Sori extending from base of sinus almost to top of segments. Spores light brown with a broad perispore wing, verrucose, mean diam. c.38 μm.

D i s t r. Old World tropics, Polynesia and Australia, becoming naturalized in the American tropics.

N o t es. There are few localised Sri Lanka specimens in herbaria, although G.Wall in his Catalogue of the Ferns Indigenous to Sri Lanka,1873 comments 'common in the Central and Southern Provinces. Very common about Galle'. In contrast Thwaites Enum.; 387,1864 states' 'Hab. Forests of the Central Province, not very common'.

S p e c i m e n s   E x a m i n e d. KANDY DISTRICT: Kandy Botanic Garden, *Schmid 886* (BM). NUWARA ELIYA DISTRICT: Dickoya, *Herb. Beddome* (K). LOCALITY UNKNOWN: *Mrs Walker s.n.*(K); *Wall 11/72*(K); *C.B.Clarke 10/77* (K); *Herb.Beddome s.n.*(K); *Thwaites C.P.1327* (BM., PDA); *Moon 120* (BM); *Gardner 39* (BM).

8. Pteris biaurita L., Sp. Pl. 2:1076.1753; Hook. Sp. Fil.2:203.1858; Thwaites Enum. Pl. Zeyl:387.1864; Holtt., Fl. Mal.2: 407 1954; Manickam & Irudayaraj, Pter. Fl. W. Ghats - S. India: 73.1991.
Type: Plumier, Desc. Pl. Amer.: t.14 (from the West Indies).

Campteria biaurita Hook., Gen. Fil.:t. 65A.1841; Beddome, Handb. Ferns of Br. India: 116.1883.
Pteris nemoralis Willd., Enum.:1073.1809.

Rhizome erect; scales lanceolate with central dark brown thick-walled cells, outer ones light brown, margins with long hairs. Stipes stramineous, 29-96 cm long, dark brown at base. Lamina ovate to triangular, 32-67 cm long, 25-45 cm wide with 5-8(-15) pairs of pinnatisect pinnae, lowest pair once branched. Pinnae 13-24 cm long, 3-7 cm wide terminating in a caudate tip, up to 24 pairs of segments per pinna; segments oblong-falcate, 2-3.5 cm long, 0.5-1 cm wide. Sinus extending to 2-5 mm of costa, both surfaces with abundant very small hairs. Veins mainly once-forked, free except for basal members which anastomose with those of adjacent segments to form a single row of costal arches; small spinules at junction of midveins and costae, absent or very infrequent elsewhere. Sori extending from sinus almost to tips of segments. Spores dark brown,verrucose, mean diam. c. 56 µm., viable and abortive spores intermixed. Chromosome numbers 'n' = 58, 2n =58, diploid apomict. and 'n' = 87, 2n = 87, triploid apomict.

D i s t r. Pantropical.

N o t es. *P.biaurita* is distinguished from all similar species in Sri Lanka by the presence of the coastal arch, although in immature plants this may not be fully developed. Both diploid and triploid apomictic forms occur in Sri Lanka and may be distinguished from one another by the form of the costal arch which is an inverted V in the diploid and is flattened and almost parallel with the costa in the triploid. The taxonomy of *P.biaurita* on a world basis is very confused and much work has to be done to clarify it. In the meantime the two forms indicated here should not be given formal taxonomic recognition.

S p e c i m e n s   E x a m i n e d. KEGALLE DISTRICT: Mawanella, *Schmid 1063* (BM); Kadugannawa, *Gardner 1280* (K), *Ballard 1096*(K, PDA). KANDY DISTRICT: Kandy, *Schmid 846* (BM); Lady Horton's Walk, Kandy, *Mrs Chevalier s.n.*,1887 (BM), *Walker T435* (BM); Hantane, *Gardner 1128* (BM); Kandy, *Alston 1091* - juvenile plants (PDA); Lady Horton's Walk. Udawattakelle Forest, Kandy, *Fraser-Jenkins 111* (BM); Roseneath Road. above Kandy, *Walker T82,T85,T94* (BM); Catchment Area,Kandy, *Walker T320, T321, T433* (BM), *Sledge 962* (K); Madugoda to Weregantota, c. 4 miles from Madugoda, *Walker T61* (BM); Hunnasgirya, near Madugoda, *Walker T101* (BM); BADULLA DISTRICT: Badulla, *Freeman s.n.* (BM); near Lunugala. S. of Bibile, *Walker T553, T554, T559* (BM); Goussa Estate,

Walker T530 (BM). RATNAPURA DISTRICT: Kirapaldeniya, road to Weligepola, *Faden & Faden 76/654* (K); NEWARA ELIYA DISTRICT: Adam's Peak, *Moon s.n.* (BM); Bogawantalawa to Boralanda road, Adam's Peak Sanctuary, *Comanor 1073* (K); near Lonach between Ginigathena and Norton Bridge, *Sledge 1064* (K); Hapugastenne, near Norton Bridge to Laxapana Falls, *Walker T223* (BM). LOCALITY UNKNOWN: *Rawson 3220* (BM); Sri Lanka,*Gardner 41* (BM); *Thwaites CP 1048* (BM., PDA); *Gardner 1128* (K); *Gardner 1130* (BM., K); *Gardner 1331*(K); *s.coll.s.n.,* Herb. Hook.,1867 (K); Sri Lanka 1240 (K); *Macrae 128* (BM).

9. Pteris argyraea T. Moore, Gard. Chron.:671,1859; Walker, Kew Bull.14: 331.1960; Manickam & Irudayaraj, Pter. Fl. W. Ghats - S. India: 78, 1991. Type from Coonoor, India in Herb. T. Moore (K).

Pteris quadriaurita Retz. var. *argentea* Bedd., Handb. Ferns Br. India: 111.1883.

Rhizome erect; scales lanceolate with dark central region of thickened cells, light brown thin-walled elsewhere with ciliate margin. Stipes 3-60 cm long, stramineous. Lamina ovate, 26-46 cm long, 18-30 cm wide, pinnate-pinnatisect, texture thin when dry, fleshy when living. Pinnae 3-6 pairs, oblong-lanceolate, 13-18 cm long, 3-4.5 cm wide, sparse fine hairs on upper and lower surfaces almost half the pinna width occupied by a central conspicuous white band; segments 2-3 m long, 0.5-0.7 cm wide, oblong-falcate, up to 25 pairs per pinna. Sinus reaching to 2-3 mm from costa. Spinules present along the length of segments. Veins free, once-forked. Sorus from base of sinus almost to apex of segment. Spores brown, verrucose, abortive and viable intermixed, mean diam. c. 53 mm. Chromosome number 'n' = 58, 2n = 58, diploid apomict.

D i s t r. Sri Lanka, Southern India and Indonesia.

E c o l. Of limited distribution in Sri Lanka and at the present apparently confined to shady forest areas near Laxapana.

N o t e s. This species appears to be uniform in Sri Lanka but is part of a small complex consisting of diploid and tetraploid sexual members together with diploid and triploid apomicts, all of which have the characteristic white band. There is also a Thwaites specimen *C.P. 399* (PDA) from Ambagaswewa termed *Pteris biaurita* var. *argyraea* which is essentially typical *P. biaurita* with campteroid venation but with a broad white band along the pinnae. This needs recollecting and investigating.

S p e c i m e n s E x a m i n e d. KANDY DISTRICT: Kandy, *Rawson 3261* (BM); Hapugustenne,near Norton Bridge, Laxapana Falls, *Walker T218, T220, T221* (BM); Laxapana, *Walker T229,T230,T231* (BM); Laxapana,

Maskeliya Valley, *Sledge 1063* (K). LOCALITY UNKNOWN: *Thwaites C.P. 3992* (PDA).

10. Pteris confusa T.G. Walker, Kew Bull.14: 329.1960. Type from Adam's Peak, Sri Lanka, *Walker T736* (BM).

Rhizome erect; scales narrowly linear with black central area surrounded by light brown cells, with irregular margin. Stipes 29-70 cm long, stramineous. Lamina ovate or widely ovate, 21-60 cm long, 16-43 cm wide, pinnate-pinnatisect; 4-10 pairs of pinnae, lowest pinnae pair once-, or rarely twice-branched (very occassionally also second pair); pinnae 10-27 cm long, 2-6 cm wide, ending more or less abruptly in a sterile caudate apex 2.5-6 cm long; 13-25 pairs of pinna segments, oblong-falcate, 1.5-3 cm long, 0.3-0.8 cm wide, apex entire, blunt, sinus extending to 2 mm from costa; spinules present on costa, rare or absent elsewhere. Veins free, once-forked, 9-15 pairs. Sori occupying segment margins except for apex. Spores light brown, verrucose, mean diam. c. 48 μm., viable and abortive spores intermixed. Chromosome number 'n' = 58, 2n = 58, diploid, apomict.

N o t e s. This species forms part of the complex centred on *P. quadriaurita*, with which it sometimes hybridizes to give fertile triploid apomictic plants whose hybrid nature is shown by a few aborted segments at the base of the pinnae and a much decreased percentage of viable spores. See Walker, Evolution 12: 82.1958. The illustration t.57 in Manickam & Irudayaraj 1991 is not of typical *P.confusa*. Here the segments are shown as being narrow and too widely spaced.

D i s t r. Sri Lanka, possibly extending into India.

S p e c i m e n s   E x a m i n e d. MATALE DISTRICT: c 1 km W. of top of pass between Rattota and Lagalla, *Fraser-Jenkins 93/128, 93/129* (BM). KANDY DISTRICT: 4 miles NE of Madugoda, *Sledge 932* (K); Lady Horton's Drive, Kandy, *Walker T313*; Roseneath Road, above Kandy, *Walker T83, T84, T87* (BM); 2 mi. from Madugoda towards Weregantota, *Walker T27* (BM); Hunnasgirya, near Madugoda, *Walker T103, T118* (BM); Ambegamuwe, *Walker T132* (BM); near Gampola, *Walker T 327* (BM); Kandy Catchment Area, *Walker T304,T322, T323, T324, T325* (BM); Corbet's Gap, *Walker T192* (BM); 5 km S of Corbet's Gap, *Fraser-Jenkins 93/275* (BM); above Talwatta, 3 km E of Kandy, *Fraser-Jenkins 93/19* (BM); Wattegama to Madulkele, *Fraser-Jenkins 93/232* (BM); BADULLA DISTRICT: Ella, *Hepper & de Silva 4792* (K); Tonacombe Estate, near Namanakula, *Walker T528* (BM); near Lunugala, S of Bibile, *Walker T558* (BM). RATNAPURA DISTRICT: Adam's Peak, *Walker T736* (holotype), *T737, T738*(BM). NUWARA ELIYA DISTRICT: Pidurutalagala, N of Nuwara Eliya, *Fraser-Jenkins 93/376* (BM). DISTRICT NOT KNOWN: Nawanagalla, *Sledge 943*(K); 2 mi. E. of Panilkanda, *Ballard 1564* (K); Watakelle Hill, above

Watakelle, *Fraser- Jenkins 93/258* (BM). LOCALITY UNKNOWN: *Thwaites C.P.1048* (PDA): as *P.biaurita, s.coll.s.n.* (PDA).

11. Pteris quadriaurita Retz., Obs. Bot. 6: 38.1791; Hieronymus, Hedwigia 55: 328. 1914. Walker, Evolution 12: 82,1958; Walker, Kew Bull. 14: 324.1960. Type: Sri Lanka, *König s.n.* Herb. Retzii, Lund.

Rhizome shortly ascending; scales lanceolate with dark central area surrounded by light brown cells and with ciliate margin. Stipe stramineous to brown,17-55 cm long. Lamina light green, thin, herbaceous, ovate, pinnate-pinnatisect, 24-40 cm long, 16-26 cm wide. Pinnae 4-8 pairs, 8-16 cm long, 2.5-3.3 cm wide, ending ± abruptly in a serrate/crenate tip 2-4 cm long, lowest pair (sometimes lowest 2 pairs) usually once-branched; segments narrowly oblong, somewhat falcate, tapering, typically with a serrate/crenate apex, 1-2 cm long, 0.3-0.4 cm wide. Sinus extending almost to costa; prominent spinules at junction of midveins and costae; smaller and irregularly distributed on midveins of segments. Veins free, once-forked. Sori extending from base of sinus to 2-3 mm from apex. Spores light brown, verrucose, mean diam. c.35 μm. Chromosome number n = 29, 2n = 58, sexual diploid.

D i s t r. Sri Lanka and S. India.

E c o l. This species tends to grow in forest on the wetter fringe of the Dry Zone with a sparse ground flora and hence relative freedom from competition.

N o t e s. The name *P. quadriaurita* has been applied by numerous authors to upwards of 60 distinct species from many parts of the World. True *P.quadriaurita* is of very limited distribution, being confined to Sri Lanka and S.India. It may be distinguished by its light green herbaceous frond, the serrate/crenate apices of the pinnae segments and the spore characters and has a less robust appearance. It hybridizes on an extreme scale with *Pteris multiaurita* to give genetically segregating, fully fertile, sexual offspring (see 14. *Pteris* x *otaria*). It can also hybridize with *Pteris confusa* (q.v.) to give rise to stable triploid apomicts. In Manickam & Irudayaraj 1991, t.55 is not of true *P.quadriaurita* as claimed.

S p e c i m e n s   E x a m i n e d. ANURADHAPURA DISTRICT: Summit of Ritigala, *s.coll.s.n.*, 22/3/05 (PDA); MATALE DISTRICT: Poengalla, Matale, *Brodie s.n.,* Dec.1859 (E); Pallegama, *Walker T702, T703* (BM). KANDY DISTRICT: Weregamtota, near Ratna Ella Falls, *Walker T396, T399, T402, T403* (BM); Lady Horton's Drive, Kandy, *Walker T309* (BM); Woods c.1km S.E. of "Gem inn ii" above Hewaheta road at Talwatta, c. 2 ½ km E. of Kandy, Alt. c. 600 m 15 Aug 1993, (2 sheets) *Fraser-Jenkins* Field-No. *6* (K); Stream gully in forest above and S.E. of "Hunas Falls Hotel" W.side of Hunnasgiriya mountain, S.S.E. of Elkaduwa, N.E. of Kandy, Alt. c.1200-1300 m., 25 Aug 1993, (2 sheets) *Fraser-Jenkins* with *Jayasekara,*

Samarasinghe & Abeysiri Field No. *32* (K); Woods c.7 km S.E. of "Gem inn ii" above Hewaheta road at Talwatta, on old road from Talwatta to Kandy, c. 3 km E. of Kandy, (via new road)Alt.c. 600 m 22Aug 1993, *Fraser-Jenkins* Field-No. *20* (K); Woods c.7 km S.E. of "Gem Inn ii" above Hewaheta road at Talwatta, on old road from Talwatta to Kandy, c. 3 km E. of Kandy, (via new road)Alt. c. 600 m 22Aug 1993, *Fraser-Jenkins* Field-No. *21* (K); near Gampola, *Walker T235* (BM); Ganoruwa Forest, *Faden & Faden 76/384* (PDA); Roseneath Road, above Kandy, *Walker T89* (BM); 2 miles from Madugoda on Madugoda to Weregamtota road, *Walker T24, T33* (BM); 2km from Kandy on old road from Talwatta to Kandy, *Fraser-Jenkins 93/21* (BM); Minipe, *Walker T77* (BM). BADULLA DISTRICT: Goussa Tea Estate, *Walker T531, T534, T535* (BM); Badulla, *Thwaites C.P.3060*, Herb. Brodie(E); *Freeman 94A, 96C* (BM); ravine near Lunaragala, south of Bibile, *Walker T549, T550, T551, T560* (BM); Namunukula, *Walker T574* (BM). LOCALITY UNKNOWN: *Walker 27* (K).

12. Pteris mertensioides Willd., Sp. Pl. 5: 394.1810; Holtt., Fl. Mal. 2: 404.1954; Piggott, Ferns of Malaysia: 227, t.685-687,1988; Manickam & Iridayaraj, Pter. Fl. W. Ghats - S. India:76.1991.

Pteris patens Hk., Sp. Fil. 2:177, t. 137. 1858; Thwaites, Enum. Pl. Zeyl: 386. 1864; Beddome, Handb. Ferns Br. India:14. 1883.
Pteris decussata J. Sm., Journ. Bot. 3: 405. 1841 (nomen nudum).

Rhizome erect; scales brown with ciliate margins. Stipes purplish up to 100 cm long. Lamina c.170 cm long, 85 cm wide, bipinnatifid with 13-20 pinnae pairs. Pinnae up to 43 cm long, 10 cm wide, with up to 50 pairs of segments per pinna; segments up to 4.5 cm long, 0.4 cm wide, confluent at base but rapidly becoming widely separated from one another up to 6mm apart, narrow oblong, tapering to a serrated tip.Sinus extending almost to costa. Veins free, c. 20 pairs per segment, once- or twice-forked, single towards apex. Sori extending to sinus but not reaching apex of segments. Spores dark brown with a wide perispore wing, spinous, mean diam. c. 34 µm.

D i s t r. Sri Lanka, S. India,Malaysia to Polynesia.

N o t es. This is a very distinctive species with its large fronds and widely spaced long narrow pinna segments. Apparently it has not been collected in Sri Lanka in recent years.

S p e c i m e n s   E x a m i n e d. KANDY DISTRICT: Hantana, *C.B.Clarke C.P. 1047* (K). LOCALITY UNKNOWN: *Gardner 42* (BM); *Thwaites C.P. 1047* (BM); *Hance 2940* (BM); *Gardner 1126* (BM); *Wall s.n.* (K); *Ligley 1829* (K); *Wall 1884* (PDA); *Gardner 1847* (K, syntype); *Macrae 896* (BM).

13. Pteris praetermissa T.G. Walker, Kew Bull.14: 327. 1960.
Type: Sri Lanka, Le Vallon Tea Estate, *Walker T343* (BM).

Rhizome erect. Stipe stramineous, 12-61 cm long; rachis stramineous or more commonly dark brown or purplish. Lamina dark green, broadly ovate to deltoid, pinnate-pinnatisect, 17-48 cm long, 14-35 cm wide. Pinnae 3-10 pairs, 8-19 cm long, 2-3.5 cm wide, lowest pair of pinnae once-branched; segments oblong with entire rounded apex, 1.2-2 cm long, 0.4-0.5 cm wide. Sinus extending almost to costa, prominent spinules present at junction of midveins and costae and prominent elsewhere on midveins. Veins free, once-forked. Sori extending from near base of sinus almost to apex; abundant long uniseriate paraphyses terminating in a brown glandular cell present. Spores dark brown, verrucose, mean diam. c.48 mm., with a finely punctate perispore wing. Chromosome number n = 29, 2n = 58, sexual diploid.

D i s t r. Sri Lanka endemic.

N o t e s. The combination of abundant uniseriate paraphyses (not as long as in *P. gongalensis*), the prominent spinules and the finely punctate perispore wing distinguish this species from others.

S p e c i m e n s   E x a m i n e d. MATALE DISTRICT: Forest stream gulley, SE of Gombaniya mountain, above Niloomally, N. of Wattegama, W. Knuckles Range, N.E. of Kandy, Alt. 1300 m., 4 Sept 1993, (3 sheets) *Fraser Jenkins* with *Jayasekara, Samarasinghe, Bandara & Bandara* Field No. *79* (K); 1 km W. of pass between Rattota and Lagalla, NE of Matale, *Fraser-Jenkins 93/128* (BM). KANDY DISTRICT: SSE of Elkaduwa, *Fraser-Jenkins 93/32* (BM); Gallebodde Estate, near Ambegamuwa, *Walker T217* (BM); Knuckles, *Walker T248* (BM); Le Vallon Tea Estate, *Walker T343*(holotype), *T362, T363* (BM); Galaha, *Schmid 1040* (BM); Hantane, *Ballard 1047* (K, PDA); Corbet's Gap, *Ballard 1072* (K); Wattekelle Hill, NE of Kallebokka, *Fraser-Jenkins 93/259* (BM); Above Nilloomally, N of Wattegama, *Fraser-Jenkins 93/79* (BM). NUWARA ELIYA DISTRICT: Forest slope above Hakgala Botanic Gardens on N.E. Side of Hakgala mountain. E. of Nuwara Eliya, Alt. c. 1700 m., 23 Sept 1993, (3 sheets) *Fraser-Jenkins* with *Jayasekara, Samarasinghe & Abeysiri* Field No. *206* (K); Norton Dam., *Walker T233, T234* (BM); On road to Ohiya, SE of Nuwara Eliya, *Fraser-Jenkins 93/443* (BM).

14. Pteris gongalensis T.G. Walker, Kew Bull. 14: 328.1960. Type: Sri Lanka, Gongala Peak, *Walker T756* (BM).

Rhizome erect; scales narrow lanceolate with dark central area surrounded by light brown cells and with irregular margin. Stipe stramineous or purple-brown, 26-84 cm long. Lamina dark green, deltoid, pinnate-pinnatisect, 19-50 cm long, 20-34 cm wide. Pinnae 4-9 pairs, lanceolate, 10-21 cm long, 2.2-4 cm wide, caudate, lowest pair of pinnae with 1 (occassionally 2) branches;

segments oblong with entire apex, 1.2-2.5 cm long, 0.4-0.6 cm wide. Sinus extending to 1 cm of costa, small spinules present at junction of midribs and costae, absent or rare elsewhere. Veins free, once-forked. Sori extending from near base of sinus to (or very close to) apex; abundant long uniseriate paraphyses 17-19 cells long present and terminating in a brown glandular cell. Spores dark brown, verrucose, viable and abortive intermixed, mean diam. c. 56 mm. Chromosome number 'n' = 87, 2n = 87, triploid apomict.

D i s t r. Sri Lanka endemic.

E c o l. in open situations on stream banks or road cuttings at elevations of 914 m.or above.

N o t e s. The combination of a deltoid frond, the almost complete absence of spinules on the midveins of segments, the presence of abundant very long paraphyses and intermixture of viable and abortive dark brown spores distinguish this species from other members of the complex.

S p e c i m e n s   E x a m i n e d. KANDY DISTRICT: Corbet's Gap, *Walker T191, T193* (BM); Knuckles, *Walker T246* (BM). RATNAPURA DISTRICT: Gongala Peak, *Walker T756* (holotype),*T757, T765* (BM). NUWARA ELIYA DISTRICT: Hapugustenne, near Norton Bridge, *Walker T219, T222, T225* (BM).

15. Pteris reptans T.G. Walker, Kew Bull.14: 325.1960. Type: Sri Lanka, Haputale Reserve, *Walker T607* (BM).

Rhizome stout, long-creeping, in type specimen (*Walker T 607*) c. 20 cm long × 3cm wide; scales with dark central area surrounded by light brown cells and with ciliate margin. Stipes 25-100 cm long, stramineous. Lamina ovate or ovate-lanceolate, 18-60 cm long, 14-37 cm wide, pinnate-pinnatisect. Pinnae 6-15 pairs, inflexed, 10-20 cm long, 2-5 cm wide, gradually narrowing towards their apex; pinnae segments 15-35 pairs, inflexed, up to 2 mm between them., oblong or subfalcate-oblong, 1.3-2.5 cm long, 0.3-0.6 cm wide, with an entire apex. Sinus nearly reaching costa. Veins free, once-forked, prominent especially on lower surface; spinules present on costae and mainly in upper half of segments. Sori reaching neither base of sinus nor apex of segments. Spores darkish brown, average diam. c. 45 mm., rugose. Chromosome number n = 29, 2n = 58, diploid sexual.

D i s t r. Sri Lanka endemic.

E c o l. Virtually confined to high elevations around Nuwara Eliya where it is common.

N o t e s. This species is very readily distinguished in the wild by its stout creeping rhizome and the marked inflection of both the pinnae and the segments. In dried specimens the most reliable features are the gradual tapering of the pinnae which lack the marked caudate tips so common in most other species of the complex, also the prominence of the veins on the underside of the frond, and the distribution of the spinules.

S p e c i m e n s   E x a m i n e d. BADULLA DISTRICT: Badulla, *Freeman s.n.* (BM). NUWARA ELIYA DISTRICT: path to World's End at Little World's End, *Fraser-Jenkins 93/518* (BM); Road to radio station, S side of Pidurutalagala, N of Nuwara Eliya, *Fraser-Jenkins 93/393, 93/394* (BM); On road to Ohiya, Horton Plains, SW of Nuwara Eliya, *Fraser-Jenkins 93/444* (BM); track below top of Ramboda Pass leading NW, *Fraser-Jenkins 93/328* (BM); Ramboda Pass, *Walker T825, T843, T844* (BM); Kandapola Forest Reserve,near Nuwara Eliya, *Sledge 1351* (K), *Walker T855, T856, T857* (BM); Hakgala, behind Botanic Garden, *Fosberg 57235* (K, PDA); Hakgala Peak, *Ballard 1129* (K, PDA); Haldummulla road, Horton Plains, *(coll. illegible) s.n.* (PDA); near Nuwara Eliya, *s.coll. s.n.*, 6/4/06 (PDA); Nuwara Eliya, *C.P.1330 Gardner* 1847 (PDA), *Gardner 1123* (CGE), *Freeman 97D,* 98E (BM), *Schmid 1346* (G); stream bank below Nuwara Eliya, *Walker T501* (BM); near Botanic Garden, Hakgala, *Ballard 1337* (K); Hakgala, road to Nuwara Eliya from Botanic Garden, *Ballard 1211* (K); Mt.Pedro, Nuwara Eliya, *Ballard 1252* (K); Horton Plains, *Schmid 1384, 1385, 1398* (BM), *Fosberg with Mueller-Dombois 50015* (PDA); Haputale Reserve, *Walker T600-T610* (Holotype *T607,* BM). LOCALITY UNKNOWN: *Wall, s.n.-* only one of specimens on sheet (E); *Gardner 1129* (BM); *Thwaites C.P.1351, C.P.1330* (PDA).

16. Pteris × otaria Bedd., Ferns S. India: t.41, t.219.1863; T.G. Walker, Evolution 12: 82, 1958; T.G. Walker, Kew Bull.14: 324.1960; Manickam & Iridayaraj, Pter. Fl. W. Ghats - S. India: 77. 1991.
Type: Malabar, S.India.

Pteris quadriaurita var. *ludens* Bedd., Handb. Ferns Br. India:111.1883.
Pteris quadriaurita forma *ludens* Thwaites, Enum. Pl. Zeyl.:386.1864.

Rhizome shortly ascending; scales lanceolate with a dark central region surrounded by light brown cells and with ciliate margins. Stipe stramineous, 15-58 cm long. Lamina ovate, 13-40 cm long, 16-26 cm wide. Pinnae 5-10 pairs, incompletely pinnatisect, 5-19 cm long, 2-3.5 cm wide, ending in a serrate caudate apex, lowest pinnae pair having at least one similar branch - frequently up to about half the pinnae have 1 or up to 5 branches; segments variable in number, occupying almost all pinnae except near rachis and apex at one extreme, to only 4 or 5 pairs at the other extreme, 1-2.3 cm long, 0.3-0.4 cm wide, oblong-falcate with serrate/crenate tip. Sinus cut almost to costa (frequently segments nearest the rachis are reduced to small rounded lobes, c.f. t.219 Bedd. 1863); spinules usually prominent at junction of midveins and costa, irregular on midveins of segments. Veins free, once-forked. Sori extending from base of sinus to 2-3 mm from apex; spores light brown, verrucose, mean diam. c. 35 mm. Chromosome number n = 29, 2n = 58, diploid sexual.

D i s t r. Sri Lanka and S.India.

E c o l. Typically a plant of disturbed habitats such as roadside banks, abandoned coffee estates, deforested areas, etc. It is one of the commonest ferns in Sri Lanka.

This is a hybrid between *Pteris quadriaurita* and *P. multiaurita.* It is fully fertile and sexual and segregates out genetically. As a consequence one can find an unbroken series involving very small morphological steps from the fully pinnatisect form of the former species to the pinnate form of the latter one (see Walker, Evolution 12:82, 1958).One of the commonest forms, as illustrated by Beddome, t.41, is intermediate in morphology between the two parental species, the pinnae being simple for the first ¼ - ⅓ of their length as in *P.multiaurita* and then pinnatisect as in *P.quadriaurita.*

S p e c i m e n s   E x a m i n e d. MATALE DISTRICT: Poengalla, Matale, *Brodie s.n.*,1864 (E); *Thwaites C.P. 1330* (E); Pallegama, *Walker T701* (BM); Road to dam from Nalanda, *Ballard 1439* (K). KANDY DIS-TRICT: Near Peradeniya to Godawela road, *Comanor 507* (K); Ganoruwa Forest, *Faden & Faden 76/384* (K), *Faden & Faden 76/385* (K, PDA); Kandapahala Korale, near Weregamtota, *Sledge 1133* (K); Peradeniya, *s.coll., s.n.*,1915 (PDA), *Petch s.n.*,7/2/1916 (K); *Thwaites CP 3040* (PDA), *Walker T422, T423, T425, T428, T429* (BM); Kandy Catchment area, *Walker T296, T298,T303, T326,T327,T404,T405,T406* (BM), *Sledge 963* (K); Lady Horton's Drive, Kandy, *Walker T307, T311, T312, T433, T437, T439, T441, T442, T444, T895, T896, T898, T900, T903* (BM); Roseneath road, above Kandy, *Walker T88, T91,T92, T93, T96* (BM); Hunnasgirya, *Sledge 975* (K); Hunnasgirya,near Madugoda, *Walker T110*(BM); road to Nawangala, 1 mile W of Madugoda, *Walker T47* (BM); Madugoda to Weregamtota road, *Walker T25* (BM); Weregamtota, near Ratna Eliya Falls, *Walker T395, T398, T400* (BM); Minipe,*Walker T68, T73* (BM); Allagala Tea Estate, *Walker T468, T469* (BM); Ambegamuwe, *Walker T134* (BM); Utuwankande, near Kandy on Colombo road, *Walker T882* (BM); near Gampola, *Walker T239, T240* (BM); Talwatta, 2 1/2 km E. of Kandy, *Fraser-Jenkins 93/6* (BM); on old road from Talwatta to Kandy, 3 km E. of Kandy, *Fraser-Jenkins 93/20* (BM); S.E. of Gombaniya mountain, N. of Wattegama, *Fraser-Jenkins 93/80* (BM). BADULLA DISTRICT: Badulla, *Thwaites CP2450*, Herb.Brodie (E); Goussa Tea Estate, *Walker T529, T533* (BM). KALATURA DISTRICT: Pelawatta, path to Pahala Hewessa, *Ballard 1531* (K). RATNAPURA DISTRICT: Belihul Oya, *Manton P433* (BM). MONARAGALA DISTRICT: S. of Bibile, *Sledge 1173* (K). LOCALITY UNKNOWN: *Thwaites C.P.1357* (K); *Walker 414* (K); *Gardner 1133* (K); *Thwaites C.P.1351* (E, BM., PDA); *Walker s.n.* (K);

Walker 1 (K); *Robinson s.n.*,1893, 1894 (K); *Beddome s.n.* - type of t.219, F.S.I. (K); *Thwaites s.n.* (K); *s.coll. s.n.* (PDA); *Thwaites 1360* (BM., PDA); *Wall CP1351-1360*, Herb.Fraser (E); *Rawson s.n.*, Herb. T.Moore (K); *Wall s.n.* (E); *Thwaites C.P.1330* (E); *Emerson s.n.*.1827 (E); *Lindley 1829* (E); *Macrae 887* (BM).

8. IDIOPTERIS

Walker, Kew Bull.14: 429-432.1957.
Type species: *Idiopteris hookeriana* (Agardh) T.G.Walker.

Terrestrial fern with long creeping rhizome. Dimorphic, subcoriaceous, free-veined, stipitate fronds. Fertile fronds with simple pinnae, some of which may bear one or a few similar branches, and with submarginal entire indusia, the spores acutely triangular, colourless. Sterile fronds shorter than the fertile with few broad, usually unbranched, simple pinnae having serrate margins and conspicuous hydathodes. Basic chromosome number = 27.

Idiopteris comprises a single species, originally described as a species of *Pteris*, from which it differs in the submarginal indusium, spore and prothallial characters and basic chromosome number. Its true affinities are obscure, possibly with the *Taenitis* alliance. Confined to Sri Lanka and S.India.

Idiopteris hookeriana (Agardh) T.G. Walker, Kew Bull. 14, 429-432,1957.

Pteris hookeriana Agardh, Rec. Sp. Gen. Pterid., 12 , 1839; Beddome t.40,
    Ferns S. India, 1863; Beddome, Handb. Ferns Br. India, 107, 1883.
Pteris cretica L. var. B Thwaites, Enum. Pl. Zeyl., 386,1864.
Type specimens: Sri Lanka, *Emerson 1827,1828* (E).

Terrestrial, long-creeping rhizome 4 cm wide; scales 3mm long, 0.5mm wide, subulate-linear, entire, dark brown, long, narrow cells; dimorphic. *Sterile* fronds shorter than fertile; stipes 15-35 cm long; lamina broadly ovate, 13-26 cm long, 11-32 cm wide, with 1-3 pairs of pinnae; pinnae oblong lanceolate, 9-14cm long, 2.5- 4cm wide, margin serrate with prominent hydathodes,terminal pinna similar to upper laterals; texture subcoriaceous. Fertile fronds, stipes 21- 68 cm long; lamina widely ovate, 13-44 cm long, 14-28 cm wide; 3-6 pairs of subopposite or alternate pinnae, 13-19 cm long, 0.7-1.7 cm wide, narrow, entire except for serrate tip, lowest pinnae with one to several branches, upper pinnae simple, with terminal pinna similar to, but longer than upper laterals; texture subcoriaceous; veins once forked, free. Sori continuous, submarginal, with entire indusium reflexed at maturity. Sporangia long-stalked with c.23 indurated cells in annulus; paraphyses similar to aborted sporangia or absent. Spores colourless, acutely triangular, papillate. Fertile and sterile fronds pink when young, turning green on maturity; chromosome number n=27, sexual diploid.

D i s t r. Sri Lanka and S.India.

E c o l. In wet forest from low altitudes to 600m+.

N o t e. Most commonly confused with *Pteris cretica* L. from which it may be distinguished in the sterile fronds by the subcoriaceous versus papyraceous texture, presence versus absence of hydathodes, regularly serrate versus irregularly spinulose-serrate margins. The fertile fronds have submarginal versus strictly marginal indusia and the spores are colourless and papillate versus brown and rugose. Fertile fronds of *Idiopteris* also have been confused with those of *Lindsaea ensifolia* Sw. The former have free veins and the indusium opens towards the costa, the latter have reticulate venation and the indusium opens towards the margin.

S p e c i m e n s E x a m i n e d: COLOMBO DISTRICT: Kottawa Forest Reserve, on bank along path, *Ballard 1538* (K, PDA); Kottawa Forest Reserve, 200 ft. alt. *Sledge 403* (K). KALUTARA DISTRICT: Hewessa, south of Moragala, jungle c100 ft.alt., *Sledge 392* (K). RATNAPURA DISTRICT: Sinharaja Forest, nr.Hedigala, *Sledge 319* (K); Hedigala,Sinharaja Forest, *Ballard 1397* (K,PDA); Adam's Peak, 2,000ft. alt., *Gardner 1242* (K); Gilimale,near Adam's Peak, *Walker T716,T717,T718,T950* (BM); Gilimale, near Adam's Peak, *Lovis as T950-T956* (BM); Gilimale, *Sledge 1251* (K); near Ratnapura, *Thwaites C.P.1329* (PDA); Gilimale Forest, track from Carney road to Kalu Ganga, 60-120m alt. *Faden & Faden 76/465* (K, PDA); Adam's Peak, 2,000 ft.alt., *Gardner s.n.*(K,PDA). GALLE DISTRICT: Kanneliya Forest Reserve, near Udagama, 500 ft. alt., *Sledge 1387* (K); Piluwala,south east of Elpitiya, *Huber 317* (PDA). LOCALITY UNKNOWN: *Thwaites C.P.1329* (BM., E, PDA); *Wall C.P.1329* (E,K); *Emerson,1826,1827*,syntypes (E); *Wight 1913,1915* (E); *Anderson,s.n.* (E); *Brodie C.P.1329* (E); *Walker s.n.* (K); *C.P.1329*, Herb.T.Moore (K); *Gardner C.P.1329* (K); *Henderson,s.n*, June,1877 (K); *C.B.Clarke C.P.1329* (K); *Trimen C.P.1329* (K); *Beddome s.n.* (K); *Col.Robinson 90a* (K); *Skinner s.n.* (K); *s.coll., s.n.* (CGE).

9. DORYOPTERIS

J. Sm., J. Bot. (Hooker) 3: 404.1841, nom. cons. Type: *D. palmata* (Willd.) J. Sm.. (= *Pteris palmata* Willd., *Doryopteris pedata* (L.) Fée var. *palmata* (Willd.) Hicken.)

A genus of 25 species occurring throughout the tropical parts of the world. *Doryopteris concolor* is often included in *Cheilanthes* (Tryon & Tryon 1981; Schelpe & Anthony 1986), but preliminary rbcl studies (Gastony & Rollo 1995) showed that the latter is polyphyletic and that *D.concolor* cannot be grouped with *Cheilanthes* s. str. but rather with species generally included with *Doryopteris*. Two sections are recognized in *Doryopteris*. Section *Lytoneuron* Klotzsch (which includes *D. concolor*) is characterized by two vascular bundles at the stipe base and free venation except for the soriferous vascular commissure that frequently occur in fertile fronds. (J.P.Roux: Conspectus of Southern African Pteridophyta. Southern African Botanical Diversity Network Report No. 13 Nov. 2001)

Several groups within the *Pteridaceae* are undoubtedly best treated as subfamilies, as in other such large families. These are the *Pteridoideae*, the *Cryptogrammoideae* (Pich.Serm.) Fraser-Jenk., comb. nov. (basionym: family Cryptogrammaceae Pich.Serm., Webbia 17: 299. 1963) and the *Cheilanthoideae* (J. Smith) W.C. Shieh. Tryon, Tryon & Kramer (in Kramer & Green (eds.), in Kubitzki: The families and genera of vascular plants 1: 230-256. 1990) have added a number of the genera here treated as belonging to the *Cryptogrammoideae* together with *Taenitis* Willd. ex Schkuhr into a subfamily *Taenitioideae* (C. Presl) R.M. Tryon. but this treatment is not adopted here.

Doryopteris concolor (Langsd. & Fisch.) Kuhn in von Decken, Reis. 3(3), Bot.: 19. 1879.
Type: Mozambique, Zambesi River, *Kirk s.n.* (K, holo.)

Pteris concolor Langsd. & Fisch., Icon. Fil.: 19, t. 21. 1810.
Pteris geraniifolia Raddi, Opusc. Sci. Biol. 3: 293. 1819.
Pellaea contracta Fée, Mém. Fam. Foug. 5, Gen. Fil.: 129. 1852.
Cheilanthes kirkii Hook., Cent. Ferns: t. 81. 1861.
Pellaea concolor (Langsd. & Fisch.) Bak., Flor. Brasil. 1(2): 396. 1870.
Pteris cheilanthoides May., Enum. Plant. Formos.: 619. 1906.
Doryopteris nicklesii Tardieu, Not. Syst. 3: 166. 1948.
Doryopteris kirkii (Hook.) Alston, Bolm. Soc. Broteriana, ser. 2a, 30: 14. 1956.
Cheilanthes concolor (Langsd. & Fisch.) R.M. & A.F. Tryon, Rhodora 83(833): 133. 1981.

Rhizome short, thin, decumbent, ± unbranched, bearing ± dense, small, thin, narrowly linear-lanceolate (up to 3.5 × 0.5 mm), long-acuminate, distally minutely toothed, pale-brown, dark-centred scales throughout. Fronds up to c. 8 in an apical tuft; stipe upright, but bent at its apex to hold the lamina horizontally, up to c. 25 cm long, longer than lamina (often over twice as long), thin, ± flattened but sometimes widely and shallowly grooved on adaxial surface, dark-purplish black, glossy, a tuft of narrow scales, resembling rhizome scales, at base, naked above; rachis similar to upper stipe but more narrowly grooved adaxially; lamina small, simple or becoming pinnate at base, deeply bipinnatifidly lobed, deltate-pentagonal (up to c. 15 cm wide and 10 cm long), usually wider than long, ± tripartite, softly herbaceous, thin, mid- or yellowish-green, entirely glabrous, bearing a lower pair of opposite pinnae or deeply pinnatifid basal lobes and a deeply pinnatifid, deltate apical portion which bears about 3-4 opposite pairs of lobes and a few smaller apical ones, laminar main-lobes joined below by small, decurrent wings of laminar tissue on both sides of rachis, which sometimes do not quite reach the basal opposite-pair of lobes (which thus become pinnae), wings the shape of a wide-based triangle, an opposite pair thus form a diamond shape (across rachis) between each pair of laminar lobes; lowest laminar lobes subdeltate, widest at their bases, up to 7 cm long, 6 cm wide, adnate to rachis at their bases, usually with a short decurrent wing of lamina on their basiscopic sides, markedly larger and more compound than the next pair up, each one similar to the rest of the upper lamina, but markedly asymmetrical with lower basiscopic pinna-lobes between 2 and 3 times as long as acroscopic ones, inserted at c. 80°, deeply pinnatifid, the lowest one or two becoming bipinnatifid, with up to c. 4 pairs of alternate pinna-lobes; upper laminar lobes opposite, lanceolate, slightly narrowed to their bases, widest at their middles, with simple, caudate, acute apices, widely attached to rachis by their bases and interconnected by small, triangular laminar wings, laminar lobes (except uppermost) in turn, deeply pinnatifidly lobed into up to c. 6 pairs of obliquely sloping, ± narrowly rounded, entire lobes; costae and costules distinct, terete, black and glossy below, ± deeply set between lamina, indistinct and brownish-green above; pinna-lobes in lowest pair of laminar lobes connected at bases by ± narrow wing of laminar tissue, usually without diamond-shaped small lobes; obliquely sloping, ± acutely pointed, or narrowly rounded, the first two basiscopic pairs in each lowest laminar lobe developing deeply cut pinnule lobes in turn. Sori form a line around laminar edge throughout, protected by continuous pseudo-indusium; pseudo-indusium formed by reflexed lamina-margin (occasionally separating into separate small lobes at each vein-ending if only weakly fertile), thin, entire, occasionally forming long, irregular lobes, pale-green, lifting, shrinking slightly, becoming pale-brown and turning back to reveal black (later brown) sporangial line on ripening. Spores small, uniform., trilete, globose, sporoderm appearing ± smooth, but patterned with minute, wide-based tooth-like ridges at higher magnifications. *Cytotype* diploid, sexual, n = 30 (with 30 bivalents at meiosis).

D i s t r. Lower-mid and mid-level regions of a wide belt of central Sri Lanka. Common. A probable south-east Asian element which has become more widespread and nearly pan-tropical. Sri Lanka; E.C. and S. India (from Orissa southwards, the Nilgiri, Annamalai and Palni Hills, southern Ghats, eastern Ghats); S. China; Taiwan; Indonesia (Java, Boeroe, Flores, Timor, Celebes, Moluccas); Philippines; New Guinea; Oceania (New Caledonia, Fiji, Austral Isles, Hawai'ian Islands, Society Islands (Bora Bora), New Hebrides, Tahiti). Also in the Mascarenes, S. and E. Africa (N. to Ethiopia) and tropical S. America.

E c o l. Occurs in open or semi-open habitats, often among bushes, between rocks, on road-banks and earth-banks or occasionally below walls, from c. 500-1100m in Sri Lanka.

N o t e. The epithets, *kirkii* and *geraniifolia*, have sometimes been used as referring to distinct varieties, or even species, but merely refer to developmental stages or forms of frond-dissection, so are not recognised taxonomically here.

S p e c i m e n s   E x a m i n e d. MATALE DISTRICT: Nalande, Matale, Herb. *Brodie C.P. 3103*, 1858(E). NUWARA ELIYA DISTRICT: $^1/_2$ km W. of Ramboda, lower Ramboda Pass, c. 3100ft., *Fraser-Jenkins Field No. 311* p.p, with *Abeysiri*, 25 Oct. 1993 (K, BM., PDA, US). LOCALITY UNKNOWN: *Hancock 30* p.p. (US); ex Herb. *Ferguson s.n.*, 30 July, Herb. *Fendler* (GH); *Thwaites C.P. 3103* (E, GH); *Hutchinson s.n.*, Dec. 1876, Herb. *Parish* (E); *Wall C.P. 3103* (E) and *s.n.* (BO 10563); *Emerson s.n.*, 1827 (E); warmer part of the island, *Naylor-Beckett 184* (E).

10. TAENITIS

Willd. ex Spreng., Anleit. Kennt. Gew. 3: 324. 1804; Holtt. in Blumea 16: 87-95. 1968 & in Kew Bull. 30: 327-343. 1975; Tryon & Kramer in Kubitzki, Fam. & Gen. Vasc. Pl. 1: 240. 1990.
Type: *T. blechnoides* (Willd.) Sw.

Rhizome creeping, solenostelic, densely covered with almost black glossy bristles which are thickened at the base and formed of a few rows of cells reducing to a single row at the apex. Stipes in 2 rows, dark and glossy, grooved on adaxial face. Fronds simple, round to elongate or simply pinnate or rarely simple but bilobed, when pinnate the terminal pinna similar to laterals; pinnae entire, usually narrowly elliptic to linear-lanceolate, mostly coriaceous, glabrous, without main lateral veins but veins forming a network of 2, 3 or more series of narrow oblique subequal areoles, without included free veinlets. Fertile fronds often with narrower pinnae or lamina than the sterile. Sporangia either confined to a continuous or discontinuous longitudinal band between the midrib and margin, or in 3 species the band in a submarginal groove or spreading irregularly along the veins and in 2 other species the fertile pinnae very narrow and covered beneath with sporangia; paraphyses very numerous, multicellular, thickened upwards; spores tetrahedral-globose, trilete.

A genus of 16 species distributed from Sri Lanka and Southern India to southern Burma, Vietnam., Hainan and throughout Malesia to Solomon Is., New Hebrides, Fiji and N. Queensland. Only one species occurs in Sri Lanka.

Taenitis blechnoides (Willd.) Sw., Syn. Fil.: 24, 220. 1806; Hook., Sp. Fil. 5: 187. 1864; Bedd., Ferns Br. Ind.: t. 54. 1865; Holtt., Rev. Fl. Malaya 2: 586, f. 346. 1954 & in Blumea 16: 89. 1968.
Type: S. India, *Klein* in Herb. Willd. 19577 (B-WILLD., holotype, BM., photo.).

Pteris blechnoides Willd., Phytographia: 13, t. 9, f. 2. 1794.

Rhizome erect with dark brown bristle-like hairs. Fronds of young plants simple but those of mature plants pinnate, 60-80 cm tall; stipes 32-45 cm long. Pinnae very variable in number width and texture, usually in 1-5(-7) pairs, opposite or alternate, linear-lanceolate to lanceolate, 15-30 cm long, 2.5-5 cm wide in case of sterile pinnae, 1-3 cm wide in fertile ones, gradually attenuate-acute at apex, cuneate at base; midrib raised beneath with rest of venation obscure but above raised, close and anastomosing, basically obliquely subparallel. Sporangia in single usually continuous linear bands 2 mm wide parallel to and on each side of midrib about $^1/_3$ to $^1/_2$ the distance from margin.

D i s t r. Sri Lanka, S. India, S. Burma, Thailand, Vietnam, Hainan and throughout Malesia to Fiji and Caroline Is.

E c o l. Disturbed primary dipterocarp forest, dense primary forest with thick leaf-litter; 0-180 m.

S p e c i m e n s   E x a m i n e d. KEGALLE DISTRICT: Kittool Galle (? = Kitulgala) *C.P.3922* (K, PDA). RATNAPURA DISTRICT: between Gilimale and Carney, *Sledge 1247* (BM., K); Forests about Galle and Ratnapura *sine collector 200* (PDA); Hewesse, *sine collector s.n.* (PDA); (note on *sine collector 200* (PDA). NUWARA ELIYA DISTRICT: Gilimale, near Adam's Peak, *Walker T 711, T 721* (BM). MATARA DISTRICT: Matara to Wilpita Forest Reserve, jungle above nursery area close to office,7 Dec 1995, *Shaffer-Fehre* with *Jayasekara 662* (K, PDA); in close proximity of former, same day, *Shaffer-Fehre* with *Jayasekara 663* (K, PDA). WITHOUT LOCALITY: *Thwaites in Herb. Beddome* (K); *Wall via Clarke* (K); *Wall s.n.* (BM).

11. CHEILANTHES

Sw., Syn. Fil. 5: 126. 1806, nom. cons.

References. H.F. Blanford 1886. The Silver Ferns of Simla and their allies, J. Simla Nat. Hist. Soc. 1(2): 13-22 (reprinted 1886 with pagination 1-12). H.F. Blanford 1888. A list of the ferns of Simla in the N.W. Himalaya between levels of 4,500 and 10,500 feet (1372-3200 m.), J. Asiatic Soc. Bengal 57(2): 294-315 et tt. 16-20 (reprinted c. 1988, New Delhi). R.C. Ching 1941. The studies of Chinese Ferns - XXXI, Hong Kong Nat. 10: 194-204. I Manton & W.A. Sledge 1954. Observations on the cytology and taxonomy of the pteridophyte Flora of Sri Lanka, Phil. Trans. Roy. Soc. (London), ser. B, 238: 150-152 (127-185). B.K. Nayar 1962. Ferns of India No. VI Cheilanthes, Bull. Natal. Bot. Gardens 68: 1-36. G. Panigrahi 1962. Cytogenetics of apogamy in *Aleuritopteris farinosa* (Forsk.) Fée Complex, Nucleus 5(1): 53-64. S.C. Verma & S.P. Khullar 1965. Cytology of some W. Himalayan Adiantaceae (sensu Alston) with cytotaxonomic comments, Caryologia 18(1): 85-106. I. Manton & S.K. Roy 1966. The cytotaxonomy of some members of the *Cheilanthes farinosa* complex in Africa and India, Kew Bull. 18(3):

553-565. R.E.G. Pichi Sermolli 1977. Tentamen pteridophyltorum genera in ordinem taxonomicum redigendi, Webbia 31(2): 336-392 (313-512). S.K. Wu 1981. A study of the genus *Aleuritopteris* Fée in China, Act. Phytotax. Sinica 19(1): 57-74. Y. Saiki 1984. Note on Ferns (2) Asiatic Species of the *Aleuritopteris farinosa* Complex, J. Phytogeog. Taxon. 32(1): 1-13. Y-Saiki 1984. Note on Ferns (4) Classification of the *Aleuritopteris farinosa* Group, J. Phytogeog. Taxon 32(2): 91-98. R.M. Tryon, A-F. Tryon & K.U. Kramer 1990. *Cheilanthoideae* in K.U. Kramer & P.S. Green (eds. of vol. 1) in K. Kubitzki (ed.), The Families and Genera of vascular plants 1.: 240-247. Wu, S.-K. 1990. *Sinopteridaceae* in Academia Sinica (eds.), Flora Reipublicae Popularis Sinicae 3(1): 117-173. Type. *C. micropteron* Sw. *Allosorus* Bernh., Schrad. Nedd J. 1(2): 5, 36. 1806. *Aleuritopteris* Fée, Mém. Fam. Foug. V, Gen. Fil.: 153. 1852. *Cheilosoria* Trevis., Atti Ist. Venete V. 3: 579. 1877. *Sinopteris* C. Chr. & Ching, Bull. Fan Mem. Inst. Biol. 4: 359. 1933. *Negripteris* Pich.Serm., Nuov. Giorn. Bot. Ital., ser. 2, 53: 130. 1946. *Leptolepidium* Hsing & S.K. Wu, Act. Bot. Yunnanica 1: 115. 1979. Misapplied names. *Pteris* L., *Notholaena* R.Br., *Cassebeera* Kaulf., *Pellaea* Link.

Terrestrial and often lithophytic ferns up to c. 50 cm high. Rhizome thin but densely surrounded by hollow, old frond-bases, short, erect, bearing small, lanceolate scales at its apex and a crown of fronds. Stipe thin, fragile, very dark purple or black, terete or somewhat flattened adaxially, bearing narrowly lanceolate, acuminate, dark-red scales at least at their bases (lowest scales usually having darker centres), or bicolorous brown scales with paler margins, depending on species, they are characteristically dispersed on the lower part or the whole stipe and sometimes rachis, they may, sometimes, bear small ± adpressed fibrils and glands as well and contain 1 to 2 separate vascular bundles towards the base; lamina varying from deeply lobed to up to four times pinnate, pinnules catodromous or andromous, veins free, often bearing dense ceraceous glands below resulting in a dense white farina on the undersurface; *costae* and costules terete, often depressed between raised laminar areas (between the venulets), naked or bearing small fibrils; ultimate segments widely adnate to stalked, symmetrical. Sori terminal on the venulets, but technically submarginally as they are protected by the folded-under segment-margin which is thin, pale to brown and ± specialised into a pseudo-indusium., the latter being continuous and uninterrupted (in species with more palmately shaped laminae), or divided into separate, but ± contiguous lobes (the lobes being *entire*, irregularly toothed or fimbriate), occasionally in shaded plants, not so abundantly fertile, the pseudo-indusium may, occasionally, be poorly developed or almost absent, like in some species outside the area. Spores tetrahedral-globose, trilete, the ± loose sporoderm is characteristically verrucose, granulose or ± papillose-echinate, depending on species. Chromosome base number $x = 28$ (in one case), 29 or 30, without any morphological distinctions between the group with 29 and that with 30.

A large genus of perhaps 260 species world-wide, especially prevalent in semi-arid zones, with centres of diversity in Central America, the Macaronesian-Mediterranean area, S. and E. Africa and especially the very rich Sino-Himalayan region. Usually occurring in rock-crevices or among rocks in open sunny areas, on open soil, stream or road-banks, or walls. Probably originally of Sino-Himalayan affinity and occurring world-wide, including Oceania, but absent from the most northerly temperate areas.

The genus contains a considerable range of morphological diversity; although it has been split up into a number of minor genera, these are probably no more than groups of related species which follow a morphological pattern which, as a whole, can be ± easily recognised as a genus. More importantly the groups, though distinctive when only a few representatives are considered, as in Sri Lanka, are not discretely separated from each other by taxonomic characters, even taken in combination as has been claimed. Thus the Sino-Himalayan *C. argentea* Kunze group is very close to the reticulate-veined *Cassebeera* (= *Doryopteris*) and even contains some non-farinous plants. The very narrow bicolorous scales of *C. bicolor* (Roxb.) Griff. ex Fraser-Jenk. contradict Ching's attempted separation of *Aleuritopteris* from *Cassebeera* as do the bicolorous scales of several other species placed in *Aleuritopteris* and as do the ebony-black stipes and axes of *C. anceps* Blanf. and other species. Another recently separated group, *Leptolepidium*, sensu S.K. Wu, cannot be clearly separated since *C. leptolepis* Fraser-Jenk. clearly forms a link between *C. bicolor*, with which it is widely confused in herbaria, and that genus.

Concerning the recently resurrected *Cheilosoria*, the farinose species *C. thwaitesii* Mett. ex Kuhn is obviously very closely related to *C. chusana* Hook, which was placed in it by S.K. Wu (Flor. Rep. Pop. Sinica 3(1): 117-119. 1990); it thus connects *Aleuritopteris* and/or *Leptolepidium* to *Cheilosoria* and thence, via *Cheilanthes chusana* Hook. and *C. Swartzii* Webb. ex Benth. to the European *C. acrostica* (Balbis) Tod. and the type of *Cheilanthes, C. micropteron*.

The various minor genera thus appear to be mainly based on degree of dissection and lamina-shape. Though they reflect certain groups of species they are not accepted here as being of sufficient taxonomic importance to warrant constitution as genera, which should be more significan₊ entities. The main groups are treated here as sections within *Cheilanthes*, which itself belongs to subfamily *Cheilanthoideae* (J. Smith) W.C. Shieh.

Very great confusion has always surrounded the identification of the farinose species within Section *Aleuritopteris* (Fée) Baker, because it requires the use of a hand-lens for the examination of scales in order to recognise them. They are widely misidentified in herbaria and in practically all the literature post 1960, especially in Indian and Japanese work on Asian species. Yet the species are remarkably constant and distinctive in their scale-characteristics and are very easy to identify thereby, much more so than in genera such as *Dryopteris* where there is far greater variation. Recent studies by Wollenweber and Fraser-

Jenk. (1995) have cleared up much of the confusion among S. Asian species and in previous Japanese chemical papers and show that the flavonoid chemistry of at least these species is consistent and differs clearly from one species to the next, even confirming the presence of hybrids with abortive spores between species based on different chromosome base numbers.

KEY TO THE SPECIES

1 Frond with white farina underneath (unless washed off herbarium-specimen by alcohol immersion), pinna-segments mostly widely attached to their axes, except for the lowest ones
  2 Scales at the stipe-base bicolorous, mid-brown with pale margins (at least towards their apices), sometimes with a ± uniform pale-brown base
    3 Scales v. narrowly linear, confined to very base of the stipe. . . . .
    . . . . **1. C. bicolor**
    3 Scales narrowly lanceolate, reaching the top of the stipe . . . .
    . . . . **2. C. anceps**
  2 Scales at the stipe-base concolorous, deep red or ± dark-brown
    4 Lamina short (to c. 10 cm tall), deltate or deltate-lanceolate, widest at the base . . . . **3. C. krameri**
    4 Lamina long (to c. 70 cm tall), ± narrowly lanceolate, gradually narrowed to base and apex
      5 Lamina narrowly lanceolate, gradually narrowed to base and apex . .
      . . . . **6. C. opposita**
      5 Lamina deltate, widest at the base . . . . **7. C. tenuifolia**
1 Frond without white farina, segments mostly narrowly attached to their axes or stalked.
  6 Stipe-base scales deep-red, ovate, pinnules deeply lobed, white farina below very dense . . . . **4. C. bullosa**
  6 Stipe-base scales dark-brown, lanceolate, pinnules shallowly lobed or un-lobed, white farina below ± faint (occasionally nearly absent). . . . .
  . . . . **5. C. thwaitesii**

Sect. *Aleuritopteris* (Fée) Baker in Hook. & Bak., Syn. Fil.: 141. 1867.

1. Cheilanthes bicolor (Roxb.) Griff. ex Fraser-Jenk., Pakistan Syst. 5: 85-120. ["1991"] 1992.
Type. From "Rohilcund" [Uttar Pradesh, N. India] (BR).

Pteris bicolor Roxb. in Griff., Calcutta J. Nat. Hist. 4: 483-520. 1844 (incl. err. "*P. discolor*").
Cheilanthes farinosa var. *tenera* C.B. Clarke & Baker, J. Linn. Soc. (Lond.) 25: 441. 1889.
Aleuritopteris kathmanduensis Ching & S.K. Wu in S.K. Wu, Act. Bot. Yunnan. 5: 167. 1983.

Aleuritopteris longipes Ching & S.K. Wu in S.K. Wu, Act. Bot. Yunnan. 5: 165. 1983.

M i s a p p l i e d   n a m e s. *Cheilanthes farinosa* (Forssk.) Kaulf., *Aleuritopteris pulveracea* (C. Presl) Fée, *Cheilanthes hancockii* Baker, *Leptolepidium dalhousiae* (Hook.) Hsing & S.K. Wu.

Rhizome very short, thin, erect, usually unbranched, bearing a tuft of small, thin, narrowly linear-lanceolate (up to c. 3 × 0.5 mm), entire, mid-brown, dark-centred scales at apex. Fronds up to c. 5 in apical crown, stipe ± upright or slightly spreading, being bent at apex to hold the lamina at c. 45° to the ground, up to c. 25 cm long, usually c.1 ½ times as long as lamina, very thin, terete, purplish-brown or ± dark reddish-brown, glossy, the very base bearing a small tuft of partly deciduous, very narrow scales, resembling rhizome-scales, becoming very scattered then absent by $^1/_3$ of the way up the stipe, naked above (reaching top $^1/_4$ of stipe in young fronds on small plants); rachis similar to upper stipe but narrowly grooved adaxially; lamina up to 16 cm wide, 20 cm long, bipinnate below, tripinnate in lowest basiscopic pinnule, deltate-pentagonal with c. 15 pairs of opposite pinnae, markedly widest at base, herbaceous, thin, mid- or yellowish-green, without scales, under surface with a greyish-, bright-, or slightly creamy-white farina of minute, sessile, ovate, globose, ceraceous glands, top surface glabrous, sunken veinlets causing wrinkled appearance particularly in exposed plants; pinnae subdeltate to elongated triangular-lanceolate, widest at their bases, up to c. 8 cm long, 7 cm wide, lower ones stalked but remainder adnate to rachis at their bases, symmetrical except lowest pair in which lower basiscopic pinnules up to c. 3 times as long as acroscopic ones, inserted at c. 60°-70°, pinnate in the lowest pinnae, where lowest basiscopic pinnules pinnatifid, with up to c. 10 pairs of alternate pinnules; costae and costules distinct, terete, dark-brown and glossy below, ± deeply inserted, indistinct and greyish-green above; pinnules or pinna-lobes anadromous at base of lowest pinna, catadromous above, adnate to costae and mostly narrowly joined at their basiscopic bases by a decurrent wing of laminar tissue, but the lowest ones on the lowest pinnae become fully separate, sloping, lowest basiscopic pinnule on first pinna markedly the longest, upper pinna-lobes with ± shallowly rounded-lobed bases, lower ones deeply pinnatifidly lobed, lobes elongated-triangular, adnate, ± narrowly or obtusely rounded. Sori marginal in continuous line around laminar area throughout, protected by a series of pseudo-indusia; pseudo-indusia, consisting of downward-curved lobes of frond-margin, specialised into short, ± semi-circular lobes of thin, white tissue with an irregular edge of small fimbriations and lacerations, positioned parallel to lamina covering sporangia before turning brown, shrinking slightly and ± lifting to reveal ripe sori. Spores trilete, globose, uniform., small, dark; sporoderm consisting of papillae often with curved-over apices or appearing slightly hooked. Cytotype diploid, sexual, n = 30 (with 30 bivalents at meiosis); reports of n = 29 may not have referred to this species.

D i s t r. Presumably at lower or lower-mid level, probably on the edge of or within the dry zone in Sri Lanka where it is evidently very rare. Possibly a south-east Asian element but with its present centre of distribution in the Indian subcontinent where it is the most common and widespread species of *Cheilanthes*, in an unusual distribution pattern. Sri Lanka; north Pakistan; India (throughout the foothills and outer ranges of the Himalaya from west to east, scattered throughout the whole of western, central and peninsular India in most of the main hill-ranges, Southern India in the Nilgiri, Annamalai and Palni Hills, eastern Ghats, southern Ghats, Assam); Nepal; Sikkim; Bhutan; Myanmar (Burma); north Thailand; Laos; a very similar plant in Sumatra may be the same.

E c o l. Occurs on open stony- or mud-banks of roads, rivers etc., below rocks and on slopes among bushes, also in walls. It requires places with good or severe drainage, often drying up and the fronds dying during the dry season. Occurs from c. 300-1700 m (in India).

N o t e. In common with almost all other species of the genus, apart from the *C. argentea* group, *C. bicolor* was previously lumped together with *C. farinosa* and the African *C. argyrophylla* (Sw.) Cordem. (= *C. pulveracea* C.Presl, nom. superfl., *C. afra* Pich.Serm.); later it was specifically referred to that species by Blanford (1884 and 1886), who evidently did not know *C. farinosa*, which is confined to Africa and Yemen and is more closely related to, but distinct from *C. anceps*. In India *C. bicolor* has been much confused, even with such distinct species as *C. leptolepis* and *C. grisea* Blanf., or with the C. and S. American *C. chihuahuaensis* (Saiki) Fraser-Jenk. (= *Aleuritopteris mexicana* Fée, non *C. mexicana* Davenp.). It was separated from *C. farinosa* by Manton & Roy (1966) on cytological grounds, but without providing a name for it. They pointed out correctly that *C. farinosa* almost certainly corresponded with their African triploid and that the present Himalayan diploid was also morphologically distinct. Fraser-Jenk. (Pak. Syst. 5(1-2): 85-120. 1992 and Bot. Helv. 102(2): 143-157. 1993, also 1997) found that the type of *C. bicolor*, as might be expected from its locality, is the same as this species and named it accordingly. *C. bicolor* is actually very distinct from *C. farinosa* in its narrow, bicolorous scales, thin, purple-brown (not black) stipe and rachis and shorter, markedly deltate-pentagonal lamina and is unrelated to that group of species.

S p e c i m e n   E x a m i n e d: LOCALITY UNKNOWN: *Henderson C.P. 2987* (E).

2. Cheilanthes anceps Blanf., J. Simla Nat. Hist. Soc. 1: 21. 1886.
Lectoype. From India: Simla region. *H.F. Blanford* (K).

Cheilanthes candida Zoll., Nat. Gen. Arch. Neerl. Ind. 2: 203. 1845, non Mart. & Gal. 1842.

Cheilanthes farinosa var. *anceps* (Blanf.) Blanf., J. Asiat. Soc. Bengal 57(2): 301. 1888.

Aleuritopteris anceps (Blanf.) Panigrahi, Bull. Bot. Surv. India 2: 321. 1961.

A. pseudofarinosa Ching & S.K. Wu ex S.K. Wu, Act. Phytotax. Sin. 19(1): 72. 1981.

Aleuritopteris interrupta Saiki, J. Phytogeog. Taxon. 32(2): 5. 1984.

Aleuritopteris javanensis Saiki, J. Phytogeog. Taxon. 32(1): 6-7. 1984.

M i s a p p l i e d   n a m e. *Cheilanthes farinosa* (Forssk.) Kaulf.

Rhizome short, thick, erect, usually unbranched, bearing a tuft of small, lanceolate, entire, thin, dark-brown, darker-centred scales at its apex. Fronds up to c. 8 in a crown; stipe spreading, up to c. 20 cm long, usually ± same length as lamina, slightly thick, terete, ebony-black, glossy, bearing ± scattered, narrowly lanceolate, bicolorous scales similar to rhizome-scales, but longer, up to c. 4 × 1 mm., the centres ± pale- to mid-brown, edges pale, sometimes base of scale ± concolorous pale-brown, apices long-acuminate, scales decreasing in size, but covering entire stipe; rachis similar to upper stipe but shallowly grooved adaxially and mostly without scales except sometimes at the base; lamina ± small to medium-sized, bipinnate below, elongated triangular (up to c. 12 cm wide and 20 cm long), widest at the base, slightly thickly herbaceous, mid-green above, without scales, with a markedly bright, snow-white, dense farina below, consisting of minute, sessile, globose, ceraceous glands, upper surface ± smooth; pinnae c. 15 pairs, opposite, elongated triangular-lanceolate, widest at their bases, up to c. 6 cm long, 5 cm wide, lower ones stalked, remainder adnate to rachis, nearly symmetrical except the lowest two pairs, where their lower basiscopic pinnules are up to c. twice as long as the acroscopic ones, inserted at c. 60°-70°, pinnate in the lowest pinnae; costae and costules distinct, terete, black and glossy below, indistinct, flat and greyish-green above; pinnules or pinna-lobes up to c. 7 pairs, anadromous at base of lowest pinna, catadromous above, wider than in *C. bicolor*, adnate to costa and mostly narrowly connected at basiscopic bases by a decurrent wing of laminar tissue, but lowermost on lowest pinnae fully separate from each other, sloping, lowest basiscopic pinnule on the lowest pinna the longest, but approached by next, pinnule-apices rounded-pointed or obtuse, upper pinna-lobes ± unlobed, lowest ones becoming well lobed with ± shallow, rounded-triangular lobes. Sori borne in a continuous line around margins of lamina throughout, protected by a series of pseudo-indusia; pseudo-indusia downward-curved green lobes of lamina-margin, with specialised edges consisting of ± long, narrow lobes of thin, pale-brown tissue divided into a small number of long, ± narrow, toothed lacerations at their edges, and lying over the sporangia, turning mid-brown and bending back to reveal sori when ripe. Spores large, uniform., trilete, mid-brown, globose, sporoderm patterned with many small, pale, partly joined, ± irregular, rounded, clear papillae. Cytotype tetraploid, sexual, n = 58 (with 58 bivalents at meiosis) from Sri

Lanka and C. and S. India; a report of n = 29 (Verma in Mehra 1961) from the Himalaya referred to *C. formosana*.

D i s t r. At lower-mid level at the edge of the dry zone in N. central and E. central Sri Lanka. Uncommon and scattered. A S.E. Asian element which is also present throughout the outer ranges of the Indo-Himalaya and just reaches S.W. China as a rarity (though widely over-reported in error for *C. formosana*). Sri Lanka; North Pakistan; India (S. Kashmir and throughout the foothills and outer ranges of the Himalaya from west to east, S. Orissa (Kashipur), Madhya Pradesh (Bailadila), Rajasthan (Mt. Abu), Maharashtra (Mahabaleshwar), Hawalbagh, Bellecherry, S. India in the N.W. Ghats, Nilgiri, Annamalai, Palni and Shevaroy Hills, Assam); Nepal; Sikkim; Bhutan; S.W. China (Yunnan, Kwangtung); Taiwan; Burma; Thailand; Indonesia (Java, Timor, Lombok, Bali, Celebes).

E c o l. Occurs on shaded mud-banks of roads, cuttings and streams with good drainage, drying up and the fronds curling up during the dry season, but often surviving and uncurling again during rain. Occurs from c. 600-1000 m in Sri Lanka.

N o t e. Like other farinose *Cheilanthes* in Asia this species has long been confused with the Afro-Arabian *C. farinosa* and of all Asian species it is the most close to it morphologically, but has bicolorous scales (slightly less so in S.E. Asia) and is not as large or dissect.

It has mainly been confused with *C. formosana* Hayata (= *C. brevifrons* (Khullar) Khullar, see Fraser-Jenkins 1997), a much commoner species widespread throughout the Himalaya, also in C. and S. India (Rajasthan, Mahabaleshwar and Concan), east to China, Taiwan, Myanmar (Burma), Thailand and the Phillipines. This is a diploid species with smaller, narrower fronds with a wrinkled upper surface and narrower, bicolorous stipe-scales, extending well up the rachis. Even Blanford's protologue of his *C. anceps* displayed a mixed concept between the two, but the bulk of his concept seems to have been the present species, which is so lectotypified here (above).

This species is difficult to separate from *C. doniana* Fraser-Jenk. & Khullar (= *C. dealbata* D. Don, non Pursh., *Aleuritopteris doniana* S.K. Wu) from the E. part of the W. Himalaya east to S.W. China, Myanmar and Thailand, which has longer fronds with less lobed and smoother pinnules, a brighter white undersurface, less bicolorous scales and narrower sori.

S p e c i m e n s E x a m i n e d. MATALE DISTRICT: Matale, *O. Moore s.n.*, Sept. 1864 (E); Laggala Division, N.E. of Matale, c. 1 km above and W. of Illukumbara, c. 10 km W. of Pallegama, E. of Laggala, dry roadside woods on slope, c. 650 m., *Fraser-Jenkins field nos. 183*, with *Jayasekara & Bandara*, 15 Sept. 1993 (BM., K, US, PDA); Dry road side woods on slope, c. 1km above and west of Illukkumbura, c. 10 km W. of Pallegama, E. of Laggala, N.E. of Matale, Alt. 650 m., 15 Sept 1993, *Fraser*

–Jenkins with *Jayasekara & Bandara* Field No. *183* (K); Lagalle, Herb. *Brodie s.n.*, May 1864 (E); Managalla, 2000 ft, *Hancock 30* (CAL, US p.p.). KANDY DISTRICT: Hingurugama [?Hingumgama], *Freeman 87c* p.p., 79 D, 80 E (BM); forest $3^1/_2$ km N. of Corbet's Gap on road to Winchfield and Mimure, N. of Hunnasgiriya, 880 m., *Fraser-Jenkins field no. 285*, with *Abeysiri & Gunawardena*, 14 Oct. 1993 (BM., K, PDA, US); between junction with Corbet's Gap road and Pahalena Ella, S.W. of Karambaketiya, N. of Corbet's Gap on road to Nitre Cave, *Fraser-Jenkins field no. 295*, with *Abeysiri & Gunawardena*, 14 Oct. 1993 (BM., PDA). NUWARA ELIYA DISTRICT: Badulla road, below Hakgala, *Sledge P. 309* (BM-Manton and Sledge box).

3. Cheilanthes krameri Franch. & Sav., Enum. Plant. Jap. 2: 212, 619. 1879. Type: From Japan: Honshu (P).

Aleuritopteris krameri (Franch. & Sav.) Ching, Hong Kong Nat. 10: 202. 1941.

M i s a p p l i e d   n a m e s. *Cheilanthes farinosa* (Forssk.) Kaulf., *C. grisea* Blanf.

Rhizome short, thin, erect, usually unbranched, bearing a tuft of small, lanceolate, entire, thin, concolorous (or sometimes black-centred), dark-reddish-black scales over its apex. Fronds up to c. 8 in a loose, apical crown; stipe sticking out ± horizontally at its base, then curved, often so that fronds hang down, up to c. 7 cm long, usually longer than lamina, thin, terete, purple or reddish-black, glossy, with scattered, partly deciduous, ovate-lanceolate, concolorous, deep-red scales (up to c. 2×1 mm), with abrupt, long-acuminate apices, scales confined to basal $^1/_4$ or less of stipe, rachis similar to upper stipe, but more black and grooved adaxially, without scales; lamina small, becoming bipinnate below, triangular-lanceolate (up to c. 6 cm wide, 10 cm long), widest at base, fragile, crispaceous-herbaceous, without scales, midgreen above, with very scattered, white farina-glands creating a very faint white powdering above, which is lost on old fronds, but a bright-white or often slightly greenish-yellowish-white farina below, consisting of minute, sessile, globose, ceraceous glands, upper surface deeply wrinkled (bullulate), the veins deeply impressed around small bulges of lamina; pinnae c. 8 opposite pairs, elongated triangular-lanceolate, widest at their bases, up to c. 3 cm long and 2 cm wide, the lowest pair stalked, others adnate to rachis at their bases, nearly symmetrical except lowest with lower basiscopic pinnules developed and up to c. three times as long as acroscopic ones, distal pinnae often slightly wider shortly below their apices than at their bases, inserted at c. 90°, becoming pinnate in the lowest pinnae; costae and costules distinct, terete, black and glossy below; indistinct, indented and greyish-green above; pinnules or pinna-lobes up to c. 6 pairs, ± alternate, wide, adnate to costa, mostly narrowly joined by a decurrent wing of laminar tissue, only lowest one on lowest pinna fully separate, ± at 90° to costa, lowest opposite pair on lower few pinnae very close to rachis, the lowest basiscopic pinnule the longest, but approached by next, lowest pinnules becoming shallowly lobed at the sides, with ± small, rounded, obtuse lobes, pinnule-apices obtuse. Sori borne in a ± continuous, or slightly broken line around margins of whole laminar area, protected by a series of pseudo-indusia; pseudo-indusia downward-curved green lobes of lamina-margin, with specialised edges forming narrow lobes of thin, pale-brown tissue divided into a number of ± long, narrow lacerations at their apices and lying over the sporangia before turning mid-brown and bending back to reveal sori when ripe. Spores small, regular trilete, dark-brown, globose, sporoderm ± smooth with very fine, dark granulosity. Cytotype diploid, sexual, n = 30 (with 30 bivalents at meiosis), from Sri Lanka (Manton & Sledge 1954).

D i s t r. At mid- to higher-mid level in central Sri Lanka. Fairly common. A. S.E. Asian element extending as far as S. India and Thailand. Sri Lanka; S. India (Nilgiri and Shevaroy Hills); Taiwan; Japan (rare); Thailand; Indonesia (Java, Lombok); the Philippines; ?Fiji (*Horne*, cited by Copeland (1929) sub *C. farinosa*).

E c o l. On damp, shaded, vertical road-banks/rocks, often in walls, c. 1000-2000 m in Sri Lanka.

N o t e. This species has been widely misidentified. Panigrahi (1960) misidentified Manton & Sledge's plant as *C. grisea*, and later misapplied the same name to several other, less closely related species including *C. formosana* and even *C. bicolor* in the Himalaya and central India. *C. krameri* belongs to a group of species with concolorous, red stipe-scales including the small, high-altitude Himalayan *C. grisea* and the larger, central Himalayan *C. platychlamys* (Ching) Fraser-Jenk., the long, narrow-fronded *C. bullosa* Kunze (below) and *C. chihuahuaensis* from the New World, which is similar to a slightly more finely dissect *C. bullosa*. Although the species in this group are very closely related and difficult to tell apart they all appear to constitute separate species. *C. krameri* in Sri Lanka maintains its small, deltate fronds, as opposed to the long narrow ones of *C. bullosa* which remain recognisably narrow even in small, immature plants. It appears to be identical to the Japanese and Taiwan populations. Plants of *C. krameri* in the Philippines often reach a larger size but appear to belong to the same species and smaller plants are indistinguishable. *C. krameri* had not previously been reported or identified from outside Japan and Taiwan.

S p e c i m e n s E x a m i n e d. KANDY DISTRICT: Hingurugama [?Hingumagama], *Freeman 78c* p.p. (BM); c. 1 km N. of and below Corbet's Gap on road to Mimure, N. of Hunnasgiriya, N.E. of Kandy, cliff, 1150 m., *Fraser-Jenkins field no. 278*, with *Abeysiri & Gunawardena*, 14 Oct. 1993 (BM., K, PDA, US). BADULLA DISTRICT: c.1km below Rendapola, c. 4 km above and west of Boragas c. 4km E. of and below Hakgala, on road from Nuwara Eliya E. to Badulla, Alt. c. 1524 m., 22Sept 1993, *Fraser-Jenkins* with *Jayasekara, Samarasinghe & Abeysiri* Field No.*192* (K); in wall by path through tea-estate, Namunakula, 1676 m., *Sledge 1219* (BM). NUWARA ELIYA DISTRICT: Nuwara Eliya, 1950 m., *Schmid 1296* (G, BM); above Ramboda Falls, *Fosberg & Sachet 53219* (US); Ambawela road, Hakgala, *Sledge P. 242* (BM-Manton and Sledge box); c. 1 km below Rendapola, c. 4 km above and W. of Boragas, c. 4 km E. of Hakgala on road towards Badulla, c. 1524 m., *Fraser-Jenkins field no. 192*, with *Jayasekara, Samarasinghe & Abeysiri*, 22 Sept. 1993 (BM., K, PDA, US). LOCALITY UNKNOWN. *Capt. Sall s.n., 1857,* leg. *Thwaites,* 1863 (BM).

4. Cheilanthes bullosa Kunze, Linnaea 24: 274. 1851. Type: From S. India: Nilgiris. *Schmid 33.* (Kleptotype fragment in PE).

Cheilanthes rigidula Wall., Num. List: no. 2175. 1828, nom. nud.
Aleuritopteris indica Fée, Mém. Fam. Foug. 5, Gen. Fil.: 154. 1852.
Aleuritopteris bullosa (Kunze) Ching, Hong Kong Nat. 10: 202. 1941.
Cheilanthes flaccida (Bedd.) Mehra & Bir, Res. Bull. Panjab Univ. 15: 10. 1964.
Aleuritopteris flaccida (Bedd.) B.K. Nayar & S. Kaur, Comp. Beddome's Handb.: 25. 1974, comb. inval., in syn.

M i s a p p l i e d   n a m e. *Cheilanthes farinosa* (Forssk.) Kaulf., *C. thwaitesii* Mett. ex Kuhn.

Rhizome short, thick, erect, usually unbranched, bearing a tuft of small, lanceolate, entire, ± thick, dark-centred, dark-reddish-black scales over its apex. Fronds up to c.10 in a loose, apical crown; stipe slightly spreading, up to c. 50 cm long, usually ± same length as lamina, thick, terete, ebony-black, glossy, bearing scattered, partly deciduous, ovate-lanceolate, deep-red scales (up to c.3 × 1.5 mm), with long acuminate apices, scales confined to basal $^1/_4$ of stipe, leaving an asperous bump when falling; rachis similar to upper stipe but grooved adaxially, without scales; lamina medium to large, bipinnate, tripinnatifid below, elongated, narrowly triangular-lanceolate (up to c.13 cm wide, 50 cm long), widest in lower ½, crispaceous-herbaceous, mid-green above, bearing many minute, green, glandular granules above (in microscope), without scales, with very scattered white farina-glands creating very faint white powdering above, which is lost on old fronds, but with a very bright-white farina below, consisting of minute, sessile, globose, ceraceous glands, upper surface deeply wrinkled (bullulate) with veins deeply impressed around small bulges of lamina; pinnae opposite, up to c. 17 pairs, triangular-lanceolate widest at bases, up to c. 6 cm long and 4 cm wide, minutely stalked, lowest acroscopic pinnule (sometimes also basiscopic one) on each pinna ± parallel to and often slightly overlapping rachis, mostly asymmetrical with lower basiscopic pinnules developed and up to c. twice as long as acroscopic ones, inserted at c. 70°-90°, pinnate; costae and costules distinct, terete, black and glossy below; indistinct, indented and greyish-green above; pinnules or pinna-lobes up to c. 10 pairs, mostly alternate, varying from narrow to wide, adnate to costa and mostly narrowly connected by a decurrent wing of laminar tissue, but lowest pair on lower pinnae fully separate, ± at right-angles to costa, lowest opposite pair on each pinna very close to rachis, lowest basiscopic pinnule on lowermost few pinnae longest, but approached by next, upper pinna-lobes shallowly lobed to ± unlobed, the lower ones lobed up to about half their depth or becoming pinnatifid, with small, ± narrow, rounded lobes, pinnule-apices usually obtuse. Sori borne in a continuous line around margins of whole laminar area, protected by a series of pseudo-indusia; pseudo-indusia downward-curved, green lobes of lamina-margin with specialised edges forming narrow lobes of thin, pale-brown tissue with an irregular, slightly erose margin, lying over sporangia before turning mid-brown and lifting somewhat to reveal sori when ripe. Spores, small, uniform., trilete, dark-brown, globose, sporoderm bearing a ± fine, dark granulosity. Cytotype unknown due to previous identifications from India being unreliable and inaccurate.

D i s t r. At higher-mid to high levels in central Sri Lanka. Common locally. An endemic Sri Lankan and S. Indian species, probably of S.E. Asian affinity. Sri Lanka; S. India (N.W. Ghats, Nilgiri, Annamalai, Palni Hills and southern Ghats). Reports from elsewhere, including China, Africa and even Australia are erroneous.

E c o l. Occurs on damp rocks above streams, or on damp slopes among bushes, from c. 1500-2300 m in Sri Lanka.

N o t e. This species is related to *C. krameri* in the group of concolorous, red-scaled species, but is consistently distinct in its much larger, very long and narrow fronds, some of the longest in the genus. A similarly long-fronded and bullulate species is *C. argyrophylla* (Sw.) Cordem., from E. Africa, Cameroun and Réunion, but this has wider segments and wider, paler scales. It is responsible for erroneous reports of *C. bullosa* from Africa. The name *C. bullosa* has been widely misapplied in Asia, like most names in the genus, and the species has been widely referred to under various other names. Nayar (1962) figured *C. bullosa* from S. India under name *C. thwaitesii* and Kunze himself included this species under both *C. bullosa* and *C. dealbata*.

S p e c i m e n s   E x a m i n e d. BADULLA DISTRICT: edge of track through forest, Namunukula, 6000 ft, *Sledge 1218* (BM); roadside rocks above West Haputale on road from E. Horton Plains down to Haputale, c. 1750 m., *Fraser-Jenkins field no. 468*, with *Abeysiri*, 27 Oct. 1993 (BM., K, PDA, US). NUWARA ELIYA DISTRICT: Nuwara Eliya, *Beckett s.n.*, Nov. 1885 (DD); Nuwara Eliya, 6000 ft, *Naylor-Beckett 285*, herb. *Moore* (CAL, GH); Nuwara Eliya, 1950 m., *Schmid 1283* (G); Nuwara Eliya, *Freeman 76A, 77B* (BM); *Le Fall s.n.*, March 1863 (UC); "Ramboddi", 5000 ft, *Gardner 1168* (BM);

Ramboda road, near Ramboda, 3200 ft, *Ballard 1131* (US); rocks and cliff by stream on S. side of ridge, c. 1-2 km along track from c. 1 km below top of Top Pass (top part of Ramboda Pass) on W. side, heading N.W. towards Maturata, W. of Nuwara Eliya, 1700-1800 m., *Fraser-Jenkins field no. 317*, with *Abeysiri & Gunawardena*, 25 Oct. 1993 (BM., K, PDA, US); steep bank in Eucalyptus forest above road, just above lower army checkpost at bottom of road up S. side of Pidurutalagala mountain, on lower slopes, N. of Nuwara Eliya, c. 1950 m., *Fraser-Jenkins field no. 377*, with *Abeysiri*, 26 Oct. 1993 (BM., K, PDA, US); south summit of Hakgala peak, 6800 ft, *Sledge 628* (Z); Hakgala mountain, behind Hakgala Botanical Garden. 2100 m., *Faden & Faden 77/15* (Z, US, GH, F, K). LOCALITY UNKNOWN: *Thwaites C.P. 2987* (BM., G, E, CAL, BO); *Dr. Mayr s.n.*, rec. 1877 (G); *Colonel Walker s.n.*, herb. *Graham.*, rec. 1946 (G); *Fraser 145* (BM., CAL); *Capt. Sall s.n.*, 1857, leg. *Thwaites*, 1863, p.p. (BM); *J. Wolsam C.P. 2987*, 23 Aug. 1936 (CAL); *Harvey s.n.*, Sept. and Dec. 1853 (CAL); *Ferguson 54(25/53)* (US, GH, MO), *39* (GH); *Boot s.n.* (GH); *Wall s.n.* (BO).

5. Cheilanthes thwaitesii Mett. ex Kuhn, Linnaea 36: 82. 1869. Lectotype: From Sri Lanka: *"Thwaites C.P. 1321"* (B); isolectotypes: "Damboul, *Thwaites C.P. 1321"* (BM., E).

Cheilanthes laxa Moore ex Bedd., Handb. Ferns Brit. Ind. Ceylon: 92. 1883.
Cheilanthes keralensis Nair & Ghosh, J. Ind. Bot. Soc. 55(1): 52. 1976.
Aleuritopteris thwaitesii (Mett. ex Kuhn) Saiki, J. Phytogeog. Taxon. 32(1): 11. 1984.

M i s a p p l i e d   n a m e. *Cheilanthes varians* Wall. ex Hook. [= *C. belangeri* (Bory) C. Chr.].

Rhizome short, thick, erect, usually unbranched, bearing a tuft of small, lanceolate, entire, dark-brown scales over its apex. Fronds up to c. 7 in an apical crown, upright; stipe upright, up to c. 12 cm long, usually less than half the length of the lamina, thick, terete, purple- or brownish-black, glossy, bearing many, ± scattered, partly deciduous, narrowly lanceolate, concolorous mid-brown, often slightly twisted scales (up to c. 4 × 1 mm) throughout most of its length, decreasing in size and becoming more scattered on the upper stipe, leaving small asperations when falling; rachis similar to the upper stipe but grooved adaxially, bearing a few linear scales towards its base; lamina medium to large-sized, bipinnatifid, bipinnate below, narrowly lanceolate (up to c. 8 cm wide and 35 cm long), widest from just above base to the middle of frond, thinly herbaceous, mid- or slightly grey-green above, without scales, at least younger fronds with a rather faint, pale-greyish-white farina below which is lost in old fronds (removed by treating herbarium-specimens with alcohol, or not drying them rapidly), consisting of minute, sessile, globose, ceraceous glands, upper surface smooth, often bearing a few small green bulbils on the lower surface of mid or upper pinna-costae from which vegetative reproduction can occur in damp conditions; pinnae up to c.15 ± alternate or subopposite pairs, elongated-triangular, widest at their bases, up to c. 4 cm long, $1^1/_2$ wide, very shortly stalked, but close to the rachis, symmetrical, inserted at c. 90°, just becoming pinnate below, costae distinct, terete, black and glossy below; indistinct and greyish-green above; pinnules or pinna-lobes c. 9 alternate pairs, ± narrow, adnate to the costa, all except lowest narrowly joined by decurrent wing of laminar tissue, ± at right-angles to the costa, upper pinna-lobes ± unlobed, lower ones with shallow, rounded, lateral lobes, apices ± narrowly rounded. Sori borne in an interrupted line around margins of mid and upper pinna-lobes, or sometimes the entire frond, protected by a series of pseudo-indusia; pseudo-indusium downward-curved lobes of lamina-margin specialised to form semi-circular flaps of thin, whitish to pale-brown tissue with an irregular, erose margin, lying over the sporangia before turning mid-brown and lifting to reveal sori when ripe. Spores small, uniform., trilete, dark-brown, globose, sporoderm nearly smooth. Cytotype diploid, sexual, n = 29 or 30 (reported with 29 or 30 bivalents at meiosis).

D i s t r. Low to lower-mid levels in the dry zone in a belt around the central highlands of Sri Lanka from the south to the east, north and north-west. Fairly common. An endemic Sri Lankan and S. Indian species, probably of S.E. Asian affinity. Sri Lanka; S. India (southern Ghats).

E c o l. Occurs on seasonally damp mud-banks of rivers and road-cuttings and in crevices of rocks where it shrivels up during the dry season, from c. 30-500 m in Sri Lanka.

N o t e. The figures and description of "*C. thwaitesii*" by B.K. Nayar (1962), are actually of *C. bullosa*, which he also treated under its proper name. Thwaites originally collected *C. thwaitesii* under the erroneous name of *C. varians* Wall. ex Hook ["Moore"], a synonym of *C. belangeri* (Bory) C. Chr., a related S.E. Asian, lowland species which reaches the outermost foot-hills of the east Himalaya and E. Nepal also S. India. This species is one of a group that form a link between the deltate, farinose *Aleuritopteris*-group species and the rest of *Cheilanthes*, including lanceolate (or deltate) non-farinose species. It was transferred to *Aleuritopteris* by Saiki (1984), presumably only because it is (partly) farinose, even though Ching had not done so. It is closely related to *C. chusana* Hook. and *C. fragilis* Hook. from S. China etc. *C. chusana*, along with the next two species and some others, has been separated by Ching and co-workers into the genus *Cheilosoria* Trev., though Pichi Sermolli (1977) had placed *Cheilosoria* as a synonym of *Pellaea*. Ching did not transfer *C. thwaitesii* into *Cheilosoria* presumably because it is farinose, which indicates the close relationship between these splinter genera.

S p e c i m e n s   E x a m i n e d. ANURADHAPURA DISTRICT: dry rocky outcrop by road to Trincomalee, near Habarana, *Sledge 876* (BM);

P. 438 (BM-Manton and Sledge box); Habarana, *Fosberg & Jayasinghe 57019* (E). KURUNEGALA DISTRICT: Kurunegala, *Thwaites* 52 p.p. (PDA). MATALE DISTRICT: Trincomalee to Kandy road, 63/3 mile-post, *Ballard 1500* (Z); Dambulla Rock, *Sledge 1365* (Z, GH); Dambulla Rock, 200-300 m., *Faden & Faden 76/554* (Z, GH, UC); Dambulla, *Thwaites C.P. 1321* (BM., E); Dambulla, *Gardner C.P. 1321* p.p. (PDA); Gonagama, *Gardner C.P. 1321* p.p., May 1856 (PDA); Elahera, *sin. coll. C.P. 1320* p.p., July 1858 (E). KANDY DISTRICT: hillside above Weragamtota, Mahaweli Ganga, 1500 ft. *Sledge 1136* (E). BADULLA DISTRICT: Kalupahane, *Thwaites 52* p.p. (PDA); Caloopana, *sin. coll. C.P. 1321* May 1856 (PDA). AMPARAI DISTRICT: Lahugula Tank, S.E. corner, large rock-outcrop, 36 m., *Mueller-Dombois & Comanor 67072557* (US, PDA); Dangama Kanda, near Komari, *Neville s.n.*, 1885 (PDA); banks of Karandi Oya, near Potuwila (Arugam Bay), *Neville s.n.*, 1885 (PDA). MONERAGALA DISTRICT: near Bibile, *sin. coll. s.n.*, June 1901 (PDA); Uva, between Bibile and Nilgala, *sin. coll. s.n.*, Jan. 1888 (PDA). HAMBANTOTA DISTRICT: Ruhuna National Park, *Schmid 1160* (BM., G); Ruhuna National Park, Block 1, Patanagala rock-outcrop area, *Mueller-Dombois 67120828* (US, PDA); W. Kinegale, *sin. coll. C.P. 1321*, Sept. 1862 (PDA). LOCALITY UNKNOWN: *Thwaites C.P. 1321* (G, BR, K, E); *Levinge 25/27* (G, NY); *Wall. s.n.*, 1884 (G); *Wall 17107* (BM); *Wall 25/26* (BO 10490); *Moon 1819* (BR); *Walker s.n.* (BM., PDA); *Trimen s.n.* (UC); *Naylor-Beckett 910* (E); *sin. coll. C.P. 1320* p.p., Herb. •*Brodie* (E); *Henderson s.n.*, Calcutta Herb., 28 March 1868 (E); *Gardner 123A C.P. 1321* (BO).

6. Cheilanthes opposita Kaulf., Enum., Fil.: 211. 1824.
Type: from "Cape of Good Hope?", erroneous locality; actually from S. India, Röttler.

Pteris elegans Poir. in Lam., Encycl. Meth. Bot. 5: 718. 1804, non *C. elegans*
    Desv. (see Alston, J. Bot.: 173. 1936).
Cheilanthes melanocoma Bory, Bélang. Voy. Bot. 2: 71. 1833.
Cheilanthes swartzii Webb. & Berth., Hist. Nat. Canar. 3(2/3): 453. 1847.
Cheilanthes mysurensis Wall. ex Hook., Sp. Fil. 2: 94-95. 1852.
Cheilosoria mysurensis (Wall. ex Hook.) Ching & Hsing [Shing], Flor. Fujian.
    1: 84. 1982 [sub "mysuriensis"].

M i s a p p l i e d   n a m e. *Cheilanthes fragrans* (L. fil.) Sw. [= *C. pteridioides* (Reichard) C. Chr.: synonym: *C. maderensis* Lowe].

Rhizome short, thick, erect, often branching, thickly surrounded by old stipe-bases, often slightly built up on a thick mass of thin, black roots, bearing a tuft of small, markedly linear, entire, dark-brown scales mixed with mid-brown hairs over its apex and a basket-like crown of fronds at its apex. Fronds up to c. 15 in a basket-like, apical crown, upright; stipe upright, up to c.

10 cm long, usually about $^1/_8$ or less the length of the lamina, thin, terete, black, or sometimes a very dark fuscous-brown when younger, glossy, bearing a tuft of acicular, mid-brown scales (up to c. 2×0.5 mm) at its base and a few scattered, similar scales further up; rachis similar to upper stipe but flattened adaxially with small, lateral ridges, without scales; lamina small to medium., bipinnatifid, narrowly lanceolate, up to c. 4 cm wide, 30 cm long, widest shortly above middle, markedly tapering from shortly below middle down to base, lowest pinnae are merely minute, simple flaps, slightly brittle, crispaceous-herbaceous, bright-green above (frequently black in poorly prepared herbarium-specimens), paler and slightly glaucous below, without scales or farina; pinnae up to c. 30 markedly opposite pairs, elongated triangular, widest at their bases, up to c. 2 cm long, 0.75 cm wide, very shortly stalked, closely approximate to the rachis, symmetrical, inserted at c. 70°-90°, deeply pinnatifid; costae distinct, terete, dark-blackish-brown and glossy below; indistinct, greyish-green above; pinna-lobes, c. 7 subopposite pairs, ± narrow-based, adnate to the costa, joined by narrow lamina-wing, inserted at ± 60°, varying from unlobed to doubly-lobed, with small, ± narrow, obtuse-tipped pairs of lobes (particularly in sterile portions of the lamina), pinna-lobe apices obtusely rounded. Sori borne separately around margins of pinna-lobes, often in pairs, usually confined to mid- and upper pinnae, protected by pseudo-indusia; pseudo-indusia reflexed lamina-margin, specialised into semi-circular flaps of thin, whiteish pale-green tissue with ± entire margins, lying over sporangia before turning pale-brown and lifting to reveal the sori when ripe. Spores small uniform., trilete, dark-brown, globose, sporoderm with very small granulations. Cytotype diploid, sexual, n = 30 (with 30 bivalents at meiosis).

D i s t r. Lower-mid levels towards the borders of the moist zone and in the surrounding dry zone around central highlands of Sri Lanka. Scattered. A S.E. Asian species extending to Sri Lanka and S. India. Sri Lanka; S. India (N.W. Ghats, Nilgiri, Annamalai and Palni Hills, southern Ghats, N.E. Ghats in S. Orissa and Andhra Pradesh); Myanmar (Burma); S. China. Reports from much of China, Taiwan, Japan and the Philippines probably all refer to *C. chusana* Hook., though some plants from Taiwan appear close to *C. opposita*, whose presence there has been rejected by Kuo (1985).

E c o l. Occurs on open, steep slopes, usually below or on rocks, sometimes on road-banks etc., from c. 400-900 m in Sri Lanka.

N o t e. Although most Indian authors have continued to use the name *C. mysurensis* for this species, following Beddome, it has long been known (Hooker 1852) that earlier names exist, of which only *C. swartzii* seems, recently, to have penetrated the subcontinent, the latter name having been given to refer to the erroneous part of Swartz's sense of *C. fragrans* (L. fil.) Sw. which included the present species. But *C. swartzii* is also illegitimate, being predated by *C. opposita*, which was not used last century or by Christensen (1905) due to its erroneous type-locality, though he commented that it could be the best name. As a misquoted locality has no nomenclatural importance, the name is used here accordingly.

C. opposita has been rather widely over-reported due to confusion with its closest relative, C. chusana Hook., from China, Taiwan, Vietnam., Korea, Japan and the Philippines, which has a less finely dissect and smaller frond with scales borne throughout the rachis. Both have been placed in Cheilosoria by Ching and co-workers (C. opposita being so placed under its synonym., C. mysurensis). The fronds of C. opposita have a most attractive, bright-green, feathery appearance.

S p e c i m e n s   E x a m i n e d. ANURADHAPURA DISTRICT: dry rocky outcrop by road to Trincomalee, near Habarana, Sledge 877 (BM., Z). KURUNEGALA DISTRICT: Kurunegala, Chevalier s.n., 1887 (BM). MATALE DISTRICT: Dambulla, sin. coll. C.P. 1320, 1859 (E); Elahera, Matale, sin. coll. C.P. 1320 p.p., July 1858 (E). KANDY DISTRICT: Hunnasgiriya, 900 m., Schmid 953 (BM., G); 5 km E. of Teldeniya, c. $^1/_2$ km W. of Moragahamula and Medamehanuwara, E. of Kandy on Hunasgiriya road, cliff by road-bend, 450 m., Fraser-Jenkins field no. 272 with Abeysiri & Gunawardena, 14 Oct. 1993 (BM., K, PDA, US). BADULLA DISTRICT: Badulla, Freeman 74A, 75B (BM). AMPARAI DISTRICT: Wadinagala Mt., 550-600 m., Bernardi 15619 (Z). RATNAPURA DISTRICT: Belihul Oya 650 m., Schmid 1222 (BM., G); Belihul Oya, Sledge P. 429 (BM-Manton and Sledge box). HAMBANTOTA DISTRICT: 9 km post, Tissamaharama towards Kataragama, Yala region, 150 m., Bernardi 14175 (Z); from Kataragama 10 km before Buttala, Paddahelakanda cave, 420 m., Bernardi 15526 (Z). LOCALITY UNKNOWN: sin. coll. C.P. 1320 (BM., E, DD); Thwaites C.P. 1320, 1853-1855 (G, E, BO); Levinge s.n. (E); McComb s.n., 1899 (GH); Henderson 25/21 (E); Carr s.n., herb. Goudie (E); low country, Hutchison s.n. (E); Anderson s.n. (E); Wall S. 190 and s.n. (E, BO (10489)).

7. Cheilanthes tenuifolia (Burm.f.) Sw., Syn. Fil.: 129, 332. 1806. Type: From India? "Plantae Leylansia. Burmann, herb. Delessert" (G).

Trichomanes tenuifolia Burm.f, Flora. Ind.: 237. 1768.
Adiantum cicutaeifolium Lam., Encycl. Meth. Bot. 1: 44. 1783.
Pteris humilis Forst., Flor. Ins. Austr. Prodr.: 79. 1786.
Pteris nigra Retz., Obs. Bot. 6: 38. 1791.
Acrostichum tenue Retz., Obs. Bot. 6: 39. 1791.
Adiantum varians Poir. in Lam., Encycl. Meth. Bot. Suppl. 1: 143. 1810.
Pteris nudiuscula R.Br., Prodr. Flor. Nov. Holland: 155. 1810.
Cheilanthes rupestris Wall., Num. List: no. 67. 1828, nom. nud.
Cheilanthes micrantha Wall., Num. no. 68. 1828, nom. nud.
Pteris gracilis Roxb. in Griff., Calc. J. 4: 508. 1844, non Michx.
Cheilanthes hispidula Kunze, Bot. Zeit.: 212. 1848.

Cheilanthes moluccana Kunze, Bot. Zeit.: 445. 1848, non Blume
Notholaena semiglabra Kunze, Farnk. Abb. Schkuhr: 59, t. 124. f. 2. 1850.
Cheilanthes javensis Moore, Ind. Fil.: 244. 1861.
Cheilosoria tenuifolia (Burm.f.) Trevis., Atti Ist. Veneto 5(3): 579. 1877.

Rhizome short, thin, small, erect, seldom-branching, bearing a tuft of small, linear-lanceolate, ± entire or slightly toothed, mid-brown scales with attenuate apices at its apex. Fronds up to c. 5 in a loose apical bundle; stipe spreading or upright in central fronds, up to c. 40 cm long, usually $1^1/_2$ times to twice as long as lamina, very thin, but a little more substantial at base in large fronds, terete, adaxially flattened, brownish-black, or black above, with a basal tuft of linear, almost fibrillose, mid- to pale-brown scales (up to c. 2 × 0.25 mm) with long-acuminate apices, and very few, similar, mostly deciduous scales a little higher up on lower stipe; rachis similar to upper stipe but widely grooved adaxially, scales absent; lamina small to medium-sized, occasionally central fronds on rhizome ± large, tripinnate to quadripinnatifid, deltate-pentagonal (up to c. 25 cm wide, 30 cm long), widest at base, fragile-herbaceous, pale- to mid-green, scales or farina absent; pinnae subopposite, up to c. 12 pairs, triangular-lanceolate, widest at their bases, up to c. 12 cm long, 10 cm wide, long-stalked, lowest pair markedly asymmetrical, lower basiscopic pinnules up to twice as long or longer than acroscopic ones, remainder ± symmetrical, inserted at c. 60°, bipinnate, becoming tripinnatifid below; costae not winged, terete, dark-brown and glossy, costules similar below but becoming less distinct and greyish-green where pinnulets become adnate and joined at their bases further up; pinnules up to c. 10 pairs, alternate, narrowly triangular-lanceolate, widest at their bases, up to c. 5 cm long and 2.5 cm wide, stalked but adnate above, oblique, varying from being deeply pinnatifidly lobed or pinnate below to shallowly lobed above, apices ± narrowly rounded; pinnulets or pinnule-lobes lanceolate, up to c. 1.5 cm long, 0.5 cm wide, mostly ± adnate, oblique, varying from unlobed to shallowly lobed, becoming deeply pinnatifidly lobed at their bases towards bases of large fronds, apices rounded. Sori in a ± continuous, or sometimes interrupted line around segments throughout lamina, naked or partly protected by a pseudo-indusium; pseudo-indusium often absent, if present a semi-specialised, deflexed, entire, interrupted, shallow flap of thin, pale-green lamina-margin, turned down loosely over sporangia until shrivelling to reveal sori when ripe. Spores ± large, uniform., trilete, mid-brown, subglobose-tetrahedral, sporoderm smooth. Cytotype tetraploid, sexual, n = 56 (with 56 bivalents at meiosis), from Sri Lanka and the E. Himalaya; an octaploid with n = c. 112 has also been reported from S. India.

D i s t r. Low to mid-levels around the borders of and within the moist zone of central Sri Lanka. Common. A widespread S.E. Asian element. Sri Lanka; India (E. Himalayan foothills in Darjeeling, Sikkim and Arunachal Pradesh; Assam., Manipur, Nagaland, Meghalaya, Tripura and Mizo; W. Bengal plains; throughout the central Indian hills in Uttar Pradesh (including in the plains), Bihar, Orissa and Madhya Pradesh; W. India in Mahabaleshwar and Goa; S. India in the N.E. Ghats, Nilgiri, Annamalai and Palni Hills, southern Ghats, E. Ghats, N.E. Ghats and Andaman Islands); C. and E. Nepal; Bhutan; Bangladesh; Myanmar (Burma); S. and S.W. China; Taiwan; Thailand; Cambodia; Vietnam; Laos; Malaya; Singapore; Indonesia (Sumatra, Java, Bali, Lombok, Soemba, Flores, Timor, Celebes, Buru, Moluccas); Philippines; New Guinea; Admiralty Islands; Australia, Tasmania; New Zealand; Oceania (Guam; Truk; New Hebrides; New Caledonia; Fiji; Society Islands (Bora Bora); Tahiti.

E c o l. Occurs on open damp mud-banks, walls and earth-slopes in fields and waste-places or on river-banks, often among grass or low bushes, from c. 300–1200 m., or more, in Sri Lanka.

N o t e. This attractive and delicate fern has been renamed in error from many different localities as can be seen from its synonymy. It has been often been confused in herbaria with other narrower-fronded species such as *C. sieberi* Kunze and *C. javanica* Kunze. It evidently contains a cytological complex whose taxonomy requires further study.

It has been treated within *Chelosoria*, along with the related Chinese and Tibetan *C. hancockii* Baker, by Ching and co-workers.

S p e c i m e n s   E x a m i n e d. TRINCOMALEE DISTRICT: roadside banks near Trincomalee, *Sledge 874* (Z, BM). MATALE DISTRICT: Lagalle Ela, 1000 ft, *Naylor-Beckett 325* (E); Beltotte, above Mr. Braybrooke's, *sin. coll. C.P. 1322*, 3 June 1862, herb. *Brodie* (E). KANDY DISTRICT: Peradeniya, *Sledge Z. 18* (BM-Manton and Sledge box). RATNAPURA DISTRICT: Bellihul Oya, *Sledge P. 105* (BM-Manton and Sledge box); on the old wall of the temple of Uggallboda, *Emerson s.n.*, 1828, herb. *Greville* (E). NUWARA ELIYA DISTRICT: Adam's Peak, N. slopes, *Sledge P. 427* (BM-Manton and Sledge box). LOCALITY UNKNOWN: *Schmid 834, 836, 1088, 1112, 1219, 1227* (G, BM); *sin. coll. C.P. 1322* (G, BM., E); *Thwaites C.P. 1322* (BO); *Wall s.n.*, 12 June 1873 (E, BO); *Wight 104B* (E).

12. PELLAEA

Link, Fil. Spec. Cult: 48, 49. 1841; R.M. Tryon et al. in Kubitzki et al. Fam. Gen. Vasc. Pl.: 243. 1990 nom. cons.
Type species: *P. atropurpurea* (L.) Link.

Rhizome erect to decumbent or short- to long-creeping, solenostelic; scales brown or black, concolorous or with darker centre. Fronds tufted; stipe chestnut to black, glabrous, hairy or scaly, terete or grooved on adaxial surface; lamina 1- 4-pinnate, mostly with an odd terminal segment; segments articulate or not thin to leathery, glabrous, pubescent, tomentose or with lacerate scales on midribs; veins free or anastomosing, without included veinlets. Sori born at vein ends or more usually submarginal forming lines usually with a continuous indusium; paraphyses mostly absent. Spores trilete, tetrahedral-globose or globose.

Variously estimated at 35-75 species mostly in S. Africa and S. America, SW. U.S.A and Mexico but extending to tropical Africa and east to India and New Zealand. Three species occur in Sri Lanka, two native which have not been seen for over a century and one introduced.

KEY TO SPECIES

1 Pinnules with venation evident in Sri Lankan material, not articulated, often trifid-hastate; fronds 2-3-pinnate (introduced) .... **3. P. viridis**
1 Pinnules with venation (other than midrib) not visible, articulated, rarely trifid; fronds simply pinnate to 3-pinnate (indigenous)
  2 Fronds simply pinnate, with stipe and rhachis densely hairy and with ± long scales; rhizome creeping with distinctly bicoloured scales; spores echinate .... **1. P. falcata**
  2 Fronds usually 2- 3-pinnate, or rarely simple in small plants, stipe glabrous to shortly pubescent; rhachis shortly densely pubescent above; rhizome ± erect with mostly concolorous scales; spores smooth ..... **2. P. boivinii**

1. Pellaea falcata (R.Br.) Fée, Mém. Fam. Foug. 5 (Gen. Fil.): 129. 1850-1852; Bedd., Handb. Ferns Br. India, 102, t. 54 (1883); Sledge in Bot. J. Linn. Soc. 84: 11. 1982.
Type: Australia, *R. Brown* (syntypes, BM).

Pteris falcata R.Br., Prodr. Fl. Nov. Holl.: 154. 1810.
Pteris seticaulis Hook., Ic. Pl. 3, t. 207. 1840. Type: Malaya, Pulo Penang,
    Lady Dalhousie (holotype, K).
Pellaea seticaulis (Hook.) Ghosh in J. Econ. Tax. Bot. 7 (3): 681.1986.
Pellaea falcata "sensu Beddome non. R.Br.", Manickam & Irudayaraj, Pterid.
    Fl. W. Ghats: 84, t. 60. 1992.

Rhizome mostly long-creeping, ± 4 mm thick; scales dense, pale brown at edges, dark brown to black in middle, lanceolate, ± 2 mm long, 0.25 mm wide. Fronds scattered, narrowly oblong-lanceolate in outline, 20–47 cm long; stipes dark brown to black, (5–)(–22) cm long, together with rhachis densely scaly and hairy; lamina narrowly oblong, 12–40 cm long, 4–8 cm wide, simply pinnate. Pinnae in 12-24 pairs, basal ones often subopposite and shortly stalked, upper ones alternate and ± sessile, oblong-lanceolate, 0.7–4(–5.5) cm long, 0.3–1.2 cm wide, usually with sparse pale brown scaly hairs on both sides particularly on midrib. Sori forming a continuous marginal line save near apex.

D i s t r. S. India, Sri Lanka, Malaya, Australia and New Zealand.

E c o l. Not known, 1500 m. Forest floor and shaded stream banks in India (fide Manickam and Irudayaraj).

N o t e. Manickam & Irudayaraj indicate this species is still found occasionally in the Western Ghats, but only very old specimens have been seen from Sri Lanka. It may be extinct.

S p e c i m e n s   E x a m i n e d. MATALE DISTRICT: East Matale, Teligama (= Telegamuwa), Oct. 1871, *C. P. 3933* (LIV). NUWARA ELIYA DISTRICT: 5 mi. N. of Hakgala, Bambarawella Falls (March 1890), *Nock s.n.* (PDA); Hakgala, *Thwaites C.P.3933* (K, PDA). WITHOUT LOCALITY: *Beddome* collection (K).

2. Pellaea boivini Hook., Sp. Fil. 2: 147, t. 118A. 1858; Bedd., Ferns S. India: 12, t. 36. 1863 & Ferns Br. India: 102, t. 53. 1883; Schelpe, Fl. Zambesiaca, Pterid.: 131. 1970; Sledge in Bot. J. Linn. Soc. 84: 11. 24. 1982; Ghosh in J. Econ. Tax. Bot. 7 (3): 684 (1986); Manickam & Irudayaraj, Pterid. Fl. W. Ghats: 85, t. 61. 1992. Type: "East coast of Africa, Nissobé" i.e. Madagascar, Nosy Bé, *Boivin* (syntypes, K).

Pteris boivini (Hook.) Thwaites, Enum. Pl. Zeyl.: 386. 1864.

Rhizome up to 1.5 cm thick; scales dense, pale brown, mostly without darker centre, lanceolate, 3–5.5(–10) mm long, 1 mm wide. Fronds tufted, (6–)24–54 cm long. Stipes black, (2–)8–30 cm long, shiny and glabrous or ± pubescent. Lamina ± triangular to ovate in outline, 5.5–24 cm long, 3.5–25 cm wide, usually 2– 3-pinnate but can be simply pinnate in very small plants; pinnae in up to 6 pairs, subopposite, up to 10 cm long, 6 cm wide; secondary pinnae in 1-6 pairs; pinnules ovate-triangular, cordate, 0.5–2.5 cm long, 2.5–8(–12) mm wide, occasionally 3-fid, coriaceous, shortly stalked, glabrous but rhachises and pinnule-stalks with stiff pale brown hairs. Sori forming a marginal line protected by thin membranous reflexed margin.

D i s t r. S. India, Sri Lanka, Zambia, Zimbabwe, S. Africa, Madagascar and Mauritius.

E c o l. Not known.

N o t e. Manickam & Irudayaraj state that it is now a very rare species in the Tirunelveli Hills, S. India. It may be extinct in Sri Lanka.

S p e c i m e n s   E x a m i n e d. WITHOUT LOCALITY. *Skinner s.n.* (K); *Thwaites C.P. 3363* (BM., PDA, K); *Mrs Walker 30* (K); *Wall s.n.* (K).

3. Pellaea viridis (Forssk.) Prantl in Engl., Bot. Jahrb. 3: 420. 1882; Sledge in Bot. J. Linn. Soc. 84: 11, 24. 1982; Schelpe, Fl. Zambesiaca, Pterid.: 133. 1970. Type: Yemen, Al Hadiyah, *Forsskål* (no material has been found).

Pteris viridis Forssk., Fl. Aegypt.-Arab.: CXXIV, 186. 1775.
Pellaea boivinii sensu Manton & Sledge in Phil. Trans. Roy. Soc. B 238: 150. 1954. non Hook.

Rhizome short-creeping, 5–10 mm wide; scales black with pale margins or admixed with concolorous pale brown scales, linear-attenuate, ± 3 mm long, entire, ciliate or not. Fronds tufted, 10–90 cm tall. Stipes chestnut brown or ± black, 2.5–40 cm long, glabrous or with spreading narrow or hair-like scales. Rhachis and secondary rhachises chestnut to blackish, glabrous, shortly pubescent or with hair-like scales. Lamina narrowly oblong, lanceolate or triangular-pentagonal in outline, (5–)13–50 cm long, (2–)6–30 cm wide, 2–3-pinnatifid to 3-pinnate. Basal pinnae often the most developed; ultimate segments narrowly oblong, lanceolate, trifid-hastate or ovate, 0.8–3.2 cm long, 0.3–2.8 cm wide, obtuse or subacute, glabrous or with hairs on venation beneath; petiolules glabrous or pubescent. petiolules glabrous or pubescent. Venation evident or obscure; var. *viridis* Fronds 2-pinnate to 3-pinnatifid; rhachis glabrous or with short unicellular hairs towards apex.

D i s t r. Arabia, eastern, central and southern Africa, Madagascar and Mascarene Is.; introduced into Sri Lanka.

E c o l. Forest and on stumps of tea bushes; 600–1500 m.

N o t e. Schelpe (loc. cit.) records it from India but I have seen no material nor is it mentioned in Dixit's Census or other recent literature. It must have been in Sri Lanka before 1923 since Freeman collected 1908-1923. Recent work on African species whilst this volume has been in press suggest, that the true var. *viridis* i.e. Forsskål's type is the same as *P. adiantoides* Willd.

S p e c i m e n s   E x a m i n e d. KANDY DISTRICT: Woods c.7 km S.E. of "Gem inn ii" above Hewaheta road at Talwatta, on old road from Talwatta to Kandy, c. 3 km E. of Kandy, (via new road)Alt. c. 600 m., 22Aug 1993, (2 sheets) *Fraser-Jenkins* Field-No. *11* (K); Kandy, *Freeman 53A, 55C, 56D* (BM). RATNAPURA DISTRICT: Sabaragamuwa, Belihul-Oya, *Schmid 1221* (BM., G); forest above Le Vallon Estate, *Sledge 1109* (K). NUWARA ELIYA DISTRICT: Hakgala, *Freeman 54B* (BM); Hakgala Estate, *Piggott 2664* (K); Mount Vernon Estate, *Schmid 1050* (BM., G); Ramboda Pass, *Sledge P177* (BM); N. slopes of Adam's Peak, *Sledge 605* (BM); *Sledge Z 26* (BM). WITHOUT LOCALITY: *Amaratunga 241* (PDA).

13. ACTINIOPTERIS

Link, Fil. Sp.: 79. 1841; Pic.Serm., Webbia 17: 1–32. 1962 & Webbia 17: 317–328.1963; R.M. Tryon et al. in Kubitzki et al., Fam. Gen. Vasc. Pl.: 239. 1990. Types species: *A. radiata* (Sw.) Link

Rhizome creeping, with tufted fronds and with linear attenuate entire rhizome-scales with or without a dark central stripe. Stipe usually stramineous. Fronds flabellate to obcuneate, dichotomously divided into linear segments, green or glaucous, with or without scales dorsally; fertile fronds usually taller than sterile fronds and sometimes differently dissected; veins free. Sori borne in a submarginal line; indusia continuous, membranous, entire.

A genus of 5 African species, 2 extending to India one of which occurs in Sri Lanka.

Actiniopteris radiata (Sw.) Link, Fil. Sp.: 80. 1841; Hook., Ic. Pl. 10: t. 975, fig. 2–3. 1854; Thwaites, Enum. Pl. Zeyl.: 386. 1864; Pic.Serm., Webbia 17: 8, fig. 1/a-f. 1962 & Webbia 17: 318, fig. 1/a-f. 1963. Lectotype: India, Coromandel, *Koenig* (Herb. Montin) (S-PA).

Asplenium radiatum Sw., Schrad., Journ. Bot. 1800, 2: 50. 1801.
Acrostichum radiatum Poir., Encycl. Méth. Bot., Suppl. 1: 128. 1810. Type:
    India, *Koenig* (? P, holo.).
Acropteris radiata (Sw.) Link, Hort. Berol. 2: 56. 1833.

Rhizome scales with a broad, thick, black shining central stripe and narrow light chestnut edges. Fertile and sterile fronds similar, drying after death, lamina reflexing downward owing to an abrupt bend in its basal part, forming a narrow V-shaped angle with stipe or even adherent to it. Fertile fronds (5)12–18(26) cm. long with stipe 3–5 times as long as lamina, more scaly in upper part than elsewhere. Lamina flabellate with edges forming an angle of 150°–180°, sometimes very broadly flabellate, angle of edges up to 270°, cut into two symmetrical halves by a clearly evident median deep notch, and consisting of a 5–6 times dichotomous branch system. Branches of the first dichotomy equal (1.8 mm.), forming angle of about 90°; branches of second dichotomy unequal, inner (3.5 mm.) longer than outer (1.9 mm.); last dichotomy located in upper half of frond. Segments of blade 32–48, narrowly linear, slightly and gradually broadening upwards, with apex divided into 2–6 hardened acute points. Basal part of lamina and adjacent upper part of stipe covered with short rounded pluricellular small hairs above and with abundant persistent scales beneath. Scales light fulvous, sometimes with a short black stripe in upper part, strongly falcate to tortuous, ovate-lanceolate, irregularly lobate to coarsely dentate at bases, tapering upwards and ending in a long hair.

D i s t r. Throughout tropical Africa to Cape Verde Is., S. & SW. Africa (Namibia) and E. to Arabia, India and Sri Lanka.

E c o l. Rocky roadside banks, cocoa plantations, crevices in walls, on ruins etc; 400-480 m.

S p e c i m e n s E x a m i n e d. TRINCOMALEE DISTRICT: Trincomalee, *Gardner* in *C.P. 1342* (PDA). KANDY DISTRICT: Teldeniya, *Faden & Faden 76/558* (K, US); Guru-oya *Jayasuriya 1913* (PDA); *Kundu 165* (PDA); Urugala, *Amaratunga 1480* (PDA); 9 mi. from Kandy on the Teldeniya road, *Sledge 1035* (K, PDA); E. of Kandy, road above the Maha Oya, *Sledge 1147* (BM). BADULLA DISTRICT: 4 mi. S. of Ella,

Dambasagedara, *Wheeler 12703* (PDA). NUWARA ELIYA DISTRICT: Hanguranketa, *de Silva s.n.* (PDA). MONERAGALA DISTRICT: Ruhuna National Park, Block I, Magulmaha Vihare ruins, *Cooray 67113012R* (PDA). WITHOUT LOCALITY: *C.P. 1342* (BM); *Gardner 1342* (K); *Robinson s.n.* (K); *Skinner s.n.* (K); Gonagama, *C.P. 1342* (LIV).

14. ADIANTUM

L., Sp. Pl. 1094. 1753; L., Gen. Pl. ed. 5: 485. 1754; Sledge, Ceylon J. Sci. (Bio. Sci.) 10: 144–154. 1973; Goudey, Maidenhair ferns in cultivation. 1985; R.M. Tryon in Kubitzki, Fam. Gen. Vasc. Pl.: 249. 1990, Lectotype species: *Adiantum capillus-veneris* L.

Small to medium rather delicate plants with erect to shortly or widely creeping densely brown-scaly rhizomes; stipe dark brown to black, polished, with 2 vascular strands united above to form a single T-shaped, lunular or trapezoidal strand. Fronds tufted or shortly spaced, pinnate to 4-pinnate or simple, the pinnae and pinnules stalked or ± sessile, the lamina with anadromous architecture and venation; segments fan-shaped, wedge-shaped, trapezoidal or dimidiate, glabrous or pilose, occasionally powdery beneath; venation usually free. Sori borne on the inner surface of reflexed marginal lobes which function as indusia ('false indusia').

A genus of over 200 species occurring in both tropical and warm temperate parts of both hemispheres but particularly in S. America; 15 are native to or naturalised in the Flora area. Several others are widely cultivated and some of these may also become naturalised. Only one reference to a vernacular name has been discovered; the general English name Maidenhair fern, mostly originally used in Britain for *Adiantum capillus-veneris*, may be used for all of them.

Apart from the species dealt with in full, several others have been grown in Peradeniya Botanic Gardens and elsewhere but have not apparently been naturalised yet. They are included in the key but of course other species may have been cultivated of which I have no record. There are now innumerable *Adiantum* cultivars that any fern enthusiast may well have in his collection. One cultivar originated in Sri Lanka and something must be said about it.

In 1884 Moore described *Adiantum fergusonii* (Gardeners' Chronicle (new series) 22: 360. 1884 & (third series) 2: 470, fig. 96. 1887). In the herbarium covers at Kew is a letter from Ferguson describing the finding of this fern at Negombo about 21 mi N. of Colombo. Ferguson was staying at the rest house and went to pay his respects to a judge, F.J. De Livera who lived in the house next door. This unknown fern was growing in a pot between *Adiantum tenerum* Willd. and *Adiantum farleyense* Moore (now considered a cultivar of *tenerum*) and presumably Ferguson requested a specimen. The plant was believed to have originated at a sale of plants in Colombo. In the literature it is usually stated that Ferguson had found it as a chance 'seedling' in his conservatory at

Colombo but his letter is quite unequivocal; later it was apparently very widely cultivated in Sri Lanka and soon in Europe and many other places. It is now considered to be a cultivar of *Adiantum tenerum* Willd. (a native of tropical America from Florida and Mexico to Panama, W. Indies, Trinidad, Colombia and possibly Ecuador and Peru). I have seen a few specimens, Peradeniya Botanic Gardens, *Schmid 907* (BM), Colombo (cult.), *Gamble* (in 1899) *27544* (BM), and Sri Lanka, *Curtis s.n.* (K). Its present status as a cultivated plant in Sri Lanka is unknown. J.G. Baker thought it was a robust variety of *Adiantum capillus-veneris* but its true origin was realized quite early. Specimens of *Adiantum tenerum* Willd. from Peradeniya Botanic Garden have been collected by *Schmid* (*902, 903, 908*).

Adiantum polyphyllum Willd., a native of Venezuela to Colombia to Ecuador and Peru, also Trinidad, has been grown at Peradeniya, and is a striking tall fern (Peradeniya Garden, *Alston 789* (PDA), *Schmid 884, 885* (BM), cultivated, *Freeman* 72A (BM)). *Adiantum macrophyllum* Sw., also from the American tropics, is according to Rev. E.V. Freeman 'firmly established in many gardens' but has 'so far not strayed beyond the gardens'; a specimen *Freeman 73a*, unlocalised, is at the BM. The taxa are included in the key but remain without a number.

KEY TO THE SPECIES

1 Fronds simply pinnate
  2 Pinnae stalked
    3 Sori elongate .... **1. A. philippense**
    3 Sori round .... **7. A. diaphanum**
  2 Pinnae sessile or subsessile
    4 Pinnae large, ovate, about 6 x 3–6 cm. with linear sori about 3 cm long
      on both outer margins (cultivated) .... **A. macrophyllum**
    4 Pinnae not as above. Fronds attenuate above and often rooting at the apex
      5 Pinnae clothed beneath with usually abundant white hooked unicellular
        hairs .... **2. A. caudatum**
      5 Pinnae without hooked hairs
        6 Rhachis hairy on both surfaces with 3–5-celled patent hairs and
          unicellular crisped hairs mixed .... **3. A. indicum**
        6 Rhachis glabrous beneath except at distal extremity...........
          .... **4. A. zollingeri**
1 Fronds not simply pinnate
  7 Fronds pedately divided
    8 Stipe and pinnules hairy .... **5. A. hispidulum**
    8 Stipe and pinnules glabrous .... **6. A. flabellulatum**
  7 Fronds bipinnate to several times divided
    9 Terminal pinna elongate and resembling laterals
      10 Stipe and rhachis glabrous, black and glossy

11 Fronds 2–3-pinnate, up to 1.2 m tall, 50 cm wide; pinnules 2–2.5 cm long; sori squarish, reniform or obovate, at tips of lobes (cultivated) . . . . **A. polyphyllum**

11 Fronds 1–2-pinnate without above characters combined

12 Pinnules up to 1 cm long . . . . **7. A. diaphanum**

12 Pinnules larger, up to 6 cm long . . . . **8. A. trapeziforme**

10 Stipe and rhachis hairy, not glossy

13 Upper margin of pinnule with one long sorus (rarely two) . . . . .
. . . . **9. A. pulverulentum**

13 Upper margin of pinnule with many sori . . . . . **10. A. latifolium**

9 Terminal pinna not elongate nor resembling laterals

14 Margins of sterile pinnules sharply serrate, the veins running into the teeth

15 Fronds bipinnate with rhachis glabrous

16 Pinnules jointed at base (articulated); black of stalklet not running into pinnule blade; plant more robust; (cultivated) . .
. . . . **A. tenerum** cv. **fergusonii**

16 Pinnules not jointed at base; black of stalklet clearly extending into pinnule; more slender . . . . **12. A. capillus-veneris**

15 Fronds 3–4-pinnate with rhachis hairy . . . . . . **11. A. formosum**

14 Margins of sterile pinnules with veins ending in the sinus between indefinite teeth or shallow crenations

17 Lowermost pinnule on upper side of pinnae overlapping the main rhachis . . . . **13. A. concinnum**

17 Lowermost pinnule on upper side of pinnae not overlapping the main rhachis

18 Short creeping rhizomes with clustered stipes; pinnules with cuneate bases; sori ± round . . . . **14. A. raddianum**

18 Long creeping rhizomes with spaced stipes; pinnules with broadly cuneate to reniform bases; sori oblong to lunate . . . .
. . . . **15. A. poiretii**

1. Adiantum philippense L ., Sp. Pl. 2: 1094 (1753)
Type: Illustration by Petiver, Gazophyll Nat. 1.4. fig.4 (1702); copies in LINN, BM etc.

Adiantum lunulatum Burm.f., Fl. Ind. 235. 1768; Thwaites, Enum. Pl. Zeyl. 387. 1864; Bedd., Handb. Ferns Brit. Ind. 82. 1883; Morton, Contr. U.S. Nat. Herb. 38: 370. 1974. Type: A. Burmann, specimen from Malabar, selected by Morton (G, lectotype).

Rhizome short, suberect or creeping with tufted fronds; scales dark brown, subulate, ± 3 mm long. Fronds erect, 30-45 cm long, simply pinnate, often bare and rooting at the tips; stipe and rhachis dark brown or blackish purple, shiny, glabrous. Pinnae up to 12 pairs, alternate, rectangular to fan-shaped,

1.2-6 cm long, 0.7-2.5 cm wide, the upper and outer margins lobulate, the lower entire, often breaking off at matuarity, glabrus; pinna-stalklets 0.2-2.5 (-4) cm long, those of lower pinnae longer. Sori narrowly oblong, 0.4-2.5 cm wide.

D i s t r. Old World tropics to Polynesia.

E c o l. Moist roadside banks, damp cliffs, sheltered places, caves; 60-900 m.

N o t e: Although Verma, Nova Hedwigia 3:464. (1961), followed by Sledge (1982) and Morton (1974), used the name *A. lunulatum* for this species on the grounds that Petiver's drawing of *A. philippense* is difficult to identify, Pichi-Sermolli (1957) had pointed out that both names apply to the present species. C.R. Fraser-Jenkins (in prep.) confirmed this and located Camel's original herbarium-specimen from Luzon, Philippines sent to Petiver, in the Sloane Herbarium, BM.

S p e c i m e n s   E x a m i n e d. ANURADHAPURA DISTRICT: Anuradhapura-Trincomalee road, 22/5–22/6 mile post, *Ballard 1483* (K, P, US). MATALE DISTRICT: Nalanda Dam., *Ballard 1441* (K, P, US). COLOMBO DISTRICT: Labugama, *Schmid 1089* (BM., G). KANDY DISTRICT: ? Wariagalla (Wariyagala), *Alston s.n.* (PDA); Madugoda to Urugala road, *Simpson 8814* (BM); Madugoda to Weragamtota, *Walker T23, T 41* (BM); Kandy-Mahiyangana road, mile post 25/11, *Faden & Faden 76/567* (K, PDA); Hunnasgiriyia, *Schmid 949* (BM., G). Haragama, *Thwaites C.P. 1323 p.p.* (PDA). BADULLA DISTRICT: Badulla Road, mile 71/6, *Ballard & de Silva 1330* (K, US). Badulla, *Freeman 57A, B & C* (BM); Way to Uma Oya, Padalu Kandura ('Pundala'), *Silva 28* (PDA); Welimada to Badulla, Hangiliella, *Sledge 775* (K); Way to Uma Oya, Padalu Kandura, *s.coll. '46'* (PDA); Near Lunugala, *s.coll. s.n.* (PDA). RATNAPURA DISTRICT: Belihul-Oya, near rest house, *Ballard 1359* (K, P, PDA, P, Z). A4 between Balangoda and Belihul-Oya, *Comanor 1105* (K, PDA). Pelmadulla, *Schmid 1215* (BM., G); Belihul-Oya, *Sledge 792* (BM). MONERAGALA DISTRICT: Etbedde, Wellawaya R., *Lewis s.n.* (PDA). NOT TRACED: Gonagama, *Gardner* in *C.P. 1323 p.p* (PDA); Gonagama, C.P.1323 (LIV). WITHOUT LOCALITY: *Gardner* in *C.P. 1323 p.p* (K); *Moon 290* (BM); *Robinson s.n.* (K); *Macrae 903* (BM). *Wall s.n.* (BM).

2. Adiantum caudatum L., Manton Pl. 2: 308. 1771; Bedd., Handb. Ferns Brit. Ind. 83 excl. fig.44. 1883; Sledge, Ceylon J. Sci. (Biol. Sci.) 10: 146, fig.2. 1973.
Type: Ceylon, Burman, Thes. Zeyl. 8, t.5, fig.1. 1737.

Rhizome short, erect; scales numerous, dark brown with pale edges, narrow, 4–5 mm long. Fronds in dense tufts, 12–30 cm long; stipes dark purple to almost black, ± densely covered with dark reddish, stiff, spreading hairs;

rhachis with shorter hairs on upper surface and longer spreading less dense hairs on shiny lower surface; lamina simply pinnate with many close subsessile oblong-trapeziform pinnae 1.5–2 cm. long, 0.6–1 cm. wide; upper pinnae gradually smaller towards apex which is often leafless and rooting at the tip. Pinnae with lower edges 4–5-lobed, each lobe again often bifid, with very prominent pleated venation, densely hairy with a mixture of long multicellular brown hairs on both sides and abundant short white hook-tipped unicellular hairs beneath; sori at apices of lobes, reflected indusium ± circular or some-what elongate, hairy. All specimens investigated cytologically appear to have been triploid and apogamous.

D i s t. Widespread in Asia, Sri Lanka, S. India, Assam., Burma, Thailand, Indochina and Malaya; other records from further afield probably refer to other species of this critical group.

E c o l. Evergreen forest, thick degraded forest, shady slopes, dry banks, sheltered boulders, rubber plantations;150–900m

V e r n. Vala-venna (S).

N o t e. A key to some of the *Adiantum caudatum* group is given by Sinha in Manton, Ghatak and Sinha, Journ. Linn. Soc. (Bot.) 60: 223–235. 1967. The main couplet 'stipe not more than 6 cm' (*caudatum* and its allies) 'stipe very long 12 to 18 cm' (*malesianum*) appears not to work.

S p e c i m e n s   E x a m i n e d. ANURADHAPURA DISTRICT: Top of Ritigala, *s.coll.* in 1905 (PDA). MATALE DISTRICT: Matale-Dambulla road, mile post 28/4, *Ballard 1436* (K). COLOMBO DISTRICT: Mt. Lavinia and many other places near Colombo, *Mrs Chevalier s.n.* (BM). KEGALLE DISTRICT: Kegalle (Kegala) road towards Kandy, *Lewalle 6775* (BM; BR). KANDY DISTRICT: Kandy, *Alston 1092* (K; PDA); cult. at Leeds Bot. Gard. ex Kandy, *Alston 17448* (BM); Kandy, Roseneath, *Ballard 1010* (K, P, PDA, US, Z); Gannoruwa Forest, *Faden 76/387* (K); Kandy, *Gardner s.n.* (K); Kandy (voucher from plant grown at Leeds Bot. Gard.), *Jarrett 116* (K); Kandy, Lady Horton's Drive, *Sledge 511* (BM); Hunnasgiryia, *Sledge 966* (K); 18 mi from Kandy, *Walker T14*, specimen grown at Kew (K). BADULLA DISTRICT: Towards Passara, *Freeman 60A* (BM); Badulla, *Freeman 62C* (BM.); Hingurugama, *Freeman 61A* (BM). AMPARAI DISTRICT: Lahugula Tank, SE. corner, *Mueller-Dombois & Comanor 67072548* (PDA). RATNAPURA DISTRICT: Belihul Oya near rest house, *Ballard 1364* (K). MONERAGALA DISTRICT: 4 mi N. of Siyambalanduwa, *Fosberg & Sachet 53072* (K, PDA, US). WITHOUT LOCALITY: *Sir & Lady Barkly* (BM); *Finlayson s.n.* (BM); *Fraser 52* (BM., K). 52 ex Heward in Herb. Gay (K); *155* (BM., K); *s.n.* in 1849 (K); *Gardner 1124* (BM). *Moon s.n;* (BM). *Robinson 105* (K); *Macrae 221. 909* (BM); *Wall s.n.* (BM); Hort. Kew ex Ceylon, *Herb John Smith* (BM); *No. 47* (PDA); *No. 47/1 p.p* (PDA); *C.P. 978* (BM; PDA).

3. Adiantum indicum J. Ghatak, Bull. Bot. Surv. India 5: 71. 1963; Manton, Ghatak & Sinha, J. Linn. Soc. (Bot.) 60: 223–234. 1967.
Type: India, Calcutta, Belgharia, *Ghatak 301* (CAL, holotype).

Adiantum incisum sensu Pich.Serm., Webbia 12: 669. 1957 p.p. non Forssk.

Rhizome short, erect; scales dark brown with light brown margins, setaceous, ± 8 mm. long, 1 mm. wide, entire. Fronds tufted, ± 35 cm long, ± erect, covered when young with deciduous white scales 5 mm. long, 1 mm. wide. Stipes dark brown, about $^1/_7$ the length of rhachis, ± 5 cm long, sparsely covered with ferruginous slightly curved hairs. Lamina ± 30 cm long, ± 3 cm wide, simply pinnate with ± 38 pairs of pinnae and naked apex ± 9 cm long rooting at tip; rhachis ferruginous hairy, less so on lower surface. Pinnae ± oblong-triangular, ± 1.5 cm long, ± 7 mm wide, subsessile, striate, lobed to ± $^1/_3$ their width, almost glabrous above, more hairy beneath with short unicellular but not hooked hairs and fewer multicellular ferruginous slightly curved ones; lobes usually 4–5 with somewhat truncate apices; one sorus per lobe. Tetraploid.

D i s t r. India; some Sri Lanka specimens seem to belong here.

E c o l. Not known.

N o t e: *T.G. Walker T126*, from Haragama Pussalamankada, 15 mi. from Kandy at 450 m. by waterfall, 18 Jan. 1954(BM) appears to be *Adiantum malesianum* Ghatak; the lowermost pinna are fan-shaped and have numerous fine white straight hairs beneath together with a few brown multicellular ones but the stipes are only 4.5 cm. long. The Walker material was brought to my attention added after the study had long been completed.

S p e c i m e n s   E x a m i n e d. ANURADHAPURA DISTRICT: A few miles from Ritigala, Galpitigala, *Ballard 1472*(K) (det.Jarrett). KURUNEGALA DISTRICT: Kurunegala Rock, *Walker T 13* (BM). POLONNARUWA DISTRICT: Polonnaruwa, *Randall s.n.* (K) (det. Jarrett). KANDY DISTRICT: Hunnasgiriya, *Walker T 116; T 119* (BM); Haragama, Pussalamankanda, 15 mi. from Kandy, *Walker T 122; T 124* (BM); Gonagampitiya c. 15 mi. from Kandy W. of Maha Oya, *Walker T 451* (BM). Near Kandy on Colombo Road, Utuwankande, *Walker T 884; T 885*(BM). BADULLA DISTRICT: Ella Pass, *s.coll. s.n.* in 1890 (? *Thwaites*) (PDA) (det. Sledge); S. of Bibile, near Lunugala *Walker T 556* (BM). LOCATION UNKNOWN: Pinburuwellegama, 4 mi.E. of Wariyapola, *Walker T 886; T 882; T 890* (BM).

N o t e. Although Sledge determined the specimen above as cf. *indicum* he does not mention the specimen in his revision of the Sri Lanka species. Further work, particularly cytological examination is required.

4. Adiantum zollingeri Mett. ex Kuhn, Ann. Mus. Lugd.-Bat. 4: 280. 1869;
Sledge in Manton, Ghatak and Sinha, J. Linn. Soc. (Bot.) 60: 234. 1967;
Holttum., Fl. Malaya 2 (ed.2): 638 (1968); Sledge, Ceylon J. Sci. (Biol. Sci.)
10: 146, fig. 3. 1973.
Type: Java, *Zollinger 2806* (L, holotype, BM photo. and iso.)

Adiantum caudatum var. *subglabrum* Holttum., Fl. Malaya 2: 600. 1954.
Types: from Kedah and Perlis but no specimens cited. *Adiantum
rhizophorum* sensu Thwaites, Enum. Pl. Zeyl. 387. 1864, non Sw.

Very similar to *Adiantum caudatum* but with rhachises densely and shortly
hairy on upper surface and glabrous on lower surface except at distal extremity; pinnae quite glabrous or rarely with sparse hairs (but no hooked ones) and
usually less deeply lobed; veins not raised on the upper surface and pinnae
not markedly striate; pinnae articulated to their stalks which in old fronds
persist as peg-like projections on the frond rhachis. Diploid.

D i s t r. S. India, Sri Lanka, Malaya (Kedah, Perlis) Thailand and Indochina
(fide Holttum), Timor, Java, Flores, ?New Guinea.

E c o l. Shady forest, rocky banks, steep rocky forested slopes, wet rocks
but also more frequent in dry zone than *A. caudatum*; 120–900 m.

S p e c i m e n s   E x a m i n e d. TRINCOMALEE DISTRICT fide
Sledge. KURUNEGALA DISTRICT: "Kornegalle" (? Kurunegala), *Mrs
Chevalier s.n.* (BM); Kurunegala Rock, *Walker T 12, T17, T 19* (BM);
Pinburuwellagama 4 mi. E of Wariapola, *Walker T 888, T 889* (BM);
Kurunegala Rock, *Sledge 931*(K). MATALE DISTRICT: Trincomalee–Kandy
road, mile post 63/3, *Ballard 1505* (K); E. of Dambulla near Village D K
Muna towards Wasgomuwa, Strict Natural Reserve, *Bernardi 14362* (G, K).
Matale–Dambulla road before mile marker 44 (A9), *Comanor 728* (K, PDA,
US); 8 mi ESE. of Dambulla, hills above Mahaweli Project, *Davidse 7417*
PDA); about 6 mi. E. of Dambulla, just E of the Kandalama Tank, Erawelagala
Mt., *Davidse & Sumithrarachchi 8078* (PDA); Nalanda, *Sledge 861* (BM).
POLONNARUWA DISTRICT: Polonnaruwa, *Randall in Thwaites s.n.* (K).
KANDY DISTRICT: Madugoda to Weragamtota, *Walker T 21* (BM);
Haragama, Pussalmankanda, 15 mi. from Kandy, *Walker T 123* (BM);
Weragamtota, *Walker T 324* (BM); Gonagampitiya, 15 mi. from Kandy W. of
Maha Oya, *Walker T 450* (BM); Ganoruwa forest, *Faden & Faden 76 /386*
(K); 30 mi. E. of Kandy, beyond Madugoda on road to Weragamtota, *Sledge
948* (K); voucher specimen from plant grown at Leeds (*Walker T 15*), *Jarrett
119* (K); N. side of Hassalaka Oya, Kandapahala towards Torapitiya, *Sledge
1130* (K); Haragama, *Thwaites in C.P. 3102 p.p* (PDA). BADULLA
DISTRICT(?): Hingurugama, *Freeman 61B* (BM). RATNAPURA DISTRICT:
Belihul-Oya, *Schmid 1215* (BM., G); Niriella, *Schmid* 1105 (BM., G).

NUWARA ELIYA DISTRICT: Nuwara Eliya to Maturata, about 4 mi. N. of latter at culvert 26/3 *Maxwell 1003* (PDA); Pussalamankada, *Sledge 984*(K). DISTRICT UNCERTAIN: Uva, *Thwaites in C.P. 3102* p.p (PDA). WITHOUT LOCALITY: '47/1' p.p (PDA); *Beddome s.n.* (BM); *Robinson 104* (K); *Thwaites CP 3102* (BM., K, PDA). *Wall s.n.* (BM).

5. Adiantum hispidulum Sw., J. Bot. (Schrad.) 1800. 2: 82. 1801 & Syn. 124, 321. 1806; Hook & Baker, Syn. Fil. ed. 2: 126. 1874; Bedd., Handb. Ferns Brit. Ind.: 86, fig. 300, frontispiece. 1883; Sledge, Ceylon J. Sci. (Bio. Sci.) 10: 148, fig. 4. 1973; W.C. Shieh, Wang-Chueng in Fl. Taiwan ed. 2, 1: 239, t. 99. 1994.
Type: probably from Australia (S, holotype).

Rhizome creeping, short and coarse with dark brown narrow entire scales about 3 mm long. Fronds close, erect, 15–45 cm tall; stipe dark red-brown to black, up to 30 cm long, slightly rough and hairy, scales brown, scattered near base, narrow; lamina ± fan-shaped, pedately divided into 10–15 branches. Pinnules asymmetrically suboblong, 0.5–1.7 cm long, 3–8 mm wide, the upper and outer margins gently rounded, finely toothed, the lower straight and entire, softly pubescent to ± glabrous; pinnule-stalklets 0.5–1 mm long. Sori 10–18, small, closely placed on upper and outer edges in notches between the lobes, 3–4(–5) mm wide, flaps circular to broadly oblong or reniform., covered with numerous small pointed brown hairs.

D i s t r. East Africa, S. India, Sri Lanka, Malesia to Polynesia and New Zealand.

E c o l. Shady roadside banks, steep earth banks, 'jungle', also tea plantations; 160–1675 m.

S p e c i m e n s   E x a m i n e d. KEGALLE DISTRICT: Kitulgala, *Schmid 1075* (BM., G). KANDY DISTRICT: Peradeniya, *Petch s.n.* (PDA); Hunnasgiryia, *Schmid 958* (BM); Road to Nawanagalla, 1 mi. W. of Madugoda, *Walker T 54* (BM). BADULLA DISTRICT: Badulla, *Freeman 66a* (BM); Ella, *Hepper & de Silva* 4791 (PDA); Namunakula, Tonnacombe Estate, *Sledge 1179* (K). RATNAPURA DISTRICT: Belihul-Oya, *Schmid 1224* (BM., G). NUWARA ELIYA DISTRICT: near Ramboda, *Ballard 1133* (K,P,PDA,US); Nuwara Eliya, *Freeman 67B* (BM); Ramboda, *Gardner 1123* (BM), *Gardner in C.P. 1325 p.p* (BM., LIV, PDA); Ramboda Pass, *Sledge 181* (BM); *Sledge 658* (BM., K); foot of Adams Peak, *van Beusekom & van Beusekom s.n.* (PDA). DISTRICT UNCERTAIN: Uva, ? *Thwaites in C.P. 1325 p.p* (BM; PDA). WITHOUT LOCALITY: *Moon s.n.* (BM); Herb. *John Smith* (BM); *Wall s.n* . (BM).

5. Adiantum flabellulatum L., Sp. Pl. 1095. 1753; Bedd., Handb. Ferns
Brit. Ind. 88. 1883; Sledge, Ceylon J. Sci (Biol. Sci.) 10: 148. 1973; W.C.
Shieh, Wang-Chueng in Fl. Taiwan ed. 2., 1: 239. 1994.
Type: China, *Osbeck**; Linnaean Herb. 1252/7 (LINN, lectotype).

Rhizome shortly creeping, erect or ascending; scales dense, yellowish brown,
very narrow, 5 mm long. Fronds tufted, ± 15 cm long; stipes nearly black, 9–
44 cm long, hairs in adaxial groove. Lamina 8–17 cm long, main rhachis short
with 2–3 alternate pinnae on each side, lower ones largest and once or twice
branched. Pinnules subrhomboidal, shortly stalked, sterile ones 1.2–1.5 cm
long, 7–9 mm wide, denticulate, subcoriaceous, glabrous; fertile ones usually
smaller, often only ± 1 cm long, shallowly lobed, usually each lobe with an
oblong or reniform sorus occupying entire apex.

D i s t r. India, Sri Lanka, Indochina, SE. China, Taiwan, S. Japan, Ryukyu
Is., Malaysia.

E c o l. Rocky banks in forest; 400–600 m

S p e c i m e n s   E x a m i n e d. BADULLA DISTRICT: Haputale,
Talipotenna Estate, *Westland s.n.* (PDA). RATNAPURA DISTRICT: W. end
of Katussagala Hill, *Faden & Faden 76/471* (PDA); Balangoda, *Gardner in
C.P. 3390 p.p* (PDA); Between Haputale and Balangoda, ? *Thwaites in C.P.
3390 p.p* (BM., PDA). WITHOUT LOCALITY: 'No. *50*' (PDA); *C.P. 3390*
(BM., LIV); *Wall s.n.* (BM).

N o t e. Although Sledge determined the specimen above as cf. *indicum* he
does not mention the specimen in his revision of the Sri Lanka species.
Further work, particularly cytological examination is required.

7. Adiantum diaphanum Blume, Enum. Pl. Java 215. 1828; Sledge, Ceylon
J. Sci. (Bio. Sci.) 10: 148, fig. 5. 1973; W.C. Shieh, Wang-Chueng in Fl.
Taiwan ed. 2, 1: 237. 1994.
Type: Java, Cheribon, 'Lingam Jattie' (Linga Jati), *Blume* (L. syntypes, K
probable isosyntypes).

Rhizome small, short, erect, with lanceolate scales to 2.5 mm long but
soon glabrous. Fronds with lamina simply pinnate, linear-lanceolate or mostly
bipinnate at base, 6–13 cm long; stipes dark brown or nearly black, polished,
3–25 cm long. Pinnules shortly stalked, ± oblong, subrhomboid or elliptic-
rhomboid, 0.8–2 cm long, 0.5–1 cm wide, membranous, usually with sparse
dark setae beneath. Sori rounded reniform., 1 mm wide, situated at the base of
incision between the lobes of the upper edge of the pinnules, glabrous.

D i s t r. Java to Fiji, New Zealand and Australia; introduced elsewhere.

E c o l. Roadside banks.

---

*The specimen in LINN may be the Osbeck specimen

S p e c i m e n s   E x a m i n e d. KANDY DISTRICT: Duckwari–Rangalla road, Girindiella turn-off, 21/12 mi. post, *Ballard 1424* (K; PDA; US). BADULLA DISTRICT: Badulla, *Freeman 68C, 69D* (BM).

8. Adiantum trapeziforme L., Sp. Pl. 1097. 1753; Sledge, Ceylon J. Sci. (Biol. Sci.) 10: 150, fig. 7. 1973.
Type: Sloane, Jamaica 1, t. 59. 1707 (lectotype, chosen by Lellinger in Proc. Biol. Soc. Wash. 89: 704. 1977).

Rhizome creeping with linear-lanceolate entire or denticulate scales pale brown at base, dark brown and shining at apex, (1–)2–3 mm long. Fronds 40–95 cm long, usually ± separated; stipe dark reddish brown to blackish purple, about half the length of the frond, glabrous but minutely tuberculate, shining. Lamina broadly ovate, tripinnate at base. Pinnae in 1–3(–5) pairs, alternate, stalked. Pinnules ± trapeziform., 2.5–5.5 cm long, shortly stalked, glabrous, the sterile margin incised-lobate, fertile margin shallowly lobed; veins extending into teeth. Sori oblong, 1–3.5 mm long, 10–15 along the upper margin of the pinnule and 3–4 on outer margin, 1–2 per lobe; indusium glabrous, dark brown.

D i s t r. Central America and West Indies.

E c o l. Shady places, forest and 'secondary jungle', secondary forest on steep slopes, gullies, erosion scars; 490–750 m.

S p e c i m e n s   E x a m i n e d. KANDY DISTRICT: Above Kandy, track from Roseneath Road, *Walker T 99* (BM); Ganoruwa forest, *Faden & Faden 76/389* (K); Mahaveli Gorge, 3 mi. below Peradeniya, *Fosberg 50671* (K, US); Peradeniya Botanic Garden, *Schmid 883* (BM); Kandy Catchment Valley, *Sledge 1096* (K); Udawatakele, *Sohmer 8040* (PDA); Peradeniya, Ganoruwa Forest, *van Beusekom & van Beusekom 1657* (PDA); Ganoruwa Hill, *Wirawan 607* (K); Ganoruwa jungle behind Botanic Garden, *Kostermans 24921* (K, L).

9. Adiantum pulverulentum L., Sp. Pl. 1096. 1753; Sledge, Ceylon J. Sci. (Biol. Sci.) 10: 150, fig. 8. 1973. Type: Plumier, Descr. pl. Amér., t. 47. 1693 (repeated in Traité foug. Amér. t. 55. 1705) based on a plant from Martinique or Hispaniola (lectotype chosen by Proctor, Fl. L. Antill. 2: 185. 1977).

Rhizome shortly creeping; scales light brown with thick dark brown walls, linear-lanceolate, 1–2.5 mm long, 0.3–0.8 mm wide, entire to denticulate, ciliate with marginal hairs. Fronds tufted, 0.3–1 m long; stipe blackish purple, about half the length of frond, shining but with small brown pectinate scales; lamina bipinnate, 25–50 cm wide; pinnae in 3–10 pairs, alternate; pinnules oblong, usually falcate, up to 1–2 cm long with usually 1 long linear pectinate sorus per segment, 5–10 mm long; indusium erose.

D i s t r. Central and South America and West Indies.

E c o l. 'Secondary jungle'; 750 m.

N o t e. A note on *Freeman 70* states it is just beginning to become naturalised but the general label gives 1908–1923 so it is not known exactly when. Probably much commoner than the specimens indicate.

S p e c i m e n s   E x a m i n e d. KANDY DISTRICT; Kandy *Freeman 70A, 70B* (BM); Peradeniya Botanic Garden, *Schmid 891* (BM); Kandy Catchment Valley, *Sledge 1093* (K).

10. Adiantum latifolium Lam., Encycl. Meth. 1: 43. 1783; Sledge, Ceylon J. Sci (Biol. Sci.) 10: 150, fig. 9. 1973.
Type: West Indies, Guadeloupe, *Proctor 20110* (A, neotype).

Adiantum intermedium sensu Hook. & Baker, Syn. Fil. ed. 2: 116. 1874 non Sw.

Rhizome long creeping much branched, slender; scales light brown with thick dark brown cell walls, linear-lanceolate, 2–3 mm long, 0.5–0.8 mm wide, entire to sparsely denticulate. Fronds separated, 30–95 cm long; stipe blackish purple, $^1/_2 - ^2/_3$ the length of frond, grooved, ± glabrous, with pectinate scales up to 3 mm long; lamina triangular to oblong, 15–40 cm wide. simply to mostly bipinnate; pinnae in 1–4 pairs, alternate, to 20 cm long; pinnules obliquely oblong-lanceolate, up to 2.5–5.5 cm long, 0.7–1.5 cm wide, acuminate at tip; sterile pinnules denticulate, glabrous. Sori several on upper and outer edges, linear, 2–5 mm long; indusium glabrous.

D i s t r. Central and S. America and West Indies.

E c o l. Steep slopes, road cuttings with almost vertical red lateritic pebbly clay, sandy ground, shady places, tea and rubber plantations; near sea level – 800 m.

N o t e. Proctor (Fl. L. Antill. 2: 186. 1977) discussed the typification of this species stating that no specimen was extant and selected *Proctor 20110* from Guadeloupe as neotype. Lamarck cites three pre-Linnean authors but the figures refer to other species.

S p e c i m e n s   E x a m i n e d. COLOMBO DISTRICT: Old Ratnapura road 10 mi. from Colombo, 11/1 mi. post, *Ballard 1434* (K, L, P, PDA, US, Z). E. of Colombo, N. of river from Kaduwala, at water intake station, *Maxwell 1033* (PDA). KEGALLE DISTRICT: Mawanella, *Schmid 1065* (BM., G); Kitulgala, *Schmid 1074* (BM., G). KANDY DISTRICT: New Peradeniya–Gadawala road at marker 5/2, *Comanor 504* (PDA); Gampola to Nawalpitiya road at marker 17/12, *Comanor 530* (K); Between Kandy and Kurunagala

near Galegedara, *Walker T 5* (BM); Delpitiya, *Faden & Faden 76/24*
(K, PDA); Peradeniya Botanic Garden, *Schmid 900* (BM); Giragama,
Amaratunga 353 (PDA). KALUTARA DISTRICT: Track to Pahala Hewessa,
Sledge 896 (BM., K).

11. Adiantum formosum R.Br., Prodr. 155. 1810 ; Sledge, Ceylon J. Sci.
(Bio. Sci.) 10: 152, fig. 10. 1973; Duncan & Isaac, Ferns & Allied Pl. Vict.
Tasm. S. Austr. 139, figs. 13.4, 13.6a, 13.9 (left).1986.
Type: Australia, New South Wales, Port Jackson, *Brown* [67] (BM., holotype,
K, isotype).

Rhizome long, creeping, scaly, branched and forming colonies; scales pale
brown, lanceolate, 2–3 mm long, 0.5 mm wide. Fronds erect, tall and robust,
0.3–1.5 m tall, (3–)4-pinnate, sometimes with up to 1000 pinnules. Stipe
purplish brown or black, shiny, mostly 30–45 cm tall, rough and scaly at the
base, glabrous above; rhachis dark brown to black, setulose-pubescent. Pinnules obcuneate to oblong or asymmetrically rhombic, 0.7–1.5 cm long, 3–7
mm wide, the upper and outer margin usually only slightly incised and mostly
sharply serrate, lower margin ± straight and entire, glabrous or hairy; pinnule-
stalklets 0.5–1.5 mm long. Sori 4–8 per pinnule, 2 per lobe, reniform., broadly
lunate or oblong, 0.5–1.5 mm wide.

D i s t r. Australia and New Zealand but now widely cultivated and
naturalised in some countries.

E c o l. Roadside banks.

S p e c i m e n s   E x a m i n e d. NUWARA ELIYA DISTRICT: Nuwara
Eliya, *Schmid 1491* (BM., G). DISTRICT UNCERTAIN: Kandy to Nuwara
Eliya, *Gunawardhana s.n.* (BM., K).

12. Adiantum capillus–veneris L., Sp. Pl. 1096. 1753; Bedd., Handb. Ferns
Brit. Ind: 84. 1883; Sledge, Ceylon J. Sci. (Bio. Sci.) 10: 152, fig. 6. 1973;
W.C. Shieh, Wang-Chueng in Fl. Taiwan ed. 2, 1: 237. 1994.
Type: a specimen from France, Montpellier in Magnol's Herbarium sent by
De Sauvages to Linnaeus (LINN 1252.9. lectotype).

Rhizome creeping, short, ± coarse; scales brown, linear-lanceolate, ± 1–3
mm long, with attenuate tips, entire. Fronds clustered, erect or spreading or
hanging, 5–50(–90?) cm tall; stipes dark brown to purplish black, mostly 5–
23 cm long, slender, brittle, smooth and polished; basal scales light brown,
shiny; rhachis and pinna-stalks very dark, slender; lamina mostly 11–23 cm
long, 2–3 pinnate. Pinnules mostly obcuneate, 0.5–3 cm long, 0.6–2.8 cm
wide, slightly asymmetrical with rounded to strongly cuneate bases and rounded
outer edges, shallowly to deeply and irregularly incised or lobed, the lobes

often long and narrow, toothed at the apex, glabrous; veins repeatedly forked, free; pinnule-stalklets very slender, 2–8 mm long. Sori 4–7 per pinnule, relatively large, at apices of lobes, oblong-elliptic, oblong or square, not curved, 1–4 mm wide, 1.5–2 mm long.

D i s t r. Almost cosmopolitan.

E c o l. Wet roadsides, crevices in dry banks, near small waterfalls; 400–960 m.

S p e c i m e n s E x a m i n e d. KANDY DISTRICT. SE. of Kandy, 1 mi W. of Marassana, *Maxwell 990* (PDA). BADULLA DISTRICT: Badulla road mi. 67/70, *Ballard 1311* (K, P, PDA, US, Z); Hakgala to Badulla road, E. of Welimada, Puhulpola, *Sledge 773* (BM., K). MONERAGALA DISTRICT: Mile post 135/11 on Koslanda–Wellawaya road, Hewelkandura, *Faden & Faden 76/587* (K, PDA). DISTRICT UNCERTAIN: Near Badulla road, P308 (cultivated at Kew) *Alston 11736* (K); Nuwara Eliya to Badulla, *Mrs Chevalier* (BM). NOT TRACED: Uy(?g)aldur(v?)a Valley *s.coll.* in 1881 (?*Trimen*) (PDA). WITHOUT LOCALITY: 'No 48' (PDA); *C.P. 1324* (BM); *Bradford s.n.* (BM); *Robinson s.n.* (K).

13. Adiantum concinnum Willd., Sp. Pl. 5: 451. 1810; Sledge, Ceylon J. Sci (Bio. Sci.) 10: 152, fig. 11. 1973. Type: Venezuela, Caracas ('Cumaná'), *Humboldt & Bonpland s.n.* (B-Willd. 20099-2, lectotype; chosen by R.M. Tryon in Contr. Gray Herb. 194: 168. 1964).

Rhizome thick, compact or shortly creeping; scales uniform brown, linear-lanceolate, (1.5–)3–4 mm long, 0.8–1 mm wide, entire. Fronds tufted, arching or pendent, 20–95 cm long; stipe chestnut-brown to purplish black, $^1/_4$ to $^1/_3$ the length of frond, shining, glabrous, with a few broad scales at base; lamina lanceolate to ovate-lanceolate, 6–25 cm wide, long-attenuate at apex, 2–3-pinnate; pinnae in 10–15 pairs, gradually reduced, alternate, glabrous, basal pinnule overlapping rhachis; pinnules obovate to oblong or rhombic-ovate, 0.7–1 cm long, cuneate at base, shortly stalked, colour of stalk running into segment base, glabrous, lobed and incised, veins running into sinuses; sterile segments denticulate. Sori rounded reniform; indusium often whitish, 1–1.5 mm wide, glabrous, thin, entire with abruptly flaring margins.

D i s t r. Central America, western S. America and West Indies introduced elsewhere and occasionally naturalised.

E c o l. Roadside banks in shady jungle; 450–1950 m (presumably Freeman's "Galla" record is from near sea level).

N o t e. Some material with rather larger pinnules has been referred to cv. *latum* Moore e.g. *Alston 1666* (K) at first named as *Adiantum capillus-veneris* but later determined as *concinnum* by F. Ballard; the PDA duplicate was similarly redetermined by Sledge. *Amaratunga 890* has been named *Adiantum tinctum* Moore but is undoubtedly identical. Moore's type could not be found at Kew so I have not put it in synonymy; Goudey has it as a cv. of *Adiantum raddianum* but his figures do not seem to differ from *Adiantum concinnum*. *Sledge 587* has been determined as cv. *multifida*, a name not given by Goudey.

S p e c i m e n s E x a m i n e d. KANDY DISTRICT: Madugoda, *Alston 1666* (K, PDA); Haragama, *Amaratunga 890* (PDA); NE of Hunnasgiriya (c. 9 mi.) near mile post 29/21 on the road to Mahiyangana, *Davidse & Sumithraarachchi 8426* (PDA); Teldeniya–Mahiyangana road beyond Urugala, *Faden & Faden 76/559* (PDA); Kandy, *Schmid 855* (BM., G). Hunnasgiryia, *Schmid 950* (BM., G); Madugoda to Weragamtota, *Walker T26; T 40* (BM); Kadugannawa, *Sledge 587* (BM); *589* (K); Madugoda, *Sledge 933*(). NUWARA ELIYA DISTRICT: Nuwara Eliya, *Schmid 1350* (BM; G). GALLE DISTRICT: 'Low country from Galla' *Freeman 64B* (BM).

14. Adiantum raddianum C. Presl, Tent. Pterid. 158. 1836; C. Presl, Abh. Königl. Böhm Ges. Wiss. IV, 5: 158. 1837; Sledge, Ceylon J. Sci. (Bio. Sci.) 10: 152, fig. 12. 1973.
Type: Raddi, Pl. Bras. Nov. Gen.–Cogn., t. 78/2. 1825.

Adiantum cuneatum Langsd. & Fisch., Ic. Fil. 23, t. 26. 1810; Willd., Sp. Pl., ed. 4, 5: 450. 1810; Raddi, Pl. Bras. Nov. Gen.–Cogn. 59, t. 78/2. 1825 non Forst., 1786. Type S. Brazil, Ilha de Santa Caterina, *Langsdorff s.n.* (LE, holotype, BM., isotype).

Rhizome short creeping, much branched, slender, with tufted fronds; scales brown, broadly lanceolate, 0.8–1.5 mm long, entire. Fronds erect or pendulous, 15–50 cm tall; stipe and rhachis dark brown, glabrous; lamina 3–4-pinnate, irregularly or dichotomously arranged. Pinnules typically obtriangular, 5–10 mm long, 4–12 mm wide, cuneate at the base but in cultivars exceedingly varied in shape and size (see note), crenate-serrate, glabrous; pinnule-stalklets brown, 1–3 mm long, glabrous. Sori round or reniform, 1.5 mm wide, situated in the marginal sinuses; indusium pale, membranous.

D i s t r. Originally South America but now very widely naturalized in many tropical areas.

E c o l. Roadside, moist banks, under tea bushes, under large boulders, often by streams; 650–2140 m.

N o t e. This species has been extensively cultivated and numerous very distinctive cultivars are available, many morphologically quite distinct from the typical form. An account of these is given in C.J. Goudey 'Maidenhair Ferns in Cultivation' 128–259. 1985. *Adiantum raddianum* c.v. *weigandii* (= *A. weigandii* Moore in Gardn. Chron II, 20: 748. 1883; Verdc. in J.E.A. Nat. Hist. Soc. 24: 39. 1962) is widely cultivated and may well occur in Sri Lanka.

S p e c i m e n s   E x a m i n e d. KANDY DISTRICT: Munagalla Tea estate ground just near road at Hwy marker 11/11 on road from Dolosbage E. to Raxawa Tea Est., *Grupe 196* (PDA); Peradeniya Botanic Gardens, *Schmid 912* (BM); Hunnasgiryia, *Schmid 961* (BM., G); Galaha, *Schmid 1028* (BM., G); Gampola, *Sledge 600* (BM). BADULLA DISTRICT, Welimada-Haputale road, Glenanone Estate, Blackwood Forest, mile post 16, *Faden & Faden 76/664* (PDA). RATNAPURA DISTRICT: Belihul Oya, *Schmid 1226* (BM., G). NUWARA ELIYA DISTRICT: Castlereagh Estate, *Alston 1857* (PDA); Horton Plains, Diyagama road, *Comanor 974* (PDA); Nuwara Eliya, *Freeman 63A* (BM) (probably the first record of this species for the island); Single Tree Mount, *Sinclair 10084* (BM., E, L, SING); Hakgala, *Sledge 636* (K); Hakgala Forest Reserve, just beyond Hakgala Garden on way up trail to Hakgala Peak, *Sohmer et al. 8506* (PDA); Ellamulla, *Waas 1931* (PDA). WITHOUT LOCALITY: Grown in propagating house at Kew ex Ceylon as "*A. aethiopicum*" *596/50* (BM); *Amaratunga 232* (PDA).

15. Adiantum poiretii Wikstr. in Vet. Akad. Handl. 1825: 443. 1826; Sledge in Ceylon J. Sci. (Bio. Sci.) 10: 154. 1973.
Type: Tristan da Cunha, *Petit-Thouars* in Herb. Juss. 1427 (P, holo).

Adiantum thalictroides Willd. ex Schltdl., Adumbr. Pl. (5): 53. 1832. Type:
   Mauritius, *Petit-Thouars* in Herb. Willd. (B-WILLD 20101, lectotype).
Adiantum aethiopicum sensu Bedd., Handb. Ferns. Brit. Ind.: 84. 1853 non L.

Rhizome creeping, slender but with thicker branches with closely spaced fronds; scales brown, lanceolate, up to 8 mm long, 0.8 mm wide, adpressed, acuminate, slightly ciliate. Fronds erect, (10–)20–60(–100) cm tall, 3(–4) -pinnate, branching irregular, with main branching subpinnate with small branches often dichotomously arranged. Stipe and rhachis dark brown or black and glabrous, save for brown lanceolate scales at base of stipe. Pinnules triangular to fan-shaped, 0.5–1.5(–2) cm long, 0.5–1.5 cm wide, usually broader than long, truncate to rounded or more rarely cuneate at base, crenate on upper and outer margins and also shortly incised, often breaking off and leaving stalklet at maturity, glabrous or in some variants with yellow powder or gland-tipped hairs beneath; stalklets 2–5 mm long, brown, glabrous. Sori curved, rounded-oblong to reniform., situated in marginal sinuses, 1–2.5 mm wide, ± 1 mm long.

D i s t r. Tropical and temperate Africa, Madagascar, Mascarene Is., India, Tristan da Cunha, Central and South America from Mexico to Uruguay.

E c o l. Presumably forest at ± 1500 m.

N o t e. Pichi Sermolli (Webbia 12: 687–696. 1957) makes out a case for considering *A. poiretii* endemic to Tristan de Cunha which, if accepted, results in *A. thalictroides* being the correct name to use for the wide-spread species. After examining Tristan material I am not convinced the differences are adequate and have preferred to follow Alston, Schelpe, Tryon and others. The matter is, however, by no means closed. Only the sheet mentioned by *Sledge* has been seen. Since this was collected in February 1881, one must assume that it is no longer present on the island. But a sterile specimen *N.D. Simpson 8827* (BM) from forest bank E. of Madugoda, 8 November1931 appears to be *A. poiretii* rather than *A. aethiopum* L. as it had been named.

S p e c i m e n s E x a m i n e d. NUWARA ELIYA DISTRICT: "Above Elgin, Dimbula" *'W.F. s.n.' (Ferguson)* (PDA) [the two localities do not appear to be close according to modern gazetteers].

S A L V I N I A C E A E

(by Jennifer M. Ide*)

H.G.L. Reichenbach, Bot. Damen, Künstler und Freunde Pflanzenw. 255. 1828
N.B. Some authors include *Azolla* in this family rather than in a family of its
own, Azollaceae.
Type: *Salvinia* Séguier (Pl. Veron. 3:52. 1754)

Small, free-floating aquatic, herbaceous plants forming extensive colonies on
still water. Annual or perennial. Plants of a single whorl of three leaves; two
lateral, green, floating leaves more than 10 mm long, and one submerged and
divided into many filiform., pendant segments usually longer than aerial leaves,
covered in multicellular filamentous hairs, and serving as a root. True roots
absent. Floating leaves folded lengthwise in vernation, flat or folded longitudinally when mature, sessile or shortly petiolate. All species readily propagate
vegetatively, additional plants remaining attached by the horizontal, freely
branching rhizome. Rhizome siphonostelic. Sporangia produced in hairy sporocarps borne singly on branches of submerged, trailing receptacles between
root-like filaments of submerged frond. Sporocarp-bearing branches arise in
pairs or alternate irregularly (both conditions may appear in the same "inflorescence"). The thin-walled, unilocular sporocarps each contain a single sorus
of either mega- or microsporangia. Microsporangia numerous, each producing
64 microspores; megasporangia few, each maturing only one megaspore. Mature sporocarps sink and decay releasing entire sporangia that rise to the
surface. Gametophytes minute, endosporic, floating.

Family of a single genus, *Salvinia*. 10 species and hybrids, mostly native to
tropical America and Africa, including Madagascar, now cosmopolitan due to
introductions, but absent from colder regions. Widespread in Old World except
SE Asia-Australasia; and in New World north to the southern United States.

All species so far examined have a chromosome base number of n=9;
ploidy levels up to heptaploid are known.

Literature:
Mitchell, D.S. 1972. *The Kariba Weed: Salvinia molesta,* Brit. Fern Gaz.
10(5): 251-252

---

* 42, Crown Woods Way, Eltham, London, U.K.

Mitchell, D.S. & P.A.Thomas 1972 *Ecology of water weeds in the neotropics.* A contribution to the International Hydrological Decade. UNESCO Paris

Kramer, K.U. (1990) in Kramer, K.U. & P.S. Green *The Families and Genera of Vascular Plants*, ed. K. Kubitzki, Vol. I, Pteridophytes and Gymnosperms. p. 256-258

K.R. Bhardwaj, 1991 *Response of Barley Crop to Salvinia manuring*, in T.N. Bhardwaja & G.B.Gena, *Perspectives in Pteridology Present and Future*, Aspects of Plant Sciences Vol. 13. Today and Tomorrow's Printers and Publishers, New Dehli [Describes field trials which showed that *Salvinia* can be used as a fertiliser to improve the crop yield of barley]

V.S. Manickam and V. Irudayaraj, *Pteridophyte Flora of the Western Ghats— South India*, B.I. Publns. New Dehli. 1992. pages 343-344

Cook, C. D.K. 1996 *Aquatic and Wetland Plants of India.* Oxford University Press, Oxford.

SALVINIA

Séguier, Pl. Veron. 3:52. 1754.

Type. *S. natans* (L.) All. (*Marsilea natans* L.)

Marsilea natans L. Species Plantarum 2. 1099. (1753)
Salvinia natans Allioni, C. Flora Pedemontana 2: 289 (1785)

Plants of still, open, exposed bodies of water of high nutritional status, although de la Sota has reported one species in Argentina, *Salvinia nuriana* nov.sp., which is adapted to oligotrophic, stagnating water (de la Sota & Cassa de Pazos, *Notes on the Neotropical species of* Salvinia *Séguier*, in R.J. Johns (ed.), *Holttum Memorial Volume*, Royal Botanic Gardens, Kew, 1997). Often cover water surface and shade out other plants below.

Salvinia molesta Mitch., Brit. Fern Gaz. 10(3) : 251. 1972.

Synonymy: *S. auriculata* auct. non Aubl.

Specimens collected before 1973 are usually identified as *S. auriculata* Aubl., but this species is not present in Sri Lanka.

Illustrations. Cook, C. D.K. 1996 *Aquatic and Wetland Plants of India.* Oxford University Press, Oxford. Good description and illustrations (fig.8, page 29).

Mitchell, D.S. & P.A.Thomas 1972 *Ecology of water weeds in the neotropics.* A contribution to the International Hydrological Decade. UNESCO Paris (figs.1-4 illustrate the arrangement of the sporocarps on the receptacle of the four species of the auriculata group of the genus, which in their vegetative state are difficult to distinguish by the inexperienced.)

Schneller, J.J. (1990) in Kramer, K.U. & P.S. Green *The Families and Genera of Vascular Plants*, ed. K. Kubitzki, Vol.I, Pteridophytes and Gymnosperms. A. in fig.130, p.257 is of *S. molesta* and not *S. auriculata*.

Sota, E.R. de la, 1995 *Nuevos sinonimos en Salvinia Sg. (Salviniaceae – Pteridophyta)* Darwinia 33(1-4): 309-313

Robust, soft to the touch, very aggressive weed, reproducing rapidly by vegetative propagation and capable of covering large areas of open water very quickly. Tends to grow in individual colonies of up to 20+ plants.

Rhizome terete, up to 1.5 mm diam. Sparsely hairy. Internodes up to 1 cm between plants. Aerial leaves borne in opposing pairs. Flat in less crowded situations or folded lengthwise along midrib when crowded. Pale green. Subsessile to shortly stalked, up to 2 mm. Orbicular to oblong. Margin entire or shallowly scalloped or fluted. Retuse to emarginate distally; truncate to cordate at base. Variable in size from 15-54 mm long × 12-24 mm wide (as measured over longest and widest parts). [Much smaller plants may be encountered (e.g.leaf dimensions approximating 10 × 8-9 mm) This may be due to the nutritional status of the water.] Upper surface covered with dense, hydrophobic, colourless, multicellular, stalked hairs ending in four branches united at their tips (resembling an egg whisk!), uniting cells brown pigmented. Arranged in neat radiating and tangential rows at lamina margins, rows becoming less distinct towards centre. Branches of hairs reduced in length towards lamina margin and may be absent. Hairs of lower (abaxial) leaf surface filamentous, of ±5 cells, with brown pigment at cross walls so that hairs in bulk appear being coloured. Dense along midrib, frequent to dense at lamina margin; centre of lamina lobes with scattered hairs or glabrous. Veins distinctly abaxial. On each side of midrib, main veins anastomose to form ± 1 row of polygonal areoles; from these, veins extend, parallel to one another, to margin of lobes; they anastomose occasionally to form long rectangular areoles, becoming shorter towards distal margins. Between the ± parallel sides of main areoles, finer anastomosing branches create 2-4 rows of polygonal sub-areoles. Submerged frond shortly stalked (7-12 mm), dividing thereafter into a cluster of 12-35 filaments. Filaments of submerged frond vary in length from c. 2 - 20 cm; length can vary within a given colony, being shorter in younger parts. Filaments covered with numerous (7-9) brown, septate hairs. Sporocarp clusters up to 12+ per plant, and 7+ cm long, emerging from centre of filaments (submerged fronds) and fused to their base. The elongated receptacle has simple lateral branches, arising ±in pairs proximally, becoming more alternate distally; first branch is longer (up to 4 mm) than subsequent ones (0.5-1.0 mm); branches can vary in length even within the same cluster; each branch bears a sessile sporocarp. As many as 37 to 55 densely hairy sporocarps per cluster borne singly on the lateral branches. The first proximal sporocarp of a cluster is megasporangiate and usually slightly smaller and more rounded than the remaining microsporangiate ones which are globose, width 1.5-2.0 mm., slightly longer than broad; all are apiculate. The elongate receptacle of sporangiate clusters bears hairs similar to those on filaments of the submerged frond.

C y t o l o g y. The species is a pentaploid hybrid (5n = 45) between two South American species, *S. auriculata* Aubl. and *S. biloba* Raddi (Mitchell, 1973), but has become naturalised in many tropical countries of the Old World. Though usually described as sterile, most sporangia being abortive, a few apparently fully developed mega-and microsporangia, are frequently found.

D i s t r. Pantropic due to the activity of man. Within Sri Lanka: central and southeastern lowland areas

E c o l. Fresh, nutrient-rich water; aggressive weed growing abundantly in fully exposed ponds, paddy fields, tanks, canals, lagoons, reservoirs, and lakes, and is increasing in polluted water with high levels of nitrogen.

N o t es. The sporocarp is equivalent to the sporangium, the sporocarp cluster,in turn, to the sorus.Referring to the sporocarp clusters as 'inflorescence', even infructescence, would confuse angiosperm- with cryptogam-terminology. Most texts report that the first two proximal sporocarps are megasporangiate; on herbarium vouchers available and examined for this study only the first sporocarp was megasporangiate. This is possibly a regional variation.

All specimens examined have been collected post-Second World War, which lends credence to the suggestion that *S. molesta* (as supposedly *S. auriculata* Aubl.) was introduced, about 1939, to disguise open bodies of water from enemy aircraft.

S. molesta can be distinguished from *S. auriculata* by its very robust growth habit, and the elongated chain of mostly male sporocarps containing many mainly empty sporangia. A few, fully developed mega- and microsporangia have been observed in their respective sporocarps but it is not known if the spores contained therein are viable.

It is not fully established if the submerged leaf acts as a root with an absorbing capacity or as a stabilising organ only.

S p e c i m e n s   E x a m i n e d. ANURADHAPURA DISTRICT: In pool, 13 July 1972, *Hepper & Jayasuriya 4653* (K). POLONNARUWA DISTRICT: Mile 48.5 Polonnaruwa/Batticaloa Road, floating on standing water, near small river, 16 Dec 1980, *Piggott 2654* (K). BATTICALOA DISTRICT: South of Kalkudah on highway A15 at mile 18, growing in pond, 20 Apr 1968, *Mueller-Dombois 68042014* (K). COLOMBO DISTRICT: Kandy Road, paddy field, 5 or 6 miles from Colombo, 29 Jan 1951, *Ballard 1572* (K); paddy field near Colombo, 23 March 1954, *Sledge 1361* (K); Muthuraja

Wela, just N. of Uswetakeliyawa, sea level, floating in canal, 30 Nov 1976, *Faden, & Faden 76/416* (K); Negombo - Blue lagoon near mouth of Dandugam Oya, 31 July 1970, *Meijer 583* (PDA); Negombo, sea level, 7 Dec 1951, *Senaratna 5492/52/H1873* (K). KALUTARA DISTRICT: Road from Kalutara (NE) to Kethena, 5-10 m., in open water at edge of paddy, 14 Feb 1968 *Comanor 1012* (K). NUWARA ELIYA DISTRICT: Kotmale Reservoir at Sangilipalama, *Jayasuriya 3312* (PDA); SW side of Gregory lake, E. of Nuwara Eliya, Alt. 1890m., 23 Sept 1993, *Fraser-Jenkins* with *Jayasekara, Samarasinhe & Abeysiri* Field No. *225*(K) ; shore of Gregory lake, *Sohmer, n Jayasuriya, Eliezer 8472* (PDA). GALLE DISTRICT: Nagoda, Wadiyahenakande Forest, 187 m., *Read 2297* (PDA).

SCHIZAEACEAE

(by D. Philcox* †)

Kaulf., Wesen Farrenkr.: [119]. 1827. Type: *Schizaea* Sm., nom. cons.

Rhizome usually short, creeping, fronds arising close together, less often erect, covered with thick, dark, septate hairs when young. Fronds variously structured, branches grading from dichotomous to pinnate; venation usually free. Sorophores (sporangia-bearing lobes) arise at the ends of veins of fertile leaflets in *Lygodium.*, in small pinnate groups at ends of fronds or branches in *Schizaea* or on special branches of the fronds in *Anemia*. Spores trilete, or monolete in *Schizaea*, surface usually sculptured.

A family of 4 genera, containing about 150 species, usually from warm or tropical countries. Three genera and 5 species occur in Sri Lanka.

KEY TO THE GENERA

1 Leaves vine-like, spreading, climbing; pinnae short- petiolulate, laciniate-lobed with lobes bordered on surface by sporangia . . . . **1. Lygodium**
1 Leaves not vine-like or spreading
  2 Leaves filiform., simple, grass-like . . . . **2. Schizaea**
  2 Leaves bipinnate or tripinnate, not grass-like . . . . **3. Anemia**

1. LYGODIUM

Sw., J. Bot. (Schrad.) 1800: 7, 106. 1801, nom. cons.
Type species: *Lygodium scandens* (L.) Sw.

Rhizome creeping, short or long, densely dark-hairy, where short, fronds produced close together, but more spaced where long. Fronds occurring in 2 rows on upper surface of rhizome. Fronds erect when young, dichotomously 1- or 2-branched, with usually palmately-lobed leaflets; when older, fronds with elongate, twining, winged, rhachises. Sterile leaflets with oblique lateral veins 1 - 3 times forked, margins entire or serrate; fertile leaflets often with contracted blades, with narrow sorophores arising from margins at ends of most veins. Sporangia mostly oblong-ovoid, shortly stalked.

---

* Royal Botanic Gardens, Kew, U.K.

A pantropic genus of about 40 species, also occurring outside the tropics in New Zealand and South Africa, and northwards to Japan, then in eastern United States to Massachusetts.

KEY TO THE SPECIES

1 Leaflets exceeding 10 cm long, sterile usually 5 - 30 × 1.5 - 2 cm., fertile narrower, 0.6 - 1.25 cm; sorophores 5 - 15 mm long... **1. L. circinnatum**
1 Leaflets never exceeding 10 cm long; sorophores up to 5 mm long
  2 Fertile leaflets 1.7 - 2.5 × 1.5 - 2 cm., sterile up to 5 cm long; sorophores 2.5 - 3.5 mm long .... **2. L. microphyllum**
  2 Fertile leaflets 1 - 6 (-10) × 0.4 - 1.25 cm (without sorophores), sterile similar in size; sorophores 3 - 5 mm long .... **3. L. flexuosum**

1. Lygodium circinnatum (Burm.f.) Sw., Syn. Fil.: 153. 1806.

Ophioglossum circinnatum Burm.f., Fl. Ind.: 228. 1768.
Type from Java (G) [not seen].

Rhizome short-creeping, densely covered with black hairs, stipes distributed closely together, densely hairy at base. Fronds climbing to c. 10 m., rhachis 2 - 5 mm diam., glabrous, shortly winged or ridged; primary branches very short, c. 2mm long, apex ± sunken, dormant, pale-brown hairy; secondary branches 2 - 8 cm long, or at times 1-dichotomously branched. Sterile leaflets cuneate at base, palmately 2 - 7-lobed with lobes 15 - 20 (-35) × 1.5 - 2 (- 3.75) cm., narrowing towards apex, entire, softly membranaceous, sparsely hairy on main veins, otherwise glabrous, sometimes fertile on upper part. Fertile leaflets similar to sterile ones but (0.2-) 0.6 - 1.2 cm wide, rarely 3 - 5-lobed, more usually in pairs at ends of ultimate branches; numerous sporangia-bearing lobes (sorophores), 2 - 9 (-14) × 1 - 1.6 mm protruding from margin. Spores verrucose. Cytology: n = 29; 2n = 116 (Manton & Sledge 1954).

D i s t r. Sri Lanka, Andaman and Nicobar Islands; (doubtfully northern India); Thailand, throughout Malaysia; Philippines; China and Hongkong.

E c o l. In lightly shaded places in primary and secondary forest in lowlands, never in very dry areas; up to 1500 m.

S p e c i m e n s   E x a m i n e d. WITHOUT LOCALITY: *Gardner* in *C.P. 1406* p.p. (PDA); *Robinson* C15 (K); February 1854, *Thwaites s.n.* in *C.P. 1406* p.p. (K); *s.coll. s.n.* in *C.P. 1406* p.p. (K, PDA). KEGALLE DISTRICT: Allagalla, 610 m., 19 Feb 1954, *Sledge 1152* (K); east of Kitulgala, 6 April 1954, *Sledge 1405* (K); Rondura, coll.'*J.E.S.*' *s.n.* (PDA); Ambagamuwa, July 1853, *Thwaites s.n.* in *C.P. 1406* (PDA). KANDY DISTRICT: Kandy, January 1870, *Hutchinson 1756* (LIV), February 1872,

Hutchinson 1757 (LIV). RATNAPURA DISTRICT: Balangoda, 28 Aug 1963, *Amaratunga 738* (PDA). MATARA DISTRICT: Deyiyandera, February 1881, ? *Trimen s.n.* (PDA).

2. Lygodium microphyllum (Cav.) R.Br., Prodr.: 162. 1810.

Ugenia microphylla Cav., Ic. Descr. Pl. 6: 76, tab. 595, fig. 2d-e. 1801. Types from the Marianna Islands and the Philippines.

Lygodium scandens Sw., J. Bot. (Schrad.) 1800: 106. 1801, excl. syn. Linn. Type not designated, but the illustration in Rumph., Amb. 6: tab. 32, fig. 2 cited by Schrader, may be considered as typifying the name.

Climber, rhizome wide-creeping, irregularly dichotomously branched, c. 2.5 mm diam., with dense, short blackish-brown hairs. Fronds climbing, 2 - 3 m., often more, long, up to 15 cm wide, rhachis up to 1.5 mm diam., glabrous; stipes c. 10 cm long, glabrescent, narrowly winged above; rhachis similar to upper part of stipe, glabrescent, narrowly winged throughout. Primary branches 4 mm or more long, apex covered with dense brown hairs; secondary branches 5 - 8 (- 15) cm long, glabrescent, narrowly winged, pinnate. Leaflets petiolulate, glabrous, 2 - 6 on each side of rhachis, 0.5 - 2 cm apart, single or pair at apex, (0.4-) 1.7 - 2.5 x (0.4) 1.5 - 2 cm., sterile often to 5 cm long, deltoid to oblong-deltoid, narrowing to subacute or obtuse apex, base subtruncate to broadly cuneate or subauricled, entire; petiolule 2 - 3 mm long. Fertile leaflets with sorophores protruding 2.5 - 3.5 mm from margin, c. 1 mm wide. Spores with raised reticulum on outer surface. Cytology: n = 30 (Manton & Sledge 1954).

D i s t r. India, Sri Lanka, Andaman and Nicobar Islands; throughout Malesia; Africa to Australia and Polynesia.

E c o l. Climbing on bushes or lower branches of tall trees at edges of secondary forests, or as a weed in open places; from sea-level up to 1300 m..

S p e c i m e n s   E x a m i n e d. WITHOUT LOCALITY: *Gardner 1404* (K); *s.coll. s.n.* (?*Gardner*) in *C. P. 1404* p.p.(K). COLOMBO DISTRICT: Colombo, October 1870, *Hutchinson 1759* (LIV); Muthuraja Wela, N of Uswetakelyawa, sea level, 30 Nov 1976, *Faden & Faden 76/415* (K). KEGALLE DISTRICT: Ambagamuwa, July 1853, *Thwaites s.n.* in *C.P. 1404* (PDA). KANDY DISTRICT: Peradeniya, 14 Feb 1915, *Petch s.n.* (PDA). BADULLA DISTRICT: near Lunugala, January 1888, *s.coll. s.n.* (PDA). KALUTARA DISTRICT: road from Kalutara to Kethena, 4 - 8 m., 14 Feb 1968, *Comanor 1010* (K). RATNAPURA DISTRICT: Sannasgama, 3 Jan 1951, *Ballard 1374* (K, PDA). NUWARA ELIYA DISTRICT: near Hatton, 15 May 1906, ?*Willis s.n.* (PDA); Bogawantalawa, c. 1425 m., 7 Feb 1940, *Worthington 788* (K, PDA).

3. Lygodium flexuosum (L.) Sw., J. Bot. (Schrad.) 1800, 2: 106. 1801.

Ophioglossum flexuosum L., Sp. Pl.: 1063. 1753.
Type from India (Herb. Hermann, BM).

Rhizome short with dense, dark-brown hairs; fronds climbing to several m. tall. Stipes closely together, 50 cm or more long, stramineous, with dark-brown base, minutely hairy or glabrescent, narrowly winged above; rhachis winged throughout, pubescent above between wings; primary branches 3 - 5 mm long, dormant, covered with pale-brown hairs; secondary branches pinnate to bipinnate, c. 10 - 20 × 7 - 12 cm., oblong to subdeltoid in outline. Sterile leaflets on lowest branches palmate, often 5-lobed, cordate, 3 - 8 (-10) cm long, 0.8 - 1.5 cm wide above lobed base, apex acute or acuminate, serrate, lower petiolulate, sessile above, most leaflets asymmetric and more or less further lobed; petiolule up to 5 mm long. Fertile leaflets mostly somewhat smaller than sterile, sorophores (1.75 -) 3 - 5 (- 8) mm long, terminating small triangular lobes. Cytology: n = 30, n = 60 (Abraham et al. 1962).

D i s t r. India (Himalaya) and Sri Lanka; throughout Malesia and Philippines; China; Australia and Pacific Islands.

E c o l. In open spaces, climbing over shrubs in mainly open forest in low country, to about 1000 m., not in shady evergreen forest.

S p e c i m e n s   E x a m i n e d. WITHOUT LOCALITY: *Ballard 1387* (K, PDA); *Gardner 1405* (K); *Mackenzie s.n.* (K); *McRae s.n.* (K); *Robinson C13* (K); *Walker s.n.* (K); *s.coll. 217* (PDA). KURUNEGALA DISTRICT: Elephant Rock, 13 Nov 1970, *Fosberg & Jayasuria 52723* (K, PDA); Marakada, 24 Oct 1979, *Klackenberg 162* (K). COLOMBO DISTRICT: Colombo, Cotta Road, August 1870, *Hutchinson 1765* (LIV); Urapola, 13 Oct 1955, *Senaratna s.n.* (PDA).KEGALLE DISTRICT: Ambagamuwa, 1854, *Thwaites s.n.* in *C.P. 1405* (PDA). KANDY DISTRICT: Peradeniya, Galaha Road, University, 560 m., 25 Oct 1967, *Comanor 490* (K); Hantane, 900 - 1200 m., *Gardner 1268* (K); by Mahaweli Ganga near Minipe, 9 Jan 1954, *Sledge 946* (K). NUWARA ELIYA DISTRICT: summit of Adam's Peak, February 1877, *Leefe s.n.* (K). MONERAGALA DISTRICT: between Bibile and Moneragala, 245 m., 22 Feb 1954, *Sledge 1171* (K).

2. SCHIZAEA

Sm., Mem. Acad. Sci. Turin. 5: 419, tab. 19, fig. 9. 1793, nom. cons.
Type species: *Schizaea dichotoma* (L.) Sm., Mem. Acad. Sci. Turin. 5: 419. 1793.

Rhizome erect, suberect or creeping, initial stem or stipe erect, slender, usually narrowly-winged towards apex, densely coarsely hairy at base as other young parts. Frond simple or dichotomously branched with lamina reduced to narrow wing, bearing one (or 2) row(s) of stomata on lower surface; 2-celled glandular hairs frequent on fronds, with basal cells persistent and forming wart-like protruberances. Sorophores at apex of frond, each with median ridge below. Sporangia almost symmetrically ovoid or ellipsoid, attached to sides of ridge, protected by reflex edge of lamina. Spores pale.

A pantropic genus of about 30 species, also occurring throughout temperate regions of the Southern Hemisphere viz. South Africa, Chile, Tasmania and New Zealand, while in the Northern Hemisphere, only occurring in North America.

Schizaea digitata (L.) Sw., Syn. Fil.: 150, 380, tab.4, fig.1. 1806; Bedd.. Ferns S. Ind.: tab. 268. 1864.

Acrostichum digitatum L., Sp. Pl.: 1068. 1753. Type. Sri Lanka, Herb. *Hermann* (BM).
Actinostachys digitata (L.) Wall. ex Hook., Gen. Fil.: tab.111. 1842; Reed, Bol. Soc. Brot 21: 130. 1947.

Rhizome very short, creeping or suberect, c. 3 mm thick, apex densely brown-hairy, 3 - 4 cm below surface of ground, bearing many crowded fronds in 2 alternate rows. Fronds 20 - 35 (-40) cm long, erect, grass-like, unbranched, base slender, triquetrous, crest winged, 2 - 4 mm wide (maximum), midrib very prominent on lower surface of winged sector, glabrous or with abundant 2-celled glandular hairs, slightly grooved above. Sorophores 5 - 18, digitate, usually 2 - 4.5 (-6) cm long, oblong-linear, attached very close together at end of frond. Sporangia very small, in 4 rows, 2 alternately on either side of midrib of lobes, appearing to completely cover lower surface of sorophore. Spores small, finely, obliquely striate, pale yellow-green. Cytology: n = 350 - 370 (Abraham et al. 1962).

D i s t r. Northeast India, Bangladesh and Sri Lanka; Thailand and Indochina; throughout Malesia except for East Java and Lesser Sunda Islands; Philippines and Fiji.

E c o l. In lightly shaded forests and estates up to 1200 m.

S p e c i m e n s   E x a m i n e d. WITHOUT LOCALITY: *Beddome s.n.* (K); *Gardner 1182* (K); 1829, *Lindley s.n.* (K); *Skinner s.n.* (K); *Wall s.n.* (E); *s.coll.* in *C.P. 3105* p.p. (E, K); 1899, *s.coll. s.n.* (E); *s.coll., s.n.* (PDA). COLOMBO DISTRICT: Pore, near Colombo, April ? 1805, *s.coll. 969* (LIV); Henaratgoda, 14 Aug 1926, *de Silva s.n.* (PDA); Gampaha, 17 Dec 1938, *Sinaratna 2573* (PDA). Kegalla District: Kittool Galle [=Kitugalla], February 1854, *Thwaites s.n.* in *C.P. 3105* p.p. (PDA). KANDY DISTRICT: Peradeniya Race Course, *s.coll.* in *C.P. 3105* p.p. (E). KALUTARA DISTRICT: Reigam

Corle, September 1856, *Thwaites s.n.* in *C.P. 3105* p.p. (PDA); Hewissa, October 1871, *Hutchinson 1744* (LIV). RATNAPURA DISTRICT: Kalatuwawa Forest near Kalatuwawa Reservoir, c. 150 m., 1 Dec 1976, *Faden & Faden 76/427* (K, PDA). GALLE DISTRICT: Panangala, 21 Jan 1951, *Ballard 1549* (K, PDA); Udugama, 91 m., 21 Jan 1951, *Sledge 906* (K).

3. ANEMIA

Sw., Syn. Fil. 6, 155. 1806; Panigrahi & Dixit, Proc. Autm. Sch. Bot.: 231. 1966, nom. cons.

Type species: *Anemia phyllitidis* (L.) Sw., Syn. Fil.: 155. 1806.

Rhizome ascending or creeping. Stipes sub-tricanaliculate above when dry, smooth below. Fronds dimorphous. Fertile frond paniculate (in our material), much-branched, arising on long, paired peduncles from base of pinnate lamina. Sporangia numerous, covering ultimate branches of panicle.

A genus of about 75 species mostly from tropical and warm countries almost wholly from the New World. One species, *Anemia phyllitidis*, occurs in Sri Lanka.

Anemia phyllitidis (L.) Sw., Syn. Fil.: 155. 1806.

Osmunda phyllitidis L., Sp. Pl.: 1064. 1753.
Type from South America.

Rhizome ascending, densely brown-pilose. Stipes 15 - 20 cm long, subquadrangular, glabrescent. Fronds (7-) 10 - 18 × (4.5-) 8 - 15 cm., membranaceous, oblong, pinnatisect; pinnae 2 - 4 (-7)-pairs, 4.5 - 7.5 × 1.5 - 2.5 cm., lanceolate, strongly veined with prominent midrib, secondary veins many, subprominent, anastomosing, long acute-acuminate, upper cuneate into obtuse or subacute, somewhat oblique base, lower broadly rounded to obliquely subtruncate; minutely crenate-serrulate to subentire, ciliate at apex, sparsely hirsute, very shortly petiolulate, terminal leaflet clearly longer-petiolulate, 1- or 2-lobed towards base. Fertile fronds longer than sterile ones, usually 2 branches arising from base of lamina, peduncle (5-) 11 - 15 cm long, slender; panicle (2.5-) 6 - 10 or more cm long, bipinnately branched, with up to 12 sporangia arranged in 2 opposite rows along ultimate branchlet. Sporangia c. 0.5 × 0.35 - 0.4 mm., subellipsoid, dehiscent by longitudinal suture. Cytology: n = 76 (Manton & Sledge 1954).

D i s t r. Central America, West Indies and tropical South America. Occurs as an established alien in Sri Lanka.

E c o l. Roadside banks, stream sides.

S p e c i m e n s   E x a m i n e d. WITHOUT LOCALITY: 1938, *s.coll. s.n.*
(K). KANDY DISTRICT: Stream gully in forest above and SE of " Hunnas
Falls Hotel" W. side of Hunnasgiriya mountain, SSE. Off Elkaduwa, NE. of
Kandy, Alt. C. 1200-1300m., 25 Aug 1993, (3 sheets) *Fraser-Jenkins* with
Jayasekara, Samarasinghe & Abeysiri Field No. *29* (K); Kandy, Roseneath,
670 m., 2 Nov 1950, *Ballard 1011* (K, PDA); Atabage, crossing of Atabage
River on Gampola to Pussellawa road, 560 m., 13 Nov 1976, *Faden & Faden
76/245* (K, PDA); below Knuckles Mountain, c. 1460 m., 30 Jan 1954, *Sledge
1074* (K); Duckwari to Rangalla, Giriendiella turnoff, 21/12 milepost, 7 Jan
1951, *Ballard 1425* (K, PDA); ?Gundumalee, 14 Nov 1961, *Amaratunga 8*
(PDA).

SELAGINELLACEAE

(by D. Philcox* †)

Willk. in Willk. & Lange, Prodr. Fl. Hisp. 1 (1): 14. 1861.
Type genus: *Selaginella* P. Beauv.

Plant herbaceous, terrestrial, epiphytic or lithophytic. Stems short or long, erect, ascending or prostrate and creeping. Rooting either at base, or by rhizophores emerging along stem., especially in creeping species. Leaves small, many, iso- or dimorphic, occupying both stems and branches, those on stem often distant, while those on branches usually contiguous; dimorphic leaves arranged in two planes: larger leaves in lower plane usually patent or suberect in two opposite rows; the smaller ones usually arranged antrorsely, appressed in two parallel rows. Strobili usually terminal on branches, tetrastichous, square or platystichous. Sporophylls iso- or dimorphic. Sporangia single in axils of sporophylls. Megasporangia, usually of 3 large and 1 small megaspore(s) per megasporangium. Microspores numerous per microsporangium. Spores trilete.

SELAGINELLA

P. Beauv., Prodr. Fam. Aetheog. 101. 1805, nom. cons. Type species: *Selaginella spinosa* P. Beauv., Prodr. Fam. Aetheog. 112. 1805, nom. illegit. [*Selaginella selaginoides* (L.) Link, Fil. Sp. Hort. Berol. 158. 1841].

N o t e. Many of the specimens from LIV cited under Naylor Beckett numbers as directed by their labelling, also bear C. P. numbers on the sheets in Thwaites' handwriting, thus clarifying that they were from Thwaites' original collections. These references on the Liverpool material must not be confused with other such 'C. P.' annotations, frequently in red ink, which were made by later students of the material implying that they had been favourably compared and matched with Thwaites' numbers.

Several collections have been made from Sri Lanka of species alien to the area, for example *S. willldenowii* (Desv.) Baker and various New World species. These may probably have escaped from one or other of the Botanic Gardens at Peradeniya, Hakgala or Gampaha.

* Royal Botanic Gardens, Kew, U.K.

KEY TO THE SPECIES

1 Leaves spirally arranged, linear-subulate, glaucous . . . . **1. S. wightii**
1 Leaves not as above, arranged in 4 rows
  2 Sporophylls isomorphic
    3 Leaves isomorphic at base of stem
      4 Smaller leaves acute or shortly aristate, with aristae much less than 1/2
      length of lamina . . . . **2. S. involvens**
      4 Smaller leaves with aristae more than 1/2 length of lamina . . . .
      . . . . **3. S. latifolia**
    3 Leaves dimorphic throughout
      5 Smaller leaves subacute, ovate-oblong . . . . **4. S. integerrima**
      5 Smaller leaves aristate or acuminate, suborbicular . . . .
      . . . . **5. S. praetermissa**
  2 Sporophylls dimorphic
    6 Sporophylls not ciliate
      7 Larger leaves ciliolate . . . . **6. S. calostachya**
      7 Larger leaves denticulate . . . . **7. S. cochleata**
    6 Sporophylls ciliate
      8 Larger leaves oblong-lanceolate . . . . **8. S. ciliaris**
      8 Larger leaves ovate . . . . **9. S. crassipes**

1. Selaginella wightii Hieron., Hedw. 39: 319. 1900; Alston, Proc. Nat. Inst. Sci. Ind. 11: 215. 1945. Dixit, Sellagin. India 33, fig. 1 A-G. 1992. Syntypes from India and Sri Lanka.

Selaginella rupestris Spring, Mem. Ac. Brux. 24, 1: 55. 1850, pro parte non typica.

Stems 4 - 12 cm long, creeping, much branched from base; primary branches ascendent, dendroid, older branches 2- to 3-pinnate. Leaves c. 2 × 0.15 mm., numerous, spirally arranged, appressed, isomorphic, long-narrow- triangular, long-acuminate, aristate, broadened and adnate at base, minutely ciliate, denticulate towards apex, herbaceous to ± papyraceous, glabrous. Strobili 4 - 8 × 1.5 - 2 mm., at apex of branchlets. Sporophylls c. 1.5 × 0.8 mm., ovate-triangular, apex acuminate, aristate, base cordate, denticulate above, eciliate to short-ciliate towards base, carinate. Megaspores c. 0.5 × 0.5 mm., irregularly flattened-circular, pale orange-yellow. Spores rugulose, orange-red.

D i s t r. Southern India and Sri Lanka; East Africa.

E c o l. Among rocks in seepage areas and in rock crevices in full sun; up to 950 m.

N o t e. Hieronymus restricted *Selaginella rupestris* Spring sensu stricto, to the North American plant, and coined the name *S.wightii* for the Asian and African element, with the above distribution.

S p e c i m e n s   E x a m i n e d: MATALE DISTRICT: Dambool [=Dambulla], 1859, *s.coll. s.n.* in Herb Fraser (E); Temple Rock, Dambulla, *Thwaites s.n.* in *C.P. 1414* p.p. in Naylor Beckett 2442 (LIV); Dambulla Hill, 16 May 1931, *Simpson 8130* (BM); Dambulla Rock, *Gardner s.n.* in *C.P. 1414* p.p. (BM), *Baker 233* (PDA), 20 Dec 1881, *Trimen s.n.* (PDA), 17 Nov 1961, *Amaratunga 61* (PDA), July 1848, *Gardner s.n.* in *C.P. 1414* p.p. (K, PDA), March 1868, *Thwaites s.n.* in *C.P. 1414* p.p.(K), May 1868, *Gardner s.n.* in *C.P. 1414* p.p. (PDA), 30 Apr 1965, *Amaratunga 850* (PDA), 18 Dec 1970, *Amaratunga 2172* (PDA), 27 March 1954, *Sledge 1367* (BM., K), 22 June 1969, *Wheeler 12006* (PDA), 15 May 1969, *Kostermans 23541* (Ḱ, PDA); Dikpatana, milepost 38/2, c. 900 m., 12 Dec 1976, *Faden, Faden & Jayasuriya 76/537* (K, PDA); Illukumbura, Pitawalapatana, near 33 km marker, 25 Oct 1995, *Shaffer-Fehre et al. 421* (K). POLONNARUWA DISTRICT: Minneri [=Minneriya], September 1885, *Trimen s.n.* (PDA). BATTICALOA DISTRICT: low hill, 16 km inland from Sittandi, c. 5 km N of Eravur, N of Batticaloa, May 1885, *Nevill s.n.* (PDA). BADULLA DISTRICT: Demodara, on A16 road, to Demodara Christ Church, 950 m., 19 Nov 1976, *Faden & Faden 76/374* (K, PDA). RATNAPURA DISTRICT: Belihul Oya, 24 Jan 1963, *Amaratunga 474* (PDA). NUWARA ELIYA DISTRICT: Horton Plains, c. 2135 m., 13 May 1963, *Waidyanatha P7* (PDA), *s.coll., s.n.* in *C.P. 1414* p.p. (HAKS); World's End, *Nock s.n.* (PDA); Hakgala Botanic Garden, 3 Aug 1897, *Willis s.n.* (PDA). MONERAGALA DISTRICT: Kotaveharagala, c. 2 km E of Wellawaya, milepost 140/1, *Faden & Faden 76/604* (PDA). . WITHOUT LOCALITY: *Moon s.n.* (BM); *Thwaites s.n.* in *C.P. 1414* (E); *s.coll. s.n.* (HAKS); *Walker s.n.* (K).

2. Selaginella involvens (Sw.) Spring, Bull. Acad. Brux. 10: 138. 1843, emend. Hieron, Hedw. 50: 2. 1911; Alston, Proc. Nat. Inst. Sci. Ind. 11, 3: 220. 1945; Dixit, Selagin. India 46, fig. 12 A-I. 1992.

Lycopodium involvens Sw., Syn. Fil.: 182. 1806. Type from Japan.
Lycopodium caulescens Wall., Cat. no. 137. 1829, nomen nudum.
Lycopodium caulescens Hook; & Grev., Enum. Fil., Hook. Bot. Misc.2: 382.
    1831. Type from Nepal.
Selaginella caulescens (Hook. & Grev.) Spring, Bull. Acad. Brux. 10: 137.
    1843.

Stem with distinct rhizome 1.5 - 2 mm diam., creeping above or just below surface, bearing sparse, isomorphic, appressed, ovate, brownish leaves, 2 - 2.5 × 0.75 - 1 mm; main stem erect, 15 - 40 cm long; lateral branches alternate, 1 - 2.5 (-4) cm distant, spreading broadly at about 45° from main axis, 3-pinnate, glabrous. Leaves uniform on main stem., appressed, laxly distributed, c. 3 × 2 mm., subcordate, base, apex subacute, minutely denticulate to subentire; leaves on lateral branches dimorphic, larger 2 - 2.25 × c.1 mm., elliptic to ovate-lanceolate, falcate above, acute, somewhat rounded-oblique at base, apparently subentire; smaller, c. 1 × 0.5 mm., ovate, appearing subcarinate, appressed, aciculate, denticulate throughout. Strobili 5 - 15 × 1.5 - 2 mm., solitary at apex of ultimate branches. Sporophylls 1 - 1.5 × 1 - 1.2 mm., uniform., ovate, strongly carinate, acuminate, minutely denticulate towards base. Megaspores laterally flattened-discoid, 0.5 mm diam., dull orange-yellow. Cytology: n = 9 (Kuriachan, Cytologia 28: 376-380. 1963); 2n = 18 (Jermy, Jones & Colden, J. Linn. Soc. Bot. 60: 150. 1967).

D i s t r. India (except North), Nepal, Bhutan, Burma and Sri Lanka; Indo-China and Malesia.

E c o l. Among wet, moss-covered rocks near streams and waterfalls, and shaded roadside banks and forest edges; up to 1525 m..

S p e c i m e n s   E x a m i n e d: MATALE DISTRICT: Lagalla, milepost 35/14 on road Rattota to Dikpatana, 1160 m., 12 Dec 1976, *Faden, Faden & Jayasuriya 76/540* (K); Dankanda Pass, October, 1861, *Brodie s.n.* (E). KANDY DISTRICT: Peradeniya, Botanic Garden, January 1954, *Schmid 927* (BM); Corbet's Gap, 1340 m., 9 Dec 1950, *Sledge 560* (BM); SE of Hunas Falls Hotel, W side of Hunnasgiriya Mountain, SSE of Elkaduwa, NE of Kandy, c. 200 - 1300 m., 25 Aug 1993, *Fraser-Jenkins et al. 27* (K); Pitawela, January 1865, *Naylor Beckett 720* p.p. (LIV); Oodewella, 1066 m., 8 Dec 1950, *Sledge 518* (K); Hantane, *Thwaites s.n.* in *C.P.985* p.p. (PDA); Midcar Estate, Knuckles Range, 5 Nov 1982, *Balasubramaniam 2894* (PDA); Knuckles area, 24 Aug 1956, *Abeywickrama 399* (PDA); Kalupahana, 23 Aug 1956, *Abeywickrama 419* (PDA); foot of Knuckles, 13 June 1926, *Silva 98* (PDA); Rangala, 5 Dec 1982, *Braggins 82/467* (PDA); Oonanagalla, Madulkelle, July 1880, ? *Trimen s.n.* (PDA); Hatale, Madulkelle, October 1887, ? *Trimen s.n.* (PDA); Galaha, 940 m., 22 Jan 1954, *Schmid 1035* (BM); Peradeniya Botanic Gardens,Octagon House, 18 Oct 1927, *Silva 258* (PDA). BADULLA DISTRICT: Glenanore Estate, Blackwood Forest, milepost 15 on Welimada to Haputale road, 1525 m., 18 Nov 1976, *Faden & Faden 76/341* (K). RATNAPURA DISTRICT: Gilimale, 8 Dec 1982, *Braggins 82/534* (PDA). NUWARA ELIYA DISTRICT: Rambodde [=Ramboda], 1866, *Thwaites s.n.* in *C.P. 985* p.p. (E, K); c. 1 km below top of Ramboda Pass towards Maturata, 1700 - 1800 m., 25 Oct 1993, *Fraser-Jenkins & Abeysiri 334* (K); Kuda Oya, near Nuwara Eliya, c. 1740 m., 28 Dec 1950, *Sledge 767* (K); Nuwara Eliya, 1950 m., 11 March 1954, *Schmid 1488a* (BM), 1490 (BM); Pelagola, milepost 35/10 on Gampola to Nuwara Eliya road, 1075 m., 13 Nov 1976, *Faden & Faden 76/257* (K); Maturata, September 1853, *Thwaites s.n.* in *C.P. 985* p.p. (PDA); Pundaloya to Dunsinane road, 1 Nov 1982, *Balasubramaniam 2879* (PDA). DISTRICT UNCERTAIN: Gongalla Rocks, March 1881, ? *Trimen s.n.* (PDA). WITHOUT LOCALITY: *Amaratunga 233* (PDA); *Gardner 9* p.p. (LIV), *10* (BM); *1186* (K); *Skinner s.n.* (K); *Thwaites s.n.* in *C.P. 985* p.p. (BM); *Thwaites s.n.* in *C.P. 985* p.p. ex Herb. Darnell (E); *Walker 15* (K), *61* (K); *Thwaites s.n* in *C.P. 985* p.p. in Naylor Beckett 720 p.p. (LIV).

3. Selaginella latifolia (Hook. & Grev.) Spring, Bull. Acad. Brux. 10: 146. 1843.

Lycopodium latifolium Hook. & Grev., Enum. Fil., in Hook. Bot. Misc. 2: 386. 1831.
Type: Sri Lanka, Adam's Peak, *Emerson s.n.* (E).

Stem 20 - 30 (-50) cm long, decumbent, rooting base, c. 1 cm broad including leaves, rarely branched below, compoundly branched above with branches spreading. Leaves dimorphic but isomorphic at or towards base of main stem; larger leaves: lamina 4 - 4.75 (-5) × c. 2.5 mm., spreading, usually contiguous, at times slightly imbricate, somewhat spaced below on main stem., ovate-rhombic, unequal-sided, distally subacute, upper margin minutely denticulate, lower entire, somewhat smaller on upper branches; smaller leaves: lamina up to 2.5 × 1.5 mm., with arista 0.75 - 1 mm long, distant on main stem., imbricate on branches, broadly obovate to ovate-elliptic, minutely denticulate or subciliate. Strobili 5 - 25 mm long, c. 2mm diam., 4-angled. Sporophylls isomorphic, c. 1.5 - 2.5 × 0.75 - 1 mm., lanceolate-ovate, apex acute, acuminate, shallowly carinate, microscopically denticulate. Megaspores c. 1 mm diam., flattened-discoid. Microspores c. 0.3 mm diam., subruminate.

D i s t r. Endemic, restricted to the area of Adam's Peak

E c o l. High altitude forest floors.

N o t e. In some ways this species appears very close to *Selaginella intermedia* (Blume) Spring, but appears to differ in habit, generally being much more erect and appearing to root only at base, larger leaves being broadly rhombic and not narrowly oblong, with smaller leaves having aristae which are much shorter than laminae. Alston, in naming the Peradeniya material in 1957, considered several of Abeywickrama's collections to represent this species, but as they show a number of characters mentioned above, they are placed here under *Selaginella latifolia*. However, these characters may, at a later date, be found to fit a much wider range encompassing both species. For the purpose of this treatment, it is felt more opportune to keep the two species as distinct and to leave any further separation or combination to a future specialist or monographer.

S p e c i m e n s   E x a m i n e d: RATNAPURA DISTRICT: "Adam's Mountain", 1828, *Emerson s.n.* (E, holotype), *s.coll. 136* (BM); Adam's Peak, 23 Feb 1950, *Abeywickrama 17* (PDA), 1525 m., March 1846, *Gardner 1276* (K, PDA), *Moon s.n.* (BM), *Walker 24* (K); near Gartmore Estate, Maskeliya, 29 Apr 1926, *Silva s.n.* (PDA); Morowak Korle, *Thwaites s.n.* in *C.P. 3284* p.p.(K); Sinharajah Forest, above Beverley Estate, Deniyaya, 690 m., 4 Apr 1954, *Sledge 1396* (K); Radella, 9 Jan 1952, *Abeywickrama 10* (PDA), *11* (PDA). NUWARA ELIYA DISTRICT: Hakgala, c. 1830 m., April 1966,

Perera s.n. (PDA). WITHOUT LOCALITY: *Baker 166* (K); *Finlayson s.n.* (BM); ex Herb. Gower (E); *Mackenzie s.n.* (K); *Skinner s.n.* (K); *Thwaites s.n.* in *C.P. 3284* p.p.(BM., E, K, LIV); *Wight s.n.* (E).

4. Selaginella integerrima (Hook. & Grev.) Spring, Bull. Acad. Brux. 10: 138. 1843; Dixit, Selagin. India 53, fig. 18 A-G. 1992.

[*Lycopodium ornithopodioides* L., Sp. Pl.: 1105. 1753, pro parte, quoad spec. Hermann (BM), sed excl. lectotyp. Dillenius tab. 66, fig. 1B.]

Lycopodium integerrimum Hook. & Grev., Bot. Misc. 2: 396. 1831. Type. India (Courtallam).

Selaginella ornithopodioides sensu Trimen, J, Linn. Soc.Bot. 24: 152. 1887; Alston, Proc. Nat. Inst. Sci. India 119 9 (3): 211-235, 1945, non (L.) Spring (1838).

Selaginella concinna Ferguson, Ceyl. Ferns 64. 1880, non (Sw.) Spring. 1838.

Stem up to 45 cm long, creeping, rooting at nodes; stems pinnately divided, branches c. 4 - 5 mm broad including leaves, spreading. Leaves dimorphic; larger leaves up to 2.5 × 1.5 mm., ovate, obtuse, base rounded to subtruncate, entire, distant on main stem below, imbricate above on branches; smaller leaves up to 1.5 × 0.8 – 1 mm., ovate, subacute to more rarely obtuse, entire, suberect, imbricate. Strobili 5 - 18 mm long, c. 2.75 mm diam., 4-angled, erect. Sporophylls 1.5 - 2 × 0.5 - 0.8 mm., ovate, acute, not acuminate, entire, carinate, apices sometimes slightly recurved. Megaspores pale orange, tuberculate.

D i s t r. Southern India and Sri Lanka.

E c o l. Stream banks, marshy areas and roadsides, in both shaded forest and in full sun.

N o t e. Alston (Proc. Nat. Inst. Sci. Ind. 11, 3: 222. 1945) stated that "The type specimen of *Lycopodium* [*Selaginella*] *integerrimun* Hook. & Grev. at Edinburgh is labelled 'Courtallum., Dr Wight', but this is probably an error because the specimens at Kew from Herb. Wight are labelled 'Colombo, Klein'." I have seen a sheet from Edinburgh similarly labelled, but there is a second collection on that sheet bearing the legend "Presid. of Madras. Dr Wight 1831, at Sri Lanka. named in Klein's handwriting '*L. ornithopoides*', leaving no doubt in my mind that this collection was made by Wight in Sri Lanka.

S p e c i m e n s E x a m i n e d: KURUNEGALA DISTRICT: near Moragala, March 1887, *Trimen s.n.* (PDA); Meegahatenne, near Moragala, 76 m., 20 Jan 1951, *Sledge 882* (BM., K). COLOMBO DISTRICT: Colombo, *Ferguson s.n.* (LIV, PDA); Labugama, 50 - 60 m., 7-8 Jan 1954, *Schmid 831* (BM), 60 m., 30 Jan 1954, *Schmid 1094* (BM); Raja Giri Hill, W of Guruwala, 25 - 65 m., 30 Nov 1976, *Faden & Faden 76/422* (K); Pantura [?=Panadura],1881, *Trimen s.n.* (PDA). KEGALLE DISTRICT: Kadugannawa,

Aug, 1884, *Ferguson s.n.* (PDA). RATNAPURA DISTRICT: Niriella, 225 m., 4 Feb 1954, *Schmid 1127* (BM); milepost 6/5, on bridge past Rassagala on Balangoda to Rassagala road, 750 m., 16 Nov 1976, *Faden & Faden 76/308* (K); track in Gilimale Forest from Carney road to Kalu Ganga, 60 - 120 m., 3 Dec 1976, *Faden & Faden 76/468* (K); Kalawana, 10 Aug 1968, *Wimalaratna s.n.* (PDA). NUWARA ELIYA DISTRICT: Castlereagh Estate, 3 Aug 1927, *Silva 1903* (PDA); Kadienlena, road from Talawakele to Nawalapitiya, 800 m., 1 Nov 1995, *Shaffer-Fehre et al. 466* (K). GALLE DISTRICT: Kottawa Forest Reserve (Arboretum), 21 Jan 1951, *Ballard 1542* (K); Pahala Hewessa, 20 Jan 1951, *Ballard 1523* (K); Hiyare, 22 Nov 1927, *Silva 226* (PDA); Katagoda, road from Galle to Akuressa, 8 Nov 1995, *Shaffer-Fehre & Jayasekara 487* (K), *488* (K); Mulukanda, Elpitiya, March 1954, *Abeywickrama 278* (PDA); WITHOUT LOCALITY: *Brandis 2420* (K); 1850, *Delessert s.n.* (BM); *Gardner 1184* (K); *Hutchison 124* (LIV); *Koenig s.n.* (BM); *Thwaites s.n.* in *C.P. 3280* p.p. (BM., E, K, PDA); 1831, *Wight s.n.* ex Herb. Greville (E); 1830, *s.coll. s.n.* (K).

5. Selaginella praetermissa Alston, J. Bot. 70: 65. 1932; Alston, Proc. Nat. Inst. Sci. Ind. 11, 3: 216. 1945; Dixit, Selagin. India 55, fig. 20A-G. 1992. Type. Sri Lanka, on rocks, Kandy, *Macrae 135* (BM holotype, K isotype).

Stem up to 50 cm long, creeping, 4 - 7 mm diam. including leaves, much branched from base; lateral branches usually further shortly pinnately branched; rooting usually in lower ?, with long, pale-green rhizophores. Leaves dimorphic; larger leaves 2 - 3.4 × 1.5 - 2 mm., spreading, spaced below, ultimately imbricate, oblong, rounded-obtuse or subacute, apically serrulate, obliquely rounded to subtruncate at base, shortly ciliate; smaller leaves c. 2 - 2.5 × 1 mm including apiculus c. 0.8 mm long, subovate-oblong, rounded to subcordate at base, entire except towards apex where serrulate. Strobili 4 - 8 (-10) × 1.5 - 2 mm., tetragonous, solitary at apex of branches. Sporophylls isomorphic, 1.25 - 1.5(-1.75) × c. 0.8 mm., ovate-lanceolate, acuminate, at times shallowly carinate. Megaspores pale- to dark-brown, verrucoid.

D i s t r. India and Sri Lanka.

E c o l. In marshes and stream sides.

N o t e. Alston (1945) discussed a specimen collected by Jenkins and purportedly from Assam residing in the Kew herbarium under the name *S. pinangensis* Spring, a Malayan plant not then found in India. He considered that this specimen was of *S. praetermissa* and probably had been wrongly localised when mounted. I have seen the Kew material and agree with Alston's identification and have similar doubts about its provenance.

S p e c i m e n s   E x a m i n e d: KURUNEGALA DISTRICT: Udagama, October 1886, *Ferguson s.n.* (PDA). COLOMBO DISTRICT: near Colombo,

Ferguson s.n. in *C.P. 3974* p.p. (K); between Colombo and Gilliemalle, *Ferguson s.n.* (PDA). KEGALLE DISTRICT: Kaduvelle, November 1883, *Ferguson s.n.* (PDA). KANDY DISTRICT: 940 m., 22 Jan 1954, *Schmid 1034* (BM); Kandy, *McRae 135* (BM holotype, K isotype); c. 3 km S of Moragalla, 20 Jan 1951, *Ballard 1516* (K, PDA); Hakmana, Feb1881, *Trimen s.n.* (PDA). RATNAPURA DISTRICT: Niriella, 225 m., 4 Jan 1954, *Schmid 1118* (BM); track in Gilimale Forest from Carney to Kalu Ganga, near Kalu Ganga, 60 - 120 m., 3 Dec 1976, *Faden & Faden 76/459* (K); towards Adam's Peak, 350 - 400 m., 7 March 1973, *Bernardi s.n.* (K). GALLE DISTRICT: Hiyare Reservoir, near Kottawa, 121 m., 1 Apr 1954, *Sledge 1381* (K); Hiniduma, 8 Nov 1995, *Shaffer-Fehre & Jayasekera 479* (K), 480 (K). MATARA DISTRICT: Deyandara, February 1881, *Trimen s.n.* (PDA). WITHOUT LOCALITY: *Koenig s.n.* (BM).

6. Selaginella calostachya (Hook. & Grev.) Alston, J. Bot. 70: 65. 1932, et Proc. Nat. Inst. Sci. Ind. 11, 3: 229. 1945.

Lycopodium calostachyum Hook. & Grev., Enum. Fil., Hook. Bot. Misc. 3: 108. 1833. Type: *Macrae s.n.* in Herb. Benth. (K)
Lycopodium macraei Hook. & Grev., Enum. Fil., Hook. Bot. Misc. 3: 108. 1833. Type: Sri Lanka, *Macrae s.n.* in Herb. Benth. (K).
Selaginella macraei (Hook. & Grev.) Spring in Bull. Acad. Brux. 10 : 232 (see p. 32). 1843.
Selaginella tenera var. *macraei* (Hook. & Grev.) Spring, Mem. Acad. Belg. 24: 242. 1850.
Selaginella zeylanica Baker in J. Bot. 23: 178. 1885. Type: Sri Lanka, *Gardner s.n.* (K).

Stems up to 10 cm long, procumbent, branches up to 4 (-8) mm diam. including leaves, spreading, rooting at nodes. Leaves dimorphic; larger up to 2 -3 (-4) × 1 - 1.8 (-2) mm., spaced or contiguous, ovate-rhombic, sides unequal, subacute or more rarely rounded, base auriculate, minutely, regularly denticulate; smaller 1.5 - 2 × c. 0.5 mm., narrowly ovate, acute, spaced or imbricate, minutely denticulate, mucronate with mucro equalling length of lamina. Strobili (2.5-) 3.5 - 5 (-11) × 1.75 - 2.5 mm., simple. Sporophylls dimorphic, large and small in same plane; larger c. 1 - 1.25 × 0.4 - 0.5 mm., oblong to ovate-lanceolate, obtuse to subacute, slightly unequal-sided, not keeled, minutely denticulate to apparently entire, spreading; smaller c. 1 - 1.25 × 0.4 mm., lanceolate, acute-acuminate, keeled, shortly mucronate, appressed. Spores bright orange.

D i s t r. Endemic.

E c o l. By roadsides and stream sides in mostly forested areas.

N o t e. The extreme maximum measurements above for strobili are represented by two collections at Kew, *Ballard 1027* and Col. *Walker s.n.* which also show a tendency towards *S. brachystachya*; because of this, they have not been included in the citations below.

S p e c i m e n s   E x a m i n e d: MATALE DISTRICT: Matale, *Thwaites s.n.* in *C.P. 1418* p.p. (E). KANDY DISTRICT: Hunnasgiriya, 900 m., 19 Jan 1954, *Schmid 988* (BM); Oodewella, 1065 m., 8 Dec 1950, *Sledge 517* (BM); Warriagala, 26 June 1927, *Alston 759* (K); Roseneath, 610 m., 3 Dec 1950, *Ballard 1021* (K, PDA); Corbet's Gap, 1160 m., 22 Jan 1954, *Sledge 1027* (K), c. 1220 m., 9 Dec 1950, *Sledge 573* (K); Duckwari to Rangalla, Girindiella turnoff, milepost 21/12, 7 Jan 1951, *Ballard 1421* (K); c. 2 km NE of Udugoda on road to Panwila, Wattegama to Madulkelle, N of Kandy, c. 650 m., 1 Oct 1993, *Fraser-Jenkins et al. 230* (K); Delpitiya, 580 m., 13 Nov 1976, *Faden & Faden 76/244* (K); Weragantota, 213 m., 10 Jan 1954, *Sledge 956* (K); Galaha, 940 m., 22 Jan 1954, *Schmid 1033* (BM); Peak Wilderness, Meriyakota, above Fairlawn Estate, 1630 m., 16 Aug 1984, *Jayasuriya et al. 2834* (PDA). BADULLA DISTRICT: between Welimada and Badulla, 915 m., 29 Jan 1950, *Sledge 778* (K). RATNAPURA DISTRICT: track in Gilimale Forest from Carney to Kalu Ganga, 60 - 120 m., 3 Dec 1976, *Faden & Faden 76/457* (K); Belihul Oya, 610 m., 2 Jan 1951, *Sledge s.n.* (K); Katussagala Hill, c. 600 m., 3 Dec 1976, *Faden & Faden 76/472* (K). Radella, 9 Jan 1952, *Abeywickrama 12* (PDA). NUWARA ELIYA DISTRICT: between Norton Dam and Hapugastenne, 915 m., 13 Dec 1950, *Sledge 595* (E). MATARA DISTRICT: Deniyaya, 550 m., 5 Feb 1954, *Schmid 1152* (BM), *1153* (BM). WITHOUT LOCALITY: *Gardner 1185* (K); *Gardner* in *C.P. 1417* (K); *Hutchison 290* (LIV), *316* (LIV); *McRae 953* p.p. (BM); *Thwaites s.n.* in *C.P. 1412* p.p. (BM), *C.P. 1417* p.p. (BM., E, K); *s.coll. s.n.* ex Herb. Miers (K).

7. Selaginella cochleata (Hook. & Grev.) Spring, Mem. Acad. Belg. 24: 121. 1849.

Lycopodium cochleatum Hook. & Grev., Hook. Bot. Misc. 2: 395. 1831.
    Type: Sri Lanka, Adam's Peak, *Emerson s.n.* (E).
Lycopodium brachystachyum Hook. & Grev., Hook. Bot. Misc. 3: 107. 1833.
    Type. Sri Lanka, *Lindley 1829* (K) [see note below].
Lycopodium ornatum Hook. & Grev., Hook. Bot. Misc. 3: 108. 1833. Type
    from Sumatra.
Selaginella brachystachya (Hook. & Grev.) Spring, Bull. Acad. Brux. 10:
    232. 1843; Alston, Proc. Nat. Inst. Sci. Ind. 11,3: 229.1945; Dixit,
    Selagin. India 90, fig. 54 A-G. 1992.
Selaginella ornata (Hook. & Grev.) Spring, Mem. Acad. Belg. 24: 259. 1849.
Selaginella stolonifera sensu Thwaites, Enum. Pl. Zeyl.: 377. 1864, non Willd.
    1810.

Selaginella brachystachya var. *ornata* (Hook. & Grev.) Baker, Fern Allies 113. 1887.

Stem usually prostrate or more rarely suberect, up to 2.5 mm thick, branched from base, stramineous; rhizophores emerging along stem. Branches up to 1 cm diam. including leaves, spreading. Leaves dimorphic, arranged in 4 rows, sparsely so on main stem., denser on lateral branches; larger leaves (1.5-) 2.5 - 5.5 × 1.5 - 2 mm., oblong, obtuse to subacute, lower margin at times upwardly curved towards apex and appearing ± subfalcate, widening to oblique base, appearing subamplexicaul, 2.5 - 5 mm apart and spreading approximately at 90° to stem; smaller leaves c. 1.5 - 2 × 0.6 - 0.8 (-1) mm., broadly ovate, aristate, with arista to c. 0.8 mm long, subappressed, all entire, becoming imbricate on lateral branches. Strobili (3-) 8 - 12(-40) × 1.5 - 2.5 mm., somewhat flattened, dorsiventral. Sporophylls dimorphic; larger sporophylls c. 1 - 1.5 × 0.8 mm. broadly triangular; smaller sporophylls c. 1 × 0.6 mm including arista c. 0.5 mm long, enfolding microspores. Microspore c. 0.5 × 0.3 mm., flattened-ellipsoid, pale cream. Spores orange.

D i s t r. Originally thought to be endemic to Sri Lanka, but three sterile collections from Southern India which could prove to be this species are housed in the Kew herbarium. The author has not seen Miller's Sumatran type of *L. ornatum* in Herb. Banks, but if this exists, it will widen the distributional range even further.

E c o l. Common among rocks on wet forest floor; up to 3000 m.

N o t e. This plant has more commonly been accepted under the name *Selaginella brachystachya* (= *Lycopodium brachystachyum* (1833)) with *S. cochleata* (= *L. cochleatum* (1831)) appearing as a distinct species. Recent study confirms Alston's views that this is probably a depauperate state of *S. brachystachya* and as such it has been included here. This necessitates the use of the earlier name *Selaginella cochleata* (Hook. & Grev.) Spring for the taxon. The holotype specimen of *Selaginella brachystachyum.*, housed at Kew and originally from Bentham's herbarium., bears the legend "Sri Lanka (McRae) Lindley 1829".

S p e c i m e n s   E x a m i n e d: MANNAR DISTRICT: Mullikanda, March 1954, *Abeywickrama 284* (PDA). MATALE DISTRICT: Lagalla, ? *Naylor Beckett 750* (LIV); Dankanda Pass, October 1861, *Brodie s.n.* ex Herb. Fraser (E). COLOMBO DISTRICT: near Colombo, ?1838, *Thwaites s.n.* in *C.P. 3974* (PDA); Labugama, 50 - 60 m., 7-8 Jan 1954, *Schmid 829* (BM); Waga, 15 Aug 1951, *Abeywickrama 14* (PDA). KEGALLE DISTRICT: Ambagamuwa, 580 m., 19 Jan 1954, *Sledge 994* (E), *Beddome s.n.* (K), November 1854, *Thwaites s.n.* in *C.P. 1412* p.p. (PDA); Kitulgala, Kelani Ganga, c.150 m., 25 Feb 1971, *Robyns 7221* (K, PDA); Kitulgala, 180 m., 6 Nov 1967, *Comanor 543* (K); Kottuwa Forest Reserve, 17 Aug 1826, *s.coll.*

s.n. (PDA), (Arboretum), 21 Jan 1951, *Ballard 1545* (K). KANDY DISTRICT: Wattakelle Hill, above Wattakelle village, NE of Kandy, 1200 - 1500 m., 1 Oct 1993, *Fraser-Jenkins et al 237* (K); Kalupahana, 22 Aug 1956, *Abeywickrama 418* (PDA), *458* (PDA); Knuckles, 5 Nov 1982, *Balasubramaniam 2894* (PDA), 24 Aug 1956, *Abeywickrama 406* (PDA), *408* (PDA); Rangala Forest, 26 Oct 1982, *Braggins 82/168* (PDA); Corbet's Gap, 1220 m., 9 Dec 1950, *Ballard 1060* (K, PDA), *1062* (K), 1340 m., 9 Dec 1950, *Sledge 558* (BM., K); Hakmana, February 1841, *Trimen s.n.* (PDA); SE of Hunas Falls Hotel, W side of Hunnasgiriya Mountain, SSE of Elkaduwa, NE of Kandy, c. 1200-1300 m., 25 Aug 1993, *Fraser-Jenkins 26* (K); Hunnasgiriya, 900 m., 18 Jan 1954, *Schmid 965* (BM); Madulkelle, October 1887, *Trimen s.n.* (PDA); Moray Estate, foothills of Adam's Peak, c. 1525 m., 18 Nov 1995, *Shaffer-Fehre et al. 542* (K). KALUTARA DISTRICT: Pelawatta, path leading to Pahala Hewessa, 20 Jan 1951, *Ballard 1533* (K); near Hedigalla, Pasdun Korle East, 76 m., 5 Jan 1951, *Sledge 811* (BM); Pasdun Korle, December 1848, *Gardner s.n.* in *C.P. 1412* p.p. (PDA); Bandureliya, 11 March 1952, *Abeywickrama 5* (PDA), *6* (PDA), *7* (PDA), January 1954, *Abeywickrama 259* (PDA), *265* (PDA), *266* (PDA). RATNAPURA DISTRICT: behind Ratnapura Resthouse, c. 60 m., 1 Dec 1976, *Faden & Faden 76/435* (K), *76/435A* (K); Carney, 290 m., 2 Feb 1954, *Schmid 1096* (BM); Niriella, 225 m., 4 Feb 1954, *Schmid 1114* (BM); c. 4.5 km SE of Rakwana , on lower N slopes of Beralagala, 2600 - 3000 m., 7 Dec 1977, *Fosberg 57310* (PDA); Patalassekanda, above Kurawita, c. 14 km N of Ratnapura, 2700 - 2800 m., 8 Dec 1977, *Fosberg 57354* (K, PDA); Katussagala Hill, 450 - 600 m., 5 Dec 1976, *Faden & Faden 76/488* (K); Kuruwita Kande, 13 Sept 1926, *de Silva s.n.* (PDA); Agars Land, Balangoda to Rasagalla road, 8 Nov 1982, *Balasubramaniam 2929* (PDA); Gilimale, 8 Dec 1982, *Balasubramaniam 82/541* (PDA); Adam's 'Mountain', 1828, *Emerson s.n.* (E). NUWARA ELIYA DISTRICT: Rambodde [=Ramboda], January 1847, *Gardner s.n.* in *C.P. 1412* (PDA), 1848, *Gardner s.n.* in *C.P. 1412* (HAKS, PDA); 1 km below top of Ramboda Pass towards Maturata, 1700 - 1800 m., 25 Oct 1993, *Fraser-Jenkins & Abeysiri 333* (K); Nuwara Eliya, September 1866, *Naylor Beckett 750* (LIV); about Nuwara Eliya, *Walker s.n.* (K); Sita-Eliya, near milepost 53/13, Nuwara Eliya to Hakgala road, 1720 m., 14 Nov 1976, *Faden & Faden 76/259* (K); Hakgala, December 1880, *Trimen s.n.* (PDA), *s.coll. s.n.* in *C.P. 1412* p.p. (HAKS); Hakgala Peak, c. 1830 m., 16 Dec 1950, *Ballard 1125* (K); Talawakelle, 1525 m., *Leigh s.n.* (BM); Horton Plains, 2270 m., 7-8 March 1954, *Schmid 1378* (BM), North Entrance, 2100 m., 28 March 1968, *Fosberg & Mueller-Dombois 50021* (K); N of Horton Plains Rest House, c. 2300 m., 3 Dec 1970, *Theobald & Krahulik 2736* (PDA), below Rest house at Ohiya Road, 2175 m., 9 July 1967, *Mueller-Dombois & Comanor 67070918* (PDA); Ohiya, 10 Jan 1952, *Abeywickrama 9* (PDA), *15* (PDA); between Horton Plains Rest house and World's End,

2175 m., 10 Feb 1968, *Comanor 950* (E, K); World's End, 2133 m., 29 Jan 1974, *Moldenke et al. 28282* (PDA); forest behind Farr Inn, 3 Dec 1970, *Fosberg & Sachet 53275* (K, PDA); Moon Plains, 1 Nov 1982, *Balasubramaniam 2866* (PDA), *Balasubramaniam* in Braggins *82/261* (PDA), 1830 m., 23 Dec 1950, *Ballard 1205* (K); Castlereagh Estate, Dikoya, 3 July 1927, *Alston 925* (K); near Gartmore Estate, Maskeliya, 29 Apr 1926, *Silva 102* (K, PDA); above Blair Athol, 3 July 1927, *Alston 1899* (PDA); Kandapola Forest Reserve, 19 March 1954, *T.G.Walker T861* (BM); Adam's Peak, *Moon s.n.* (BM), 1525 m., *Gardner 1275* (K), 1830 m., 14 Dec 1950, *Sledge 611* (K); Pidurutalagala, c. 2135+ m., 6 Apr 1959, *Sinclair 10104* (E, K); between Norton Dam and Hapugastenne, 1066 m., 13 Dec 1950, *Sledge 595A* (K). GALLE DISTRICT: Deniyaya, Sinharajah Forest, above Beverley Estate, c. 800 m., 4 Apr 1954, *Sledge 1398A* (K); Hedigala, Sinharajah Forest, 5 Jan 1951, *Ballard 1389* (K); Haycock Mountain, Hinidum Pattuwa, 228 m., 22 Jan 1951, *Sledge 913* (K); Hiniduma, 45 m., 21 Sept 1946, *Worthington 2247* (BM); Hiniduma Pattuwa, Kanneliya Forest Reserve, 9 Nov 1995, *Shaffer-Fehre & Jayasekera 496* (K), 152 m., 2 Apr 1954, *Sledge 1383* (E, K). DISTRICT UNCERTAIN: Manwelle, February 1881, *Trimen s.n.* (PDA). LOCALITY UNKNOWN: *Ferguson s.n.* (PDA); *Finlayson s.n.* (BM); 1849, *Fraser 142* (BM., E, K); *Gardner 1185* p.p. (K), *1412* (K), *1413* (K); *Hutchinson 295* (LIV); *Mackenzie s.n.* (K); 1894, *Rippon s.n.* (BM); *Thwaites s.n.* in *C.P. 1412* (BM., E, K, PDA); March 1836, *s.coll. s.n.* ex Herb. Wight (K); *s.coll. s.n.* ex Herb. Gower (E); *s.coll. s.n.* (K).

8. Selaginella ciliaris (Retz.) Spring, Bull. Acad. Brux. 10: 231. 1843; Alston, Proc. Nat. Inst. Sci. Ind. 11, 3: 227. 1945; Dixit, Selagin. India 79, fig. 41 A-G. 1992.

> *Lycopodium ciliare* Retz., Obs. 5: 32. 1789. Type. Sri Lanka, *Koenig s.n.* (K fragment, LUND holotype).
> *Lycopodium belangeri* Bory in Belanger, Voy. Bot. 2: 12. 1833. Types from Mahe and Malabar.

Stem up to 2 - 8 (-15) cm long, prostrate, pinnate-branched, branches spreading, 3.5 - 4.5 mm wide including leaves, lower leaves very compound, rooting in lower 1/4 with thin, wiry rhizophores. Leaves dimorphic, membranous, lateral 2 - 2.5 × 0.75 - 1 mm., ovate-oblong, unequal-sided, cordate at base, apex obtuse or subacute, patent, shortly ciliate; smaller leaves 1 - 1.5 × 0.5 - 0.75 mm., lanceolate, more or less equal-sided, acute-acuminate with very short arista c. 0.1 mm long, not long-aristate, shortly ciliate or not subappressed, imbricate, median vein prominent beneath. Strobili 5 - 15 × 2.5 - 3 mm., simple, quadrangular. Sporophylls dimorphic, larger up to c. 2 × 0.85 mm., sublanceolate-ovate, subobtuse or subacute, base rounded to subcordate, ciliate especially towards base, arranged in same plane as smaller leaves; smaller c. 1.5 - 0.5 mm., lanceolate, acuminate, obscurely carinate, markedly ciliate. Megaspores pale fawn. Cytology: 2n = 18 (Jermy et al. loc. cit. p. 149. 1967).

D i s t r. India, Burma and Sri Lanka; Philippine Islands and China to Northern Australia.

E c o l. On rocky outcrops in shade or partial shade.

N o t e. Many authors have referred this plant to *S. proniflora* (Lam.) Baker, an erect species with long-aristate, smaller leaves and not recorded from Sri Lanka.

S p e c i m e n s   E x a m i n e d: JAFFNA DISTRICT: Soranpattu, on road from main Jaffna to Kandy road to Talalady, c. 5 m., 23 Jan 1977, *Faden & Faden 77/226* (K). TRINCOMALEE DISTRICT: Nillavelly [=Nilaveli], March 1867, *Glenie s.n.* in *C.P. 3978* (K, PDA). KURUNEGALA DISTRICT: Kurunegala, August 1868, *Thwaites s.n.* in *C.P. 3975* p.p. (K, PDA). MATALE DISTRICT: Pallegama, 182 m., 3 March1954, *Sledge 1228* (BM., K). COLOMBO DISTRICT: Colombo, *Ferguson s.n.* (PDA); Raja Giri Hill, just W of Guruwala, 25 - 65 m., 30 Nov 1976, *Faden & Faden 76/425* (K). KANDY DISTRICT: Kadugannawa, August 1884, *Ferguson s.n.* (PDA); Kallebokka, *Thwaites s.n.* in *C.P. 3979* p.p. (K); Wattekelle, Peradeniya, September 1868, *Thwaites s.n.* in *C.P. 3979* p.p. (PDA). WITHOUT LOCALITY: *Gardner 1185* p.p. (K); *Koenig s.n.* (K fragment, LUND, May); *Thwaites s.n.* in *C.P. 3975* p.p. (E, K); *s.coll. s.n.* ex Herb. Gower (E).

9. Selaginella crassipes Spring, Monogr. Lyc. 2: 243. 1850. Type: Sri Lanka, *Walker s.n.* (K)

Selaginella atroviridis sensu Thwaites, Enum. Pl. Zeyl. 377. 1864, non Spring, 1838.

Stems 5 - 25 cm long, erect, rooting at lower nodes, compound; branches alternate, up to 5.5 mm diam. including leaves. Leaves dimorphic; larger leaves distantly spaced, 2 - 2.5 (- 3) × 1 - 1.5 (-2) mm., ovate, acute, unequal-sided, upper base subtruncate, lower cuneate, margin microscopically ciliolate, often revolute when dry; smaller leaves up to 1 × 0.75 mm., ovate or ovate-lanceolate, ciliolate, mucronate with mucro less than ½ length of leaf. Strobili 3.5 - 16 × 2.5 - 3 mm., simple. Sporophylls dimorphic; larger sporophylls c. 1.25 × 0.8 mm., in same plane as smaller sporophylls, rhombic-lanceolate, acute or slightly acuminate, not carinate, subciliate, especially towards base, spreading, ± recurving; smaller sporophylls c. 0.75 mm long, ovate-lanceolate, acuminate, not markedly carinate, shortly ciliate, acumen c. ? length of sporophyll. Megaspores pale yellowish-fawn; microspores orange.

D i s t r. Sri Lanka and Southern India.

E c o l. Rocky banks and stream sides, mostly in forest areas; up to 1000 m..

S p e c i m e n s   E x a m i n e d: MATALE DISTRICT: Damboul[=Dambulla], March 1868, *Naylor Beckett 2458* (LIV); a. 2 km W of Laggala, E side of pass Rattota to Laggala, c. 1000 m., 15 Sept 1993, *Fraser-Jenkins et al. 179* (K). COLOMBO DISTRICT: Labugama, January 1885, *Trimen s.n.* (PDA), 50 - 60 m., 7-8 Jan 1954, *Schmid 830* (BM). KEGALLE DISTRICT: Ruanwella, November 1883, *Ferguson s.n.* (PDA); Gallebodde, 1065 m., 26 Jan 1954, *Sledge 1045* (BM). KANDY DISTRICT: Mount Pleasant, 24 Oct 1982, *Braggins 82/114* (PDA); Galaha, 940 m., 22 Jan 1954, *Schmid 1034A* (BM); Knuckles Range, 5 Nov 1982, *Braggins 82/289* (PDA); Le Vallon Estate, 30 Dec1982, *Braggins 83/11* (PDA). KURUNEGALA DISTRICT: Ellaboda Kande, 24 March 1919, *Lewis & Silva s.n.* (PDA). KALUTARA DISTRICT: Badurelliya, road to Kalawana, 10 Nov 1995, *Shaffer-Fehre & Jayasekera 516* (K). RATNAPURA DISTRICT: Karawita Kanda. 13 Sept 1926, *Alston s.n.* (PDA); Ratnapura, 4 Jan 1951, *Sledge 807* (K); Adam 's Peak, 1525 m., *Gardner 1274* (K, PDA); near Adam's Peak, 16 Nov 1927, *de Silva 44* (PDA); Katussagala Hill, c.600 m., 3 Dec 1976, *Faden & Faden 76/470* (K). NUWARA ELIYA DISTRICT: Rambodde[=Ramboda], ?*Gardner s.n.* in *C.P. 1418* p.p. (HAKS); near Talawakelle, 1524 m., *Leigh s.n.* (BM); Maskeliya, March 1883, *Trimen s.n.* (PDA); Rasamalayi, near Gartmore Estate, 29 Apr 1926, *de Silva s.n.* (PDA). GALLE DISTRICT: Elpitiya, March 1954, *Abeywickrama 282* (PDA); Hiniduma, March 1881, *Trimen s.n.* (PDA); near Hiniduma, Kalubovitiya, towards Dawalagama, 888 m., 26 Oct 1975, *Bernardi 15482* (K).DISTRICT UNCERTAIN: Hammathawa, 11 Feb 1929, *de Silva s.n.* (PDA). WITHOUT LOCALITY: 1819, *Moon s.n.* (BM); *Finlayson s.n.* (BM); *Hutchison 312* (LIV); *Skinner s.n.* (K); *Thwaites s.n.* in *C.P. 1412* p.p. ex Herb. Hance 15777 (BM), *C.P. 1417* p.p. (E, LIV), & in *C.P. 1418* p.p. (BM., K); *Walker 40* (K), *s.n.* (K); *Wight 1900* (E); *s.coll. s.n.* (HAKS).

KEY TO SPECIES AS GIVEN IN ALSTON'S MANUSCRIPT AT BM

1 Leaves spirally arranged; plant homophyllous; strobili sharply square . . . .
. . . . **S. rupestris**
1 Leaves dimorphic, in two planes with those of upper plane smaller than those of lower
  2 Sporophylls isomorphic
    3 Plants small, procumbent, rooting at nodes
      4 Larger leaves not entire, mucronate

5 Larger leaves obtuse or subacute, upper margin microscopically denticulate; mucro of smaller leaves 1/2 length of leaf . . . . **S. radicata**
5 Larger leaves acute, both margins denticulate; mucro of smaller leaves 1/3 length of leaf . . . . **S. acutifolia**
4 Larger leaves entire, obtuse or subacute; smaller leaves acute or acuminate . . . . **S. integerrima**
3 Plants larger, suberect, lower part rooting
6 Leaves of lower part of stem isomorphic, distant . . . . **S. caulescens**
6 Leaves of lower part of stem dimorphic, more crowded . . . . **S. latifolia**
2 Sporophylls dimorphic
7 Smaller sporophylls in the same plane as the smaller leaves . . . . **S. ciliaris**
7 Smaller sporophylls in the same plane as the larger leaves
8 Larger leaves of the main stem obtuse or subacute
9 Smaller leaves over 1/2 length of the larger leaves; larger leaves ovate, ciliate . . . . **S. proniflora**
9 Smaller leaves less than 1/2 length of the larger leaves; larger leaves more or less rhombic
10 Larger leaves subentire or irregularly denticulate, oblong-rhombic
11 Larger leaves 2 - 3 times as long as broad, usually subcontiguous; stem over 5 mm., usually 1 cm diam. including the leaves . . . . **S. brachystachyum**
11 Larger leaves less than twice as long as broad
12 Leaves distant; stem up to 00 mm diam. including leaves . . . . **S. tenera**
12 Leaves subcontiguous; stem up to 00 mm diam. including the leaves . . . . **S. calostachya**
10 Larger leaves regularly denticulate, ovate-rhombic, twice as long as broad, subacute; stem up to 4 mm diam. including the leaves . . . . **S. macraei**
8 Larger leaves of the main stem acute, ovate, margin recurved; plant suberect . . . . **S. crassipes**

THELYPTERIDACEAE
(by Monika Shaffer-Fehre*)

Ching ex Pic.Serm.

Research since 1940 has supported the distinct family status of Thelypteridaceae. Distinguishing features of this family of terrestrial ferns are rhizome scales which bear acicular hairs on their margins and often on their surface as well, stipes with two vascular bundles at their base which coalesce distally into one deeply concave strand, an indument of acicular hairs on the adaxial surface of the main axis and the pinnae; sinuses between pinna lobes are usually closed at their base by a translucent membrane.

Venation patterns had the dominant rôle in the works of Beddome: The Ferns of Southern India, The Ferns of British India (1863), The Ferns of British India (1870), followed by a supplement (1876) and by The Handbook of Ferns of British India, Ceylon and the Malay Penninsula (1883). Accordingly thelypteroid ferns were grouped with *Lastrea* or *Nephrodium* depending on their free or connivently anastomosing venation; this division still has an echo in modern keys (see below).

Characters used in Holttum's classification are: shape of fronds, pinnules, the indument: hairs on the upper surface of rachis and costae non-glandular to antrorsely curved; hairs on the lower surface of rachis and costae, presence/absence of the indusium., shape of sori, and, for the first time, setae and/or glandular hairs on the sporangium and its stalk and also spore ornamentation. The interesting research on gametophytes of the Thelypteridaceae of Sri Lanka, Tigershiold (1989-90), and the (mostly unpublished) spore research using scanning electron microscopy and gene sequencing, at the Royal Botanic Gardens, Kew (the latter two on Thelypteridaceae other than from Sri Lanka) are the most recent criteria used.

Taxa of the Thelypteridaceae often thrive in disturbed environments on slopes or in ditches along waysides and among stones by or in water courses. Distinct groups of taxa can be recognized according to their altitudinal as well as geographical distribution. Most taxa grow at altitudes between 500 -1500m., from and above 1250m and between 1500 and 2200m

---

* Royal Botanic Gardens, Kew, U.K.

Representatives of the family Thelypteridaceae make up about 8% of the fern flora of the world. Following Christensen's (1911) suggestions, Ching established the family Thelypteridaceae in 1940. Soral patterns similar to those of *Dryopteris*, led taxa of current Thelypteridaceae to be associated with the Dryopteridaceae in classifications of the 19th century.

According to Holttum's classification (1981), accepted by Sledge (1981), the Old World is home to 23 genera of Thelypteridaceae and from among these 17 genera occur in Sri Lanka; Malesian species make up the richest section of the family in the Old World. In his culminating work 'Thelypteridaceae' (1981) Holttum includes an interesting discourse on affinities between Thelypteridaceae and Cyatheaceae.

For Sledge (1981) the Thelypteridaceae were the last family in his series of monographs on Sri Lankan ferns. Here descriptions of several new species are first published. It is therefore meet to take full account of the genus- and species descriptions of this highly experienced author and his texts have been used in part.

Literature, Books: Beddome, A Handbook to the Ferns of British India, Ceylon and the Malay Peninsula. 1883 (with supplement 1892). Holttum, Thelypteridaceae. Flora Malesiana, Series II –Vol.1(5): 334-560. Manickam & Irudayaraj, Pteridophyte Flora of the Western Ghats-South India. 653 p. 1992

Journals: Holttum., Blumea 19: 17-52. Studies in the family Thelypteridaceae, 3. A new system of genera in the Old World.1971. Manton .& Sledge Phil. Trans. Roy. Soc. London, B, 238: 127-185. 1954. Sledge, Bull. Br. Mus. nat. Hist. (Bot.) 8(1): 1-54. The Thelypteridaceae of Ceylon. 1981. Sledge, Bot. J. Linn.Soc. 84: 1-30 An annotated check-list of the Pteridophyta of Ceylon. 1982. Tigerschiold, Nord. J. Bot. 8 (6): 639-648. SEM of gametophyte characters and antheridial opening in some Ceylonese species of Thelypteridaceae. 1989. Tigerschiold. Nord. J. Bot. 9 (4): 407-412. Dehiscence of antheridia in thelypteroid ferns1989. Tigerschiold,. Nord. J. Bot. 9 (6): 657-664. Gametophytes of of some Ceylonese species of Thelypteridaceae 1990.

KEY TO THE GENERA

1 Ferns growing predominantly among rocks in or by river beds (oya),spores trilete . . . . **15. Trigonospora**
1 Ferns growing predominantly in different habitats, spores monolete
  2 All veins free, excurrent veins always absent
    3 Fronds bi-pinnate or deeply bi-pinnatifid, sori ex-indusiate . . . .
    . . . . **12. Pseudophegopteris**
    3 Sori round and indusiate or elongated and ex-indusiate
      4 Fronds bi-pinnate, tri-pinnatifid . . . . **5. Macrothelypteris**
      4 Fronds pinnate or bi-pinnatifid
        5 Pinnae pinnatifid with lobed pinnules, veins forked, not reaching margins . . . . **1. Metathelypteris**

5 Pinnae lobed to pinnatifid with entire pinnules, veins simple, reaching margins
  6 Sori elongate, exindusiate, pinnae shallowly lobed . . . .
  . . . . **14. Stegnogramma**
  6 Sori round, indusiate, pinnae pinnatifid
    7 Undersurface of pinnae with scattered glands and hairy costae
    . . . . **2. Parathelypteris**
    7 Undersurface of pinnae eglandular
      8 Costae with conspicuous, round, pale scales . . . .
      . . . . **16. Thelypteris**
      8 Costae without scales . . . . **11. Pseudocyclosorus**
2 Some veins free, excurrent veins with connivent anastomoses always present
9 Under surface of pinnae with hooked hairs, indusia minute . . . .
. . . . **7. Amauropelta**
9 Under surface without hooked hairs, indusia otherwise
  10 Proliferating buds present on rachis . . . . **9. Ampelopteris**
  10 Proliferating buds absent from rachis
    11 1-22 free pinna pairs, 4-8 connivent veins form excurrent vein, excurrent vein mostly discontinuous at pinna middle . . . .
    . . . . **10. Pronephrium**
    11 12-30 pinna pairs, all excurrent veins continuous
      12 Up to 9 pairs of auricles below pinnae, lamina covered in pustules when dry . . . . **3. Pneumatopteris**
      12 Fewer auricles, pustules absent from dry lamina
        13 4-8 pairs of auricles, costa raised adaxially and abaxially
        . . . . **13. Sphaerostephanos**
        13 Few or no reduced pinnae on stipe, costa slightly raised adaxially, distinctly raised and rounded abaxially
          14 Sori & glands confined to pinna lobes, positioned not lower than sinus . . . . **6. Amphineuron**
          14 Sori & glands not confined, extending below sinus
            15 Lower surface of all vascular structures (i.e. excluding lamina) densely covered in long acicular or capitate hairs; few pairs of most distal pinnae abruptly reduced . . . . **4. Cyclosorus**
            15 Lower surface of vascular structures and lamina variously but never densely covered in indument; tip of lamina reduced more gently . . . .
            . . . . **8. Christella**

1. METATHELYPTERIS

(H. Itô) Ching, Acta phytotax sin. 8:305. 1963.

Caudex usually erect; fronds small, pinnate with deeply lobed pinnae or bipinnate with adnate pinnules , lowest pinnae not or little reduced; upper surface of costae not grooved; veins free, often forked, not reaching the margins; lower surface of pinnae with unicellular, acicular and/or short capitate hairs also short multicellular reduced scales; sori indusiate, sporangia without hairs, spores dark with thick wings or raised bands.

Type species: *Metathelypteris gracilescens* (Blume) Ching [based *on Aspidium gracilescens* Blume]

Metathelypteris flaccida (Blume) Ching, Acta phytotax. sin. 8:306. 1963. Sledge, Bull. Br. Mus. nat. Hist. (Bot.), 8 (1): 8-9. 1981.

Aspidium flaccidum Blume, Enum. Pl. Jav.: 161. 1828. Type: Java, Burangrang, Blume.
Lastrea flaccida (Blume) T. Moore, Index fil.: 92 (1858). Bedd., Ferns South. India: t.250. 1864; Handb. Ferns Brit. India: 244. 1883.
Nephrodium flaccidum (Blume) Hook., Sp. fil. 4: 133, t.263. 1862. Syn. fil.: 274. 1867.
Illustrations Spec. Fil. 4:133, t. 263. 1862. (as *Nephrodium flaccidum* Hook.)

Manickam & Irudayaraj, Pter. Fl. Western Ghats, t. 129. 1992.

Rhizome erect (long-creeping in var. **repens**, 2 mm in diam.). Fronds tufted, numbers depending on size of caudex, (evenly spaced ± 0.5cm. in var. *repens*), soft herbaceous, glabrous, hairs restricted to vascular structures, mid-dark green. Stipe green, stramineous when dry, 29cm long, with ± dense cover of short, soft, pale hairs, particularly in groove, black base 1-1.5-(2.5)cm long, sparsely scaly, scales with a few acicular hairs. Lamina 20-40 × 7.5-15cm lanceolate - oblong-lanceolate with acuminate tip, pinnate to almost bipinnate with c. 20 pinnae up to 8.5cm long, 3.5cm wide (basiscopic width 2cm). Pinnae 1-2 basal pinnae reduced, others subopposite becoming alternate at tip or subopposite throughout; pinnae at base of lamina with peduncle to 0.75mm., sessile at centre, towards tip gradually assuming size of pinnules, here pinnatifid to a narrow wing, forming acuminate tip, narrow wing decurrent on rhachis to c. (5)-6-(8) pinna pair. Pinnules parallel to rhachis, but gradually assuming oblique angle to costa,(in var. *repens* pinnules less lobed and not spaced along pinna axis), 14-18 pinnule pairs below attenuate tip of pinna, proportionately longest near middle of basiscopic side, oblong with four lobed segments and a blunt to subacute apex, margins fringed with short hairs. Rachis prominent on both sides (not grooved above in var. *repens*) with short 2-3-septate hairs throughout. these are more dense towards tip. Costae densely hirsute (0.3 mm) above and with well spaced acicular hairs below, occasionally with hooked hairs. Costules and veins too with shorter hair above, acicular hairs. Veins forking once or twice, thickened at tip, not reaching margin. Sinus membrane not apparent. Sori medial, up to 1mm in diam., on acroscopic branch of vein, when there are 3 veins at base of pinnule, sorus also on basiscopic vein. Indusium c. 0.75 mm., fringed with acicular hairs. Spores monolete, translucent 0.04 mm long. Cytology: n= 31, 35, 36.

D i s t r. Throughout India, in Sri Lanka, south China, Japan; Malesia, Solomon Islands, one species in Sao Thome and Madagascar.

E c o l. In higher parts of interior (1500-2200 m), frequent by streams and in damp ground in forests.

N o t es. Stipe of young crozier hairs and scales throughout entire 9cm length; scales with strong dark brown cell walls and beset with acicular and with gland-tipped hairs.

S p e c i m e n s   E x a m i n e d. KANDY DISTRICT: Peacock Hill, Pussellawa, *Beckett 71* (BM); Corbet's Gap, secondary jungle, 1320m., 9 Dec. 1950, *Sledge 569* (BM). BADULLA DISTRICT: Tonacombe Estate, Namunukula, 1350m., 23 Feb. 1954, *Sledge 1180* (BM); Tangamalle Forest Sanctuary near Haputale, cloud forest in deep shade of bamboo, 1450m., 5.12.1995, *Shaffer-Fehre with Jayasekara & Samarasinghe 642, 643, 655* (K). NUWARA ELIYA DISTRICT: *Moon,* Adams Peak (BM); in woods at Nuwara Eliya, Sept. 1844, *Gardner 1152* (CGE; K); same locality: *Mrs. Chevalier* (BM). Herb. John Smith, anno 1866, 2 sheets, *n. coll. 1152*; Watakelly Hill, forests around Nuwara Eliya, terraneous, *Wall 44/83*, before 1873 (K); Adams Peak, 14 Feb. 1908, *Mathew* (K); Adam's Peak, north slope at 1950m 14 Dec. 1950, *Sledge 622* (BM); jungle between Pattipola and Horton Plains, 1950m., 20 Dec. 1950, *Sledge 672* (BM); Moon plains 1676m damp ground in jungle 23.12. 1950, Ballard 1202 (K); Moon plains, Nuwara Eliya, wet ground in secondary jungle, 1800m., 23 Dec. 1950, *Sledge 718* (BM); road between Hakgala and Nuwara Eliya, 1620m., 27 Dec. 1950, *Ballard 1257* (K); Hakgala by jungle stream., 1650m., 27 Dec. 1950, *Sledge743* (BM); Hoolankande, edge of jungle, c.1370m., 20. Jan. 1954, *T.G. Walker T148* (BM); Horton Plains, 2270m., 7-8 March 1954, *Schmid 1371* (BM); by Ramboda Pass - Maturata track, 1890m., 17 March 1954, *Sledge 1352* (BM); near Ambawela junction on Hakgala-Pattipola road, on banks of streams and in roadside ditches, 2 Jan. 1977, *Faden 77/20* (K).

var. **repens** Sledge, Bull. Br. Mus. nat. Hist. (Bot.), 8 (1): 8. 1981. Type Sri Lanka, *Sledge 1323* (BM., K)

S p e c i m e n s   E x a m i n e d.: BADULLA DISTRICT: Tangamalle Forest Sanctuary near Haputale, cloud forest in deep shade of bamboo, 1450m., 05.12.1995, *Shaffer-Fehre with Jayasekara & Samarasinghe 642, 643, 655*

(K). NUWARA ELIYA DISTRICT: Kandapola, nr. Nuwara Eliya, 1800m., 19 March 1954, *Sledge 1323* (BM., holotype; isotype K!). Horton Plains, road to World's end on roadside banks in forest, ± 2090m., 15 Nov. 1976, *Faden 76/284* (K). LOCALITY UNKNOWN: *Thwaites* C.P. *3802, 1365* (BM.,CGE, K, P, PDA); *Thwaites* C.P. *1365* (BM.,CGE, K); *Freeman, A238, B239, C240, D241* (1908-1923 /pres. 1932) (BM); 1950m., 11 March 1954, *Schmid 1506* (BM); *Walker* (K, P, PDA); *Wall* (K); *Robinson 158* (K).

2. PARATHELYPTERIS

(H. Itô) Ching, Acta phytotax, sin. 8 : 300 p.p. 1963.

Small ferns with slender creeping rhizomes; fronds pinnate with deeply pinnatifid pinnae, decrescent or not; veins free, reaching margins, costae grooved adaxially; pinnae abaxially with sessile, spherical glands, often with slender, septate hairs; sori indusiate, capsules eglandular, without setae. Spores opaque with narrow, irregular wing. About 10 species: mostly from warmer parts of mainland Asia, to New Guinea, Solomon and Philippine Islands, and Japan. *P. beddomei* (Baker) Ching is the sole representative in Sri Lanka.

Type species: *Parathelypteris glanduligera* (Kunze) Ching [based on *Aspidium glanduligerum* Kunze]

Parathelypteris beddomei (Baker) Ching, Acta phytotax. Sin. 8 : 302. 1963. Holtt., Blumea 19 : 32. 1971. Holtt., Flora Malesiana ser.II,1 (5) 1982 "1981". Sledge, W.A. Bull. Br. Mus. nat. Hist. (Bot.) 8(1):11-12. 1981.

Nephrodium beddomei Baker in Hook. & Baker, Syn. fil. : 267. 1867. Type: Sri Lanka, Nuwara Eliya, *Thwaites 1287* (K!).
Lastrea beddomei (Baker)Bedd., Ferns Brit. India, Corr.: 11. 1870. Handb Ferns Brit. India : 239 fig. 122. 1883.
Thelypteris beddomei (Baker) Ching in Bull. Fan. meml. Inst. Biol. (Bot.) 6 308. 1936. Holtt., Rev. Fl. Mal. 2 : 240. 1955.
Lastrea gracilescens sensu Bedd., Ferns South. Ind.: 38, t.110. 1863, non *Aspidium gracilescens* Blume nec *Nephrodium gracilescens* (Blume) Hook.
Aspidium gracilescens sensu Thwaites, Enum. Pl. Zeyl.: 391.1864, non Blume 1828.

Illustrations. Beddome, Ferns South. India t. 110. 1863 (as *Lastrea gracilescens*). Bedd., Handb. Ferns Brit. India, fig. 122. 1883 (as *Lastrea beddomei*). Holtt.., Fl. Mal. Ser II 1(5): 372, fig. 5, a. 1982 "1981".

Rhizome wide-creeping, 2mm in diam., scales scarce 0.5-2mm long. Fronds ± regularly spaced 0.5 - 1cm apart, stiff. Stipe thin, to 0.1cm in diam., with adaxial groove, softly pubescent, pale green to stramineous. Texture of lamina herbaceous to subcoriaceous; lamina deeply pinnate-pinnatifid, narrowly to moderately lanceolate, almost fusiform., greatest width almost at centre of blade between 4 - 9 cm; between 30-40 pinnae, 3.5 × 0.7cm., with abruptly reduced, remote pinna pairs at base 0.2 × 0.1cm; pinnae free, minutely pedicellate 0.1-0.3 mm; basiscopic pinna contour often with shallow depression due to shortening of almost one third of pinnules near pinna base. Pinnules (12-16)-18 per pinna; almost free near costa, distinguished by larger size and, as other larger pinnules, by frequently serrate margin; only c. 4 ultimate pinnae, and pinnules respectively, fused into tip. Rachis stramineous. Costa with minute hairs in groove, white acicular hairs abaxially on costa and costules. Veins free, 5-7 reach margin. Sinus membrane distinct. Sorus round, submarginal; arises first near sinus of basiscopic pinnules, viz. on acroscopic then basiscopic side only then on acroscopic pinnules; initial sori tend to be much larger. Indusia small, glandular. Spores opaque with a narrow, irregular wing. Cytology: n=27, 31 [c.36].

D i s t r. Southern India, Sri Lanka , Sumatra, Java and widely throughout Malaysia on mountains to New Guinea, Taiwan, Philippines and southern Japan.

E c o l. In swampy places in the highest parts of the Central Province: frequent in the vicinity of Nuwara Eliya. At altitudes between 1500-2200m.

N o t e. shallowly rooted , primary roots, covered regularly in moss, secondary and tertiary roots very thin, smooth and short-twisted.

S p e c i m e n s   E x a m i n e d. NUWARA ELIYA DISTRICT: Banks of streams in open places, Sept. 1844, *Gardner 1141* (CGE); Nuwara Eliya 1829m., 1853, *v. Fridau 626* (GZU); Nuwara Eliya environs, very common in swampy places, *Wall 44/47* before 1873; Sita Eliya, patana near river, 1740m., Oct. 1897, *Pearson 226* (CGE); Nuwara Eliya, April 1899, *Gamble 27585* (K); Nuwara Eliya, 9 May 1906, *Matthew* (K); Nuwara Eliya, *Freeman A231,B 232,C 233* (BM); near Hakgala, on earth bank near path, 1670 m., *Holttum S.F.N. 39194*, 27.12.1950; Nuwara Eliya, 24-27 Feb. 1954, *Bonner 1298 & 1363*; Nuwara Eliya, Feb. 1954, *Schmid 1298, 1363* (BM); Ramboda Pass , c. 1900m., 17. March 1954, on path, *T. Walker T837* (BM); Ramboda Pass at c. 1980m., 2. Jan. 1977, *Faden 77/34* (K); between Pattipola and Horton Plains, marshy ground by stream., 1800m., 20 dec. 1950, *Sledge 668* (BM); Ramboda Pass - Maturata track, c. 1900 m., 17 March 1954, *Sledge 1315* (BM); Nuwara Eliya, entrance, just past highest point in road, among scrub and bamboo on slope above road, 1850m., 31.10.1995, *Shaffer-Fehre with Jayasekara & Samarasinghe 446* (K); Kadienlena, road from Talawakele to Nawalapitiya at 11/6 km marker, steep rock face, moist with trickling water, 800m., 01.11.1995, *Shaffer-Fehre with Jayasekara & Samarasinghe 466* (K). LOCALITY UNKNOWN: *Thwaites C.P. 1287* (BM; CGE; K; P; PDA); *Randall 3241. Rawson 3003.*

3. PNEUMATOPTERIS

Nakai, Bot. Mag.Tokyo 47: 179. 1933. Emend. Holtt. in Blumea 19: 42 (excl. *Pseudocyclosorus*) 1971; op. cit. 21: 293. 1973.

Caudex usually erect, rarely creeping; rhizome scales broad, thin with marginal hairs; fronds pinnate,usually large, decrescent, with shallowly to deeply lobed pinnae, the lobes with cartilaginous margins; usually several pairs of basal pinnae reduced either abruptly or gradually; aerophores on reduced and lower pinnae distinct, ± swollen; stipe, lamina never conspicuously hairy; veins anastomosing in most species, free in some; lamina between veins ± pustular when dry, sessile spherical glands never present; sori usually indusiate, sporangia often bearing short, club-shaped glandular hairs, stalks with a 2-4 celled hair with enlarged terminal cell; spores pale with many small ± quadrate wings of irregular shape, hence appearing spinulose. Cytology n=36.

About 75 species; mainly in Malesia with a few species in Africa and the Mascarene Islands; mainland Asia from southern China southwards throughout Malesia to Australia (northern Queensland) and New Zealand; in the Pacific Islands from Fiji and Samoa to Hawaii. (Text: Sledge)
Type species: *Pneumatopteris callosa* Blume (Nakai) [based on *Aspidium callosum* Blume] Sledge, Bull.Br. Mus. nat. Hist. (Bot.) 8(1): 41-42. 1981. Manickam and Irudayaraj, Pteridophyte Flora of the Western Ghats-South India. 1991.

Pneumatopteris truncata (Poir.) Holtt.., Blumea 21: 314. 1973.
Polypodium truncatum Poir., Encycl. Meth. 5: 534. 1804.
Type: Brazil [*s.coll.*] (P).

Aspidium abruptum Blume, Enum. Pl. Jav.: 154. 1828. Type: Java, *Kuhl & van Hasselt* (L).
Nephrodium abruptum (Blume) J. Sm. in Hooker's J. Bot. 3:411. 1841. Hook., Spec. Fil. 4:77.1862.
Aspidium eusorum Thwaites, Enum.Pl.Zeyl.:391. 1864. Type: Sri Lanka, *Thwaites C.P. 3064* (K!) (Isotype BM!).
Nephrodium eusorum (Thwaites) Bedd., Ferns Brit. India: t.130. 1866.
Nephrodium truncatum sensu Bedd., Handb. Ferns Brit. India: 280. 1883, non (Gaudich.) C. Presl

Illustration Beddome, Ferns of Brit. India t. 130. 1866 (as *Nephrodium eusorum*).

Rhizome erect, to 4cm in diam., scales ovate 5 × 5 mm., pale brown, apex acuminate, margin entire. Fronds tufted, 90-240cm long. Stipe 28-80 cm., adaxial groove, rounded below, pale, stramineous when dry, can be to 1cm in diam. just below lamina; initially puberulous, then glabrous, a few acicular hairs in groove. Along stipe 3 or more minute pinnule lobes of laminar tissue at distances of between 4-8 cm (0.4- -1.5 cm long, ca 4mm wide). Lamina elongate-deltoid with ca 20-28 pairs of ascending (up to 40°) pinnae, tip of lamina pinnatifid with strongly falcate base, tip entire; laminar tissue pustulate throughout, comparable to goose pimples, diam. of pustules 0.05 to 0.15 mm. Pinnae on lower lamina up to 5cm apart, cuneate base slightly contracted, ± pedunculate, sessile and less distant towards tip of lamina; pinnae to 30 cm long, 3cm wide with up to 50 truncate to falcate to rounded lobes, cut from to almost ½ to costa, incisions decreasing again, becoming denticulate on shoulder, abruptly reduced below, acuminate 2.5-3 cm long pinna tip with entire margin; margin of basal basiscopic pinna lobe rounded, at an angle with rhachis, acroscopic lobe closely parallel to rhachis, occasionally underlying it, truncate, may be 1/5-1/6 longer than next lobe. Rhachis pale, with adaxial groove, abaxially round, from c. 4 mm in diam. at lamina base to 0.5 mm towards tip. Aerating tissue, cushion -like, at junction of rhachis with costa. Costa pale, raised and with groove above, rounded below. Costules raised only above, parallel and 3-5 mm apart. Veins up to 10 pairs, pinnate, two basal pairs of veins anastomosing; third pair joining sinus which has small, free process; excurrent vein below deep sinus occasionally interrupted just below outer, second pair of veins. Sinus membrane distinct. Sori 6-7,almost medial, parallel to costa. Indusia glabrous. In Sri Lankan specimens sporangial stalks lack glandular hairs (cf. Manickam et al. 1991); glands were seen in specimen *T 724*. Spores monolete, 0.05 × 0.04 mm., colourless, transparent; exine with irregular, broad-based spines.

D i s t r. Southern and north-east India, Sri Lanka , southern China, western Malesia and the Phillipine Islands.

E c o l. In forests of the interior to 1500m (rare according to Sledge).

S p e c i m e n s   E x a m i n e d. MATALE DISTRICT: Rattota, on road along Mid Car Tea Est. at 25/9 km marker by gentle waterfall over granite, forest with good light penetration, 1100m., 25.10.1995, *Shaffer-Fehre* with *Jayasekara* and *Ekanayake*, 408(mixed colln.) (K). KANDY DISTRICT: Hantane Range, in forests, July 1844, *Gardner 1104* (BM; CGE; K); Kadugannawa, shady forest, Oct. 1846, *Gardner 1252* (CGE; K); Oodawella, 1870, leg. *Randall* in herb. Rawson *3244* (BM); Lady Horton's Walk, Kandy, 600m., 24 March 1954, *Sledge 1355* (BM); Wooded stream below road, c. 2 km N.E. of Udugoda on road to Wattegama, N. of Kandy, Alt. c. 650 m., 1.Oct. 1993, *Fraser-Jenkins with & Bandara 234* (K). CENTRAL PROVINCE: not uncommon in the forests here, 600-1500m., terraneous, *Geo. Wall 44/194* before 1873; Gallebodde, by stream in jungle, 600m., 26 Jan. 1954, *Sledge 1046* (BM). KALUTARA DISTRICT: Yattupatha on road from Palawatta to Neluwa at 47/48km marker, among other ferns beside culvert, at top of boulder-

strewn, wooded valley, 10.11.1995, *Shaffer-Fehre* with *Jayasekara, 519* (K).
RATNAPURA DISTRICT: between Gilimale and Carney, jungle 150m.,
9 March 1954, *Sledge 1250* (BM); same locality and date, *T.G. Walker, T
724;* Sinharaja Forest above Beverley Estate, Deniyaya, 900m., 12 March
1954, *Sledge 1277* (BM). LOCALITY UNKNOWN: Thwaites C.P. 3064(CGE;
K; P; PDA); *Beddome* (K); 1887, *Wall*(P); *Mrs Chevalier* (BM); *Macrae*
(CGE); *Bradford* (CGE); *Walker* (K).

4. CYCLOSORUS

Link , Hort.Reg. Berol. 2: 128. 1833. Holttum, Blumea 19: 27. 1971. Sledge
Bull. Br. Mus. nat. Hist. (Bot.), 8 (1): 12-13. 1981.
Type species: *Cyclosorus gongylodes* (Schkuhr) Link [based on *Aspidium
goggilodus*]

Rhizome long-creeping, growing on wet ground; fronds pinnate, pinnate-
pinnatifid, the lower ones not or very little reduced; thin, flat scales present on
costae abaxially; upper surface of costae grooved; basal veins anastomosing,
the next pair passing to sides of sinus; abaxial surface of pinnae usually with
acicular and/or capitate hairs, sessile, red glands; sori indusiate, sporangium
stalk bearing multicellular gland-tipped hairs, capsules eglandular, spores
muricate. Cytology n=36.

Three species: pan-tropical and subtropical. One species in Sri Lanka.

Cyclosorus interruptus (Willd.) H. Itô, Bot. Mag., Tokyo 51: 714. 1937
nomen tantum. *Pteris interrupta* Willd., Phytographia 1: 13, t.10, fig.1. 1794.
Type: Southern India, *Klein* (B).

Aspidium goggilodus Schkuhr, Krypt. Gew. 1: 193, t.33c. 1809. Type from
   Guyana.
Cyclosorus gongylodes (Schkuhr) Link, Hort. Reg. Bot. Berol. 2: 128. 1833.
   Holtt., Rev. Fl. Malaya 2: 261, fig. 148. 1955.
Nephrodium propinquum R.Br., Prodr. Fl. Nov. Holl.: 148. 1810. Hook.,
   Spec. Fil. 4: 49. 1862. Bedd., Ferns South. India: 32, t. 89. 1863. Type:
   Australia, *Banks* (BM).
Aspidium propinquum (R.Br.) Thwaites, Enum. Pl. Zeyl.: 391. 1864.
Nephrodium unitum sensu Bedd., Handb. Ferns Brit. India: 268. 1883, non
   Polypodium unitum L. 1759.

Rhizome long-creeping 4 (-8) cm in diam.; tip and stipe bases covered in
brown stout, small-celled triangular scales with curved base, 2 × 1 mm to 5 ×
1.5 mm. Fronds ± evenly spaced, to 105 cm tall. Stipe ± rust brown, 44-72 cm
long, 3-4 mm in diam., basal 4-5 cm almost black, few scales similar to those
on rhizome. Lamina lanceolate to elongate-deltoid, 30-60 cm long, 22 cm

wide, with 12-25 pairs of pinnae, lowest pair not or scarcely reduced; texture chartaceous. Pinnae mostly opposite but can be opposite / alternate on same rhizomes; mainly sessile, occasionally with c. 0.5 mm long pedicel near lamina base, here slightly cuneate; pinnae at shallow angle of 15° (-45°), at centre of lamina pinnae distant by as much as their own width, more remote towards base; pinna margin with up to 30 blunt-pointed lobes, cut ? to costa; their lateral-or forward-pointing mucronate tips result from recurved lamina, lax hairs on margin, at maturity small yellow glands on underside of lamina. Rhachis with adaxial groove, densely hairy, a few scales in lesser, lateral grooves and abaxially concentrated at base of costa 1 × 0.3 to 2.5 × 0.5 mm. Costa grooved and hairy above, abaxially pale, prominent, similar scales persisting throughout its length; scales small, auriculate with long, acuminate tip, oval with same tip or narrowly lanceolate all with few, regularly spaced, stout, unicellular hairs along margin, their minute circular basal attachment and their cell pattern notable. Costule visible adaxially, abaxially costule and excurrent vein with small, white hairs, translucent orange glands on costule, only occasionally on veins. Veins 6-8 all slightly curved go free to margin, only lowest pair joining to form excurrent vein; acroscopic second vein mostly connivent with excurrent vein, second basiscopic vein, clearly separated, goes to margin above lowest point of sinus. Sinus membrane distinct in larger specimens. Sori medial, absent from lowermost veins; soral pattern restricted to free lobe of segment; occasionally sorus appears on point of fusion of lower veins at base of excurrent vein, so connecting sinuous pattern of sori; when mature sporangia fill entire lobe. Paraphyses arise on sporangium stalk and from receptacle. Indusium round 0.3 mm at tip of lobe, increasing to 0.5 mm on second vein, ± glabrous with very few small persistent hairs. Spore monolete, oval to reniform., 0.075 × 0.03 mm., dull grey, spinulose-tuberculate.

D i s t r. Tropical and subtropical Africa, Madagascar; India, Sri Lanka, China, Japan, the Phillippines, Malesia to New Zealand; Polynesia; tropical America including Brazil.

E c o l. Open, marshy places at low elevations in western and southern parts of Sri Lanka. Exclusively low altitude fern; the only other thelypteroid ferns also confined to low elevations are *Ampelopteris prolifera*, *Trigonospora glandulosa* and *T. zeylanica*. Sri Lankan specimens abaxially glabrous.

S p e c i m e n s   E x a m i n e d. MATALE DISTRICT; Tamankaduwa, 1893, *Nevill* (PDA); Sanasgama; (64-11), 3 Jan. 1951, in ditch by road side, *Ballard 1375* (4 sheets); *Barkley 1416* (BM).COLOMBO DISTRICT: Colombo, *Randall* 1870, 1871, *Rawson 3245* (BM). KALUTARA DISTRICT: Calture,(Kalutara ?) *Macrae 217* (CGE, E, K); Poruwadadanda, Ingiriya Horana road, sea level, above low ditch at road side, in humus along cultivated land, in deep shade of hedge, 7.11.1995, *Shaffer-Fehre* with *Jayasekara, 468* (K). RATNAPURA DISTRICT: Near Ratnapura, marshy ground by road, 3 Jan. 1951, *Sledge 798* (BM); road towards Pelmadulla, marshy habitat disturbed by past open gem mining, in open daylight, 25.11.1995, *Shaffer-Fehre* with *Jayasekara & Samarasinghe, 572* (K); grassy plain of Morningside estate, 900m., 26.11.1995, *Shaffer-Fehre* with *Jayasekara & Samarasinghe, 607* (K). GALLE DISTRICT: Galle, *Freeman 262* (BM); marsh near Bentota, 19 Jan. 1951, *Sledge 880* (BM); Bentota, sea level, 19 Jan 1951; in coconut grove by roadside, marshy ground - sterile, fertile fronds in open, outside shade of palms, abundant, *Ballard 1508* (2 sheets) (K); Hiniduma, sea level, water meadow with bushes (Lauraceae) 8.11.1995, *Shaffer-Fehre* with *Jayasekara,* 469 (K). LOCALITY UNKNOWN: Thwaites C.P. 1705 (BM., CGE, K, PDA); 1839, *Mackenzie* (K); *Skinner* (K); *Ferguson* (US).

5. MACROTHELYPTERIS

(H.Itô) Ching, Acta phytotax sin. 8: 308. 1963.
Type species: *M. oligophlebia* (Baker) Ching

Caudex short creeping or erect; scales at base of stipe ± thickened at base , marginal and surface hairs acicular or capitate; fronds bipinnate-tripinnatifid with ± adnate pinnules, lowest pinnae not or little reduced; rhachis scales narrow, the base thick and acicular hair-tip, hairs on lower surface of frond long, slender, multicellular; sori small, indusiate, indusium very small, sporangia usually bearing capitate hairs near annulus, spores with a ± winged perispore.

Nine species: from Mascarene islands throughout warmer parts of Asia eastwards from Japan to Australia (Queensland); Pacific islands. An account of the species has been given by Holttum 1969.

Macrothelypteris torresiana (Gaudich.) Ching, Acta phytotax. sin. 8: 310. 1963.

Holtt., Blumea 17: 25-32. 1969. Holttum, Thelypteridaceae. Fl. Males., Ser.II, 1(5): 331-560. 1982 "1981". Sledge. W.A. Bull. Br. Mus. nat. Hist. (Bot.), 8 (1): 6-8. 1981. Bull. Br. Mus. nat. Hist. (Bot.) 8 (1): 1-54. 1981. Mannickam., V.S. & Irudayaraj, V. Pteridophyte Flora of the Western Ghats: 172,1992. Type: Mariana Is., Gaudich. (P).
Polypodium tenericaule Hook. Hooker's J. Bot. Kew Gard. misc. 9: 353. 1857. Type: China, Alexander (K!).
Lastrea tenericaulis (Hook.) T. Moore, Index fil.: 99. 1858. Bedd. Handb. ferns Brit. India : 266. 1883.
Nephrodium tenericaule (Hook.) Hook., Sp. fil. 4: 142, excl. t. 269. 1862. p.p. *Aspidium tenericaule* (Hook.) Thwaites, Enum. pl. zeyl.: 393. 1864.
Lastrea setigera sensu Bedd. Ferns S. India. Correct. p.i. 1864. p.p. [errore 'L. flaccida' in t. 99. 1863.] non *Cheilanthes setigera* Blume.
Nephrodium setigerum sensu Hook.& Baker, Syn. fil.: 284. 1867. p.p. non Baker.

Illustrations: Spec. Fil. 4:142, t. 296. 1862. (as *Nephrodium tenericaule* (Hook.) Hook.)

Manickam., V.S. & Irudayaraj, V. Pter. Fl. Western Ghats t. 131. 1992.

Rhizome short-creeping. Fronds tufted, 28 - 180 cm long. Stipes glaucus when fresh, stramineous when dry, 12cm - 120cm long, base persistent 1.5 - 2 cm., dark grey-brown, with dense coat of scales, acicular hairs; scales ± 5-10 mm long 1mm wide at base, tip strongly attenuating, brown, both sides and margin with white, acicular hairs, glabrous above scales, with few scars of former scales, larger stipes with dense felty hairs in groove. Lamina to 45 cm long × 23 cm wide, deltoid-ovate, softly herbaceous to sub-coriaceous, bipinnate to subtripinnatifid in lower portion of large fronds, dark green. Pinnae recorded as 27 × 13 cm and 35 × 16.5 cm suggesting larger dimensions of lamina still. Pinnae and pinnules decurrent on rachis or costa respectively, rather than sessile or stalked. Pinnules towards tip of frond may be entire, borne almost at 90° to costa or slightly falcate, those towards base of frond have a variously pronounced dissected margin, serrated pinnules may be oblong in shape or widening, paddle-like, towards apex. Rachis stramineous, rounded abaxially, groove of stipe intruding at base with felt of short hairs towards tip with short, appressed hairs adaxially, acicular hairs abaxially, base persistent. Costae adaxially not grooved; costa and costules stramineus, adaxially more apparent due to cover of short, appressed hairs; abaxially rounded, prominent with 1.5 mm long multicellular, white acicular hairs. Veins go to margin, may fork, marked on both sides with one or more acicular hairs. Sinus membrane not apparent. Sori medial, 1mm diam., round, restricted to lower 2/3 of pinnules; where sori unequally distributed, preferentially on acroscopic side of pinnule; soriferous part of pinnule covered in minute (0.05-0.15 mm) stalked, yellow glands, red in dried specimens. Indusia fugaceous, supposedly with ascicular hairs, but very dificult to see. Spores monolete 0.05mm long, translucent, surface structure irregularly banded. Cytology: n=31.

D i s t r. *Macrothelypteris torresiana* found in the tropics of the Old World from Madagascar to India, Sri Lanka and Malaysia, even to north-eastern Australia. Adventive in the New World.

E c o l. *Macrothelypteris torresiana* is one of the commonest thelypteroid ferns in Sri Lanka. Frequent in the interior, in open grassy places or in light shade, from sea level to 1750m.

N o t es. *Macrothelypteris torresiana* varies widely in size, in texture from softly herbaceous to almost coriaceous and in comb-like to soft, lobed dissection; sometimes both forms on same frond, if so, the latter form towards tip of large frond. The subtripinnately dissected fronds, and long, septate, hyaline hairs on abaxial side of pinnae are sufficient to distinguish it from all other thelypteroid ferns. In the field it is easily recognized by its glaucous stipes.

Two collections show teratological deformations: *Gardner 1150* is a mature frond in which tips of pinnae are split into three ends; in *Chevalier* the base of pinnae is deformed, the pinnules being dwarfed and deformed. *W.F.*[Ferguson] *134* (PDA) mentions wide creeping rhizome and "very peculiar glaucous colour".

S p e c i m e n s   E x a m i n e d. MATALE DISTRICT: Rattota on road along Mid Car Tea Est. at 25/9km marker, by gentle waterfall over granite, forest with good light penetration, 1100m., 25.10,1995 *Shaffer-Fehre* with *Jayasekara & Ekanayake, 409* (K). KEGALLE DISTRICT: 'Manawella' (Mawanella?), 160m., Sab. Province, Jan. 1954, *Schmid 1067* (BM). KANDY DISTRICT: Kaduganawa, margins of forests, Oct. 1846, *Gardner 1222* (CGE;K); Corbet's Gap, roadside through secondary jungle, 1290m., 9 Dec. 1950, *Sledge 572* (BM); Kadugannawa (62 mile- post) on Colombo Kandy road, on damp in rubber plantation, Dec. 12 1950, *Ballard 1092* (K); between Madugoda and Weragamtota, roadside bank through jungle, 750m., 9. Jan.1954, *Sledge 949* (BM); near Urugala on Kandy-Mahiyangana road, c. 650m., 24 Dec. 1976, *Faden 76/561* (K). BADULLA DISTRICT: Badulla, *Freeman A 259, B 260, C261* (BM); near Badulla, roadside bank, 29. Dec. 1950, *Sledge 780* (BM); foot path to Nanumukula Peak, along bank, in shade under tree 1800m., 04.12.1995, *Shaffer-Fehre with Jayasekara & Samarasinghe, 627* (K). RATNAPURA DISTRICT: Opanaki, Kelani Valley near Ratnapura, bushy roadside bank, 5 Jan. 1951, *Sledge 801* (BM). NUWARA ELIYA DISTRICT: Ramboda, on shady banks, June 1845, *Gardner 1150* (BM., CGE,K); Ramboda Pass, 960m., 17 Dec. 1950, *Sledge 656* (BM); between Hakgala and Ambawela, 1650m., 25. Dec. 1950, *Sledge 788* (BM); Ramboda Pass c. 950m., 2 Jan.1977, *Faden 77/37* (K); Entering Horton Plains from Patipola side, steep cut above road, disturbed 170m., 01.11.1995, *Shaffer-Fehre with Jayasekara & Samarasinghe, 457* (K). LOCALITY UNKNOWN: *Thwaites* C.P. *1286* (BM.,CGE,K, P, PDA); *Thwaites* C.P. *1365* (PDA); Central Province, 450-900m common in shaded places, terraneous, *Wall 44/139*, before 1873; *Walker* (K); *Robinson* (K); *Mrs Chevalier* (BM); *Alston E71* (PDA); *Ferguson 134* (PDA)

6. AMPHINEURON

Holttum, Blumea 19: 45. 1971. Blumea 23: 205. 1977. Fig. 19.
Sledge. W.A. Bull. Br. Mus. nat. Hist. (Bot.), 8 (1):1- 54. 1981.

Caudex erect, decumbent or long-creeping; fronds bipinnatifid. Basal pinnae narrowed at bases, not reduced in size; basal veins free or joining to form short excurrent vein to base of sinus between adjacent pinna segments, the rest free; lower surface of pinnae bearing short, acicular hairs and commonly also subsessile, yellow glands frequent; sori usually confined to lobes of pinnae; indusiate; indusia often glandular, sporangium not bearing hairs or glands near the annulus,usually bearing a short, gland-tipped hair; spores dark, irregularly tuberculate or with irregular thick,branched ridges. Cytology n=36.

About 12 -15 species; one widespread in Africa, Mascarene Islands and south-east Asia, Malesia and Melanesia.

D i s t r. From E. Africa to SE Asia and Australia (Qld), in Pacific to Tahiti, c. 12 species of which two in Sri Lanka.

KEY TO THE SPECIES

1 Indusia studded with glands round their margin. At least 7 of distal veins on both sides of costule usually fertile, pinnatisect,margins cut ? to costa, sori submarginal . . . . **1. A. opulentum**
1 Glands absent from margin of indusia, sori restricted to tips of pinnule lobes, rarely more than 5 distal veins on basiscopic side fertile, pinnatifid, margins cut ¼ to ½ to costa, sori marginal . . . . **2. A. terminans**

1. Amphineuron opulentum (Kaulf.) Holttum, Blumea 19:45 (1971). Sledge. W.A. Bull. Br. Mus. nat. Hist. (Bot.), 8 (1): 27. 1981.

Aspidium opulentum Kaulf., Enum. fil. Chamisso: 238 (1824). Type: Guam., Chamisso (LE).
Aspidium extensum Blume, Enum. Pl. Jav.: 156. 1828.
Nephrodium extensum (Blume) T. Moore, Index fil.: 91 (1858). Bedd. Handb. Ferns Brit. India: 269. 1883.
Nephrodium punctatum Parish ex Bedd., Ferns Brit. India : t. 131. 1866.
Dryopteris extensa (Blume) Kunze, Rev. Gen. Pl. 2: 812. 1891. Type: Burma, Moulmain, *Parish* (K).
Aspidium ochthodes sensu Thwaites, Enum. Pl. Zeyl.: 392. 1864. non Kunze.

Rhizome creeping. Fronds shortly spaced to c. 1.5 m tall. Stipes 60 cm or longer, 4(base) to 2 mm in diam., scaly along 11cm of base; scales 3-5 mm long, 0.2 mm wide, subpubescent particularly in adaxial groove. Lamina to 90 cm long, deltoid, dark purplish, pinnae near top seldom exeeding base of terminal pinna, lower pinnae not decrescent. Pinnae opposite (more likely at base) to alternate, to 39 cm long, 2.6 cm wide, at 90° to rhachis or ascending (c. 60°), pedicillate to sessile; between 20 to more than 60 pinnules per pinna, the 2-3 cm acuminate tip of pinna ± fertile. Pinnules narrow, cut to 5/6 of their length, 2-3 mm gap between pinnules, lower surface with yellow glands on veins. Rhachis with adaxial groove, minutely pubescent, abaxially less so, prominent, rounded. Costa adaxially minutely hirsute, groove in lower half of its length, abaxially prominent, minutely pubescent. Costules, veins adaxially minutely hirsute, abaxially encrusted with minute, yellow glands. Sinus membrane distinct. Sori marginal, round, up to 13, confined to elongate lobes, a velvety band. Indusia showy in immature specimens, persistent, with small yellow glands around margin. Spores monolete.

D i s t r. East Africa, Seychelles, Southern India, Sri Lanka , Burma, Thailand, Malesia, Australia (N. Qld.), New Caledonia eastwards to Tahiti and Marquesas.

E c o l. Widely distributed but not very frequent in forests at all altitudes up to 1250m.

N o t e. The steeply angled large fronds of a deep purple lend the plant a very striking appeance.

S p e c i m e n s E x a m i n e d. KANDY DISTRICT: heights above Kandy, Nov.1829, *Col. Walker* (K); Kandy catchment, in secondary jungle, 750 m., 4 Feb.1954, *Sledge 1094* (BM); Peradeniya, near School of Agriculture, roadside bank, 16 Feb.1954., *T. Walker T430* (BM); Lady Horton's Walk, Kandy, in secondary jungle above river c. 600m., 16 Feb. & 24 March 1954, *Sledge 1142, 1354* (BM). MONARAGALA DISTRICT: Hewelkandura on Koslanda -Wellawaya road, c. 400m., 26 Dec.1976, *Faden 76/593* (K). LOCALITY UNKNOWN: *Thwaites C.P. 975* (BM; CGE; K; P, in part; PDA); *C.P. 990, Trimen* (K); *Gardner 1362* (K); *Gardner 1106* (K; BM; CGE); *1839, Mackenzie* (K); Wall (K; PDA); *Ferguson* (PDA; US); *Robinson 161* (K).

2. Amphineuron terminans (Hook.) Holttum, Am.Fern. J. 63 : 82. 1973. Sledge. W.A. Bull. Br. Mus. nat. Hist. (Bot.), 8 (1): 28. 1981.

Nephrodium terminans Hook., Spec.fil. 4: 73. 1862. Bedd. Ferns S. India: t.90. 1863. Type: Kumaun, *Wallich 386* in Herb. Hook. (K!)
Nephrodium pteroides sensu J. Sm., Cat. Cult. Ferns: 54. 1857. Baker Syn. fil.:289. 1868. p.p. Bedd., Handb. Ferns Brit. India: 296. 1883. non *Polypodium pteroides* Retz.

Rhizome long-creeping, 4-5 mm in diam., branching not observed but seems probable, due to scars. Fronds to 1.50 m long, fairly evenly spaced at 4-6cm intervals. Stipes to 60 cm long, 0.7(base) -0.2 cm in diam., pale brown, slightly reddish-brown for basal 3cm., deep adaxial groove densely filled with elongate-lanceolate scales (10-15 × 2 mm), at reduced density over 10cm of stipe base. Lamina pinnate-pinnatifid, to 50 cm long, 56 cm wide, 42 pairs of pinnae beneath pinna-like tip, 3-4 top pinnae exeeding base of terminal pinna to form obtuse tip; up to four pairs of abruptly reduced, minute pinnae lobes below main part of lamina; texture herbaceous, pale green. Pinnae slightly pedicillate (0.7-1 mm), ascending (60), lower down in frond with cuneate base 1.2cm wide near rachis, at centre 2.5 cm wide, pinnule width varying appropriately; towards top of frond pinna base obliquely truncate, lowest basiscopic pinnule rounds off away from rhachis. Pinnules 4-5mm wide, lobes oblique, sightly falcate, in excess of 50 pinnule lobes, dissected to their middle.

Rhachis adaxially grooved abaxially rounded, prominent, minutely pubescent in groove, elsewhere glabrous. Costa strigose hairy above, smooth, prominent, rounded below. Costule and veins strigose-hairy above and below. Veins 10-12 pairs, lowest two pairs of adjacent groups form excurrent vein the third pair may contribute to excurrent vein or merge sideways with sinus. Sinus membrane prominent. Sori restricted to four or five vein endings in extreme pinnule lobe, vein endings in fertile lobes not occupied by sori, encrusted with minute yellow glands. Indusia, showy in immature specimens, with small glands on margins, surfaces glabrous or with a few hairs. Spores monolete.

D i s t r. Southern India, Sri Lanka and Burma to China (Hainan Dao); throughout Malesia to New Guinea and Australia (N. Qld.).

E c o l. In Sri Lanka widespread and not uncommon in forests in the west and centre.

S p e c i m e n s   E x a m i n e d. CENTRAL PROVINCE: Common in forests, terraneous, below 900 m., *Wall 44/164,* before 1873. MATALE DISTRICT: Dry roadside woods on slope, c. 1km above and W. of Illukkumb ra, N.E. of Matale. Alt c. 650 m., 15.Sept. 1993, *Fraser-Jenkins* with *Jayase ara & Bandara 181*(K). KANDY DISTRICT: Hantane Range, in forests, July 1844, *Gardner 1106* (CGE; K; P); common at Kandy, *Mrs Chevalier s.n.* (BM); Lady Horton's Walk Kandy, c. 600m., 11 Dec. 1950, *Sledge 582* (BM); Kadugannawa, amongst undergrowth below Hevea trees near roadside c. 300m., 12 Dec. 1950, *Sledge 584* (BM); Hunnasgiriya, c. 870m., 16 Jan. 1954, *Sledge 965* (BM). BADULLA DISTRICT: Rawana Ella Falls, Ella - Wellawaya road, Ella Pass, c. 775 m., 18 Nov. 1976, *Faden 76/371* (K); Road to Wellawaya from Beragala, near Naketiya, before Diyaluma waterfall, 06.12.1995, Alt. 700m., in deep shade of overhanging trees on boulder slope above gully, with low scrub, Cerbera bushes, *Shaffer-Fehre* with *Jayasekara & Samarasinghe 661*(K). MONARAGALA DISTRICT: ravine south of Bibile, 450m., 22 Feb. 1954, *Sledge 1172* (BM). HAMBANTOTA DISTRICT; Heneratgoda, in jungle, 13 July 1927, *de Silva s.n.* (PDA); LOCALITY UNKNOWN: *Thwaites C.P. 990* (BM; CGE; P; PDA); *Robinson C 151* (K); *Bradford* ex herb. Hance (BM); *Randall* in herb. Rawson *3220* (BM); *Ferguson* (PDA,US).

7. AMAUROPELTA

Kunze, Farnkr. 1 : 86, 109, t.51. 1843.
Sledge. W.A. Bull. Br. Mus. nat. Hist. (Bot.), 8 (1):1- 54. 1981.

Type species: *Amauropelta breutelii* Kunze

Caudex mostly erect although a few species with creeping rhizomes; fronds bipinnatifid , decrescent below, attenuate upwards, aerophores often present at bases of pinnae ± swollen; veins simple, free, basal ones passing to margins above base of sinus, sessile glands sometimes present on lower surface, short, stipitate, often coloured hairs and acicular or uncinate hairs also present; sori usually supramedial, indusia small, often glandular or hairy, sometimes absent; sporangia short-stalked bearing neither hairs nor glands; spores wingless with very fine raised reticulum.

About 200 species mostly from tropical and subtropical America from Mexico to Chile and northern Argentina; eight species in Africa, Madagascar and Mascarene Islands and one in Hawaii.

A. hakgalensis is the only species recorded from Asia. See note below: Remarks.

Amauropelta hakgalensis Holttum in Sledge, Bull.Br. Mus. nat. Hist. (Bot.) 8(1) : 9, fig.1. 1981.

Holttum., Thelypteridaceae. Fl. Males., Ser.II, 1(5): 331-560. 1982 "1981".
Fraser-Jenkins, New species syndrome in Indian pteridology and the ferns of Nepal, pge 253. 1997.
Type: Sri Lanka, slopes above Hakgala Botanic Garden, 1670m., *Holttum S.F.N. 39169* (SING, holotype n. v.; K, isotype).

Illustrations W.A. Sledge, Bull. Br. Mus. nat. Hist. (Bot.), 8 (1) fig. 1, A-C. 1981.

Caudex short, erect. Fronds tufted, slender. Stipe 9-13 cm long, adaxial groove, pale brown, hairs all round, patent ± 1mm long; few scales at base 5.3 mm × 1mm., with some capitate hairs, scales decreasing upwardly, none beyond stipe. Lamina 18-40 cm long, ovate –lanceolate ± 16 pairs of alternate to opposite pinnae curving gently towards blunt apex, surface minutely hirsute (0.3 mm); largest pinnae near centre of lamina 5 cm long, 1.6cm wide at base (M. S.-F., 448, K); tip of lamina of up to 11 tiny, fused pinna pairs with crenate to simple margins; lowest 3-5 pinna pairs of lamina strongly decrescent, remote, without auricles, pinnae subsessile, truncate base, deeply pinnatifid, interstices ± 5mm., with ± 12 blunt pinnules, lowest acroscopic pinnule often slightly longer. Rhachis pale brown, with continuing adaxial groove, densely covering hairs of varying length, some hooked, some capitate. Costa prominent, grooved adaxially, fringed on both sides with acicular hairs 0.3-1 mm long. Costule slightly prominent adaxially, with acicular hairs; in dried specimens marginal fringe of hairs recurved. Veins, 2-3 pairs, simple, free, basal ones passing to margins above base of sinus. Sinus membrane not apparent. Sori supramedial, on lowest one or two basal acroscopic veins. Indusium barely visible. Spores with fine, two-dimensional reticulum. Cytology: n=29.

D i s t r. Endemic, restricted to Hakgala / Nuwara Eliya area.

E c o l. In undergrowth or exposed on slopes.

N o t e. Following an in depth discussion Sledge (1981) comes to the conclusion that the plant may be endemic or might possibly be introduced, but that it cannot be referred to any other known species of *Amauropelta*. Fraser Jenkins (1997) suggests that the taxon might have close affinities to, or indeed may be the African species *Thelypteris bergiana* Schlecht. finding both to be "almost identical" with "the frond morphology matching exactly in every respect".

S p e c i m e n s   E x a m i n e d. KALUTARA DISTRICT: Poruwadadanda, Ingiriya- Horana road, at 5/16km marker, at road side above low ditch, on border of cultivated land in deep shade of hedge, sea level, 7.11.1995, *Shaffer-Fehre with Jayasekara 468* (K). NUWARA ELIYA DISTRICT: shrubby slopes above Hakgala Botanic Garden, 1670m., 23 Dec. 1950, *Holttum S.F.N. 39169* (SING, holotype; K, isotype); same locality and date, cult. Kew, *Manton P. 220* (BM). Hakgala Bot. Gd. and slopes of Hakgala Mt., 1720-1820m., 14.Nov. 1976, *Faden 76/272* (K); Nuwara Eliya, nr Forest Department, slope above road, shallow soil on gneiss, 1700m., 31.10.1995, *Shaffer-Fehre, Jayasekara & Samarasinghe 448 & 467* (K); entering Horton Plains from Pattipola, 2km marker, steep cut (c. 2 m) above road, shade below Eucalyptus plantation, 1700m., 01.Nov 1995, *Shaffer-Fehre, Jayasekara & Samarasinghe 458* (K).

8. CHRISTELLA

H. Lév., Fl. de Kouy-Tcheou : 472. 1915. emend. Holttum, Taxon 20: 533. 1971. Blumea 19: 43. 1971. Wood, Bot. J. Linn. Soc., 67 (1) 191-202, t. 1-4. 1973. Holttum, Kew Bull. 31: 293. 1976. Sledge, W.A., Bull. Br. Mus. nat. Hist. (Bot.) 8(1):1-54. 1981. Holttum, Fl. Malesiana, Ser. II, Vol. I(5): 331-560 (550-560). 1982 "1981".

Lectotype:*Christella parasitica* (L.) H. Lév.; *Polypodium parasiticum* L., designated by Holtt.

Caudex erect to wide-creeping; fronds pinnate, decrescent with 1-5 pairs of lower pinnae gradually reduced, or rarely the lowest pinnae deflexed but not or scarcely reduced; pinnae pinnatifid, lobed or crenate; aerophores not conspicuous; costas, costules, veins and lamina surface bearing erect, acicular hairs, sometimes small capitate hairs beneath occasionally also orange-coloured glands of varying size; sori indusiate, "sporangia lacking setae or glands near annulus but always bearing unicellular, elongate, glandular hairs on the stalks of sporangia" (according to Sledge). The present author examined all Sri Lankan species under the microscope and found unicellular elongate hairs with glands only once and glands near annulus several times. "Capitate hairs beneath" were always dubious. "Anastomosing veins" becomes here 'veins joining'( ) because only a straight line of fusion,shallow or at slightly steeper angle are involved. Spores with incomplete wings verrucose. Sledge mentions 2 types of spore sculpture for Christella (see also Wood). Therefore ornamentation is always given. n=36. Taxa of this genus are variable; according to

Sledge variability are partly due to hybridization. Hybrids are readily produced experimentally and therefore occur most likely in nature as well.

About 50 species, mainly tropical and subtropical regions of Asia, less in Africa and America. Only *Christella parasitica* and *C. dentata* are common and widespread. The distribution of *C. papilio* and *C. hispidula* is not well known; they are infrequent but probably occur in many different habitats with suitable conditions. *Christella subpubescens* and *C. meeboldii* are at present known only from one and two gatherings respectively. Although both may have been overlooked through confusion with other species, it is more likely that the absence of specimens in all collections seen indicates their rarity. *Christella zeylanica* has not been regathered during the 20th century (Sledge 1981).

KEY TO THE SPECIES

1 Lowermost pair of pinnae not or hardly shortened, but often deflexed ....
.... **1. C. parasitica**
1 One or more pairs of lower pinnae distinctly shortened
  2 Rhizome erect, fronds tufted
    3 One pair of veins joining: pinnae hairy beneath...... **3. C. hispidula**
    3 ½- 2 pairs of veins joining : pinnae subglabrous beneath ....
.... **6. C. papilio var. papilio**
  2 Rhizome creeping, fronds spaced
    4 Pinnae subentire, crenate or very shallowly lobed ... **7.C. zeylanica**
    4 Pinnae pinnatifid
      5 Five or more pairs of basal pinnae shortened ....
.... **6. C.papilio var. repens**
      5 Up to four pairs of basal pinnae shortened
        6 1-1½ pairs of veins fusing below the sinus membrane; pinnae lobed
          about half way to costa .... **2. C. dentata**
        6 At least some pinnae with two pairs of veins fusing below the sinus
          membrane; pinnae lobed less than ½ way to costa
          7 Pinnae glabrous or nearly so beneath, except on costa ....
.... **4. C. subpubescens**
          7 Pinnae hairy beneath .... **5. C. meeboldii**

1. Christella parasitica (L.) H. Lév. Flora Kouy-Tcheou: 475. 1915.

Polypodium parasiticum L., Sp. Pl. 1090. 1753. Type: Canton, *Osbeck* (S-PA, in herb. Sw.). Holttum., Rev. Fl. Mal. 2:281, fig.162. 1955.
Nephrodium molle Hook., Sp.Fil. 4: 67. 1862p.p.; Bedd., Ferns S. India: t.84. 1863. Handb. Ferns Brit. India: 277. 1883p.p. and Handb. Suppl.: 76. 1892p.p., non *A. molle* Sw..
Aspidium procurrens Mett. in Annls Mus. Bot. Lug.-Bat. 1:231.1864. Type: Java, *Zippelius* (L: 908, 335-152).

Nephrodium procurrens (Mett.) Bakei, Syn. Fil.: 290. 1867. Bedd., Handb. Ferns Brit. India: 278. 1883; Handb. Suppl.: 67. 1892. p.p.

Nephrodium didymosorum Parish ex Bedd., Ferns Brit. India, t.200. 1866. Type: Burma, Moulmein, Parish.

Nephrodium molle var. *didymosorum* (Parish ex Bedd.) Bedd., Handb. Ferns Brit. India: 279.1883.

Cyclosorus didymosorus (Parish ex Bedd.) Nayar & Kaur, Companion Beddome: 68.1974.

Nephrodium tectum Bedd., Handb. Suppl.: 79. 1892, excl. King spec. Type: Singapore, *Wallich 394* (K; W).

Aspidium molle sensu Thwaites, Enum. Pl. Zeyl.: 391.1864 p.p., non Sw.. 1801.

Illustrations Sledge. W.A. Bull. Br. Mus. nat. Hist. (Bot.), 8 (1): 30. fig. 5A.1981.

Rhizome long creeping, 5 mm in diam., moderate cover of brown, narrow lanceolate scales with curved base 8-10 mm long, 1mm wide. Fronds to 55 cm long, well spaced. Stipe to 30 cm long 1-2 mm in diam., adaxial groove, at base a few dark brown scales, similar to rhizoidal scales, above base weakly hirsute with soft, white acicular hairs. Lamina deltoid, 16-18 or more sessile pinnae at shallow angle or ± horizontal, not reduced at base, this character distinguishes it from all other Sri lankan species of *Christella*; lamina moderately hairy adaxially, between veins with scattered short hairs and subsessile, colourless, gland-like capitate hairs, abaxially soft spreading acicular hairs up to 1mm long. more dense, apex pinna-like cut 9/10 to costa at base, tip crenate. Pinnae up to 12.5 cm long, 0.8-1.9 cm wide, slightly contracted towards truncate base or parallel-sided, 1/5 of length forming ± abruptly acuminate tip with entire or crenulate margin; pinna pinnatifid cut 2/3 to 4/5 to costa with 18-26 blunt-pointed to rounded, slightly curved lobes; basal acroscopic lobe largest, sometimes with wavy margin;. Rhachis 1-2 mm in diam., stramineous, adaxially grooved, hairy, abaxially rounded, prominent, moderately hirsute with acicular hairs. Costa adaxially grooved, hirsute, abaxially rounded, densely hirsute. Costules pale above, hairy on both sides, with translucent orange glands below, lowest acroscopic pinna lobe sligthly incurved towards rachis. Veins up to 7 pairs, pale above, with few translucent orange glands below; first pair joining sends very short excurrent vein to sinus, all other veins go free to margin. Sinus membrane distinct. Sori medial. of two different placements in fertile frond viz.: the didymosoral condition (see also *C. hispidula*) with sori restricted to first vein or to first vein and second acroscopic vein, i.e. a pattern of two or three sori in a line along costa. the second pattern with usual 4 (-6) pairs of sori, each 1-1.2 mm in diam., arising in a medial position along costule of individual segment. Indusia round, persistent, with tuft of long white hairs at centre. Spores monolete, oval, 0.2 × 0.1mm., grey, ornamentation of short unconnected dark ridges, partial wing, less than ½ at equator. (Sledge: "n=72 tetraploid".)

D i s t r. India, Sri Lanka, S. China, S. Japan, through S E Asia to Taiwan, New Caledonia Australia (Queensland) and east to Polynesia.

E c o l. In open or lightly shaded, grassy or bushy places, roadside banks; at all altitudes up to 1800m. In Sri Lanka very common in the Western, Central and Southern Provinces.

N o t e. Some specimens which may be *C. parasitica* but share a few characteristics with *C. dentata* appear to form a varietal group. They are distinct by the top-half of each pinna rising in very regular fashion, of a candelabra-like image or that of a scimitar in individual pinnae, here referred to as falcate or "scimitar form" e.g. (*Fraser-Jenkins 113* (K), *Shaffer-Fehre 618* (K))

S p e c i m e n s E x a m i n e d. MATALE DISTRICT: Bambaragalla, 750m., 12 Dec. 1976, *Faden 76/530* (K); Stream gulley in Midland Estate, N.E. of Matale, Alt. c. 900 m., 15. Sept. 1993, *Fraser-Jenkins* with. *Jayasekara & Bandara 119* (K); Rattota, on road along Midland Tea Estate, secondary montane forest, in coarse sand by granite rock, in shade, 750m., 25.10.1995, *Shaffer-Fehre with Jayasekara & Ekanayake, 400* (K); Rattota, on road along Mid Car Tea Estate, cardamom plantation, in deep shade, 1000m., 25.10.1995, *Shaffer-Fehre with Jayasekara & Ekanayake,* 402 (K); Bambaragala, road along Midland Tea Estate, steep slope with thin red top soil, secondary montane forest above750m., 25.10.1995, *Shaffer-Fehre with Jayasekara & Ekanayake, 425-427* (K). KANDY DISTRICT: Peradeniya, 7 Feb. 1914, *Petch* (PDA); Roseneath, Kandy, edge of path in secondary jungle, 660 m., 28 Nov. 1950, *Ballard 1000* (K); Nawanagalla, bushy ground by road through jungle, 1110m., 8 Jan.1954, *Sledge 942* (BM); Galaha, 940 m., 22Jan. 1954, *Schmid 1041* (BM). Corbet's Gap, 1200m., 7 Jan. 1951, *Sledge 844* (BM). Jungle at Hunasgiriya, 870m., 16 Jan. 1954, *Sledge 977* (BM). Panilkande, 600m., 5 April 1954, *Sledge 1401* (BM); Kandy, 720m., 9 April 1954, *Sledge 1412* (BM); Forest by "Lady Horton's Walk" N. part of Udawattakelle forest Alt. 650 m., 14. Sept. 1993, (scimitar-form) *Fraser-Jenkins* with *Jayasekara 113* (K); Denipitiya, Ramboda, Top Pass, in crevices of sheer rock face, 880m., 31.10.1995, *Shaffer-Fehre with. Jayasekara & Samarasinghe, 440,* (K). BADULLA DISTRICT: Badulla road from Nuwara Eliya, 29 Dec. 1950, *Ballard 1313* (K); Near Muppane, Kumaradale Estate, 600 m., 18Mar. 1928, *R.M. Alston* (BM); Road from Badulla to Passara, from 3M post turn to Nanumukula, peak, disturbed road-side bank, 900m., 03.12 1995, *Shaffer-Fehre with Jayasekara & Samarasinghe, 618* (scimitar form) (K). NUWARA ELIYA DISTRICT: road from Rendapola to Ambewala, in exposed soil beside boulder, grassy overhang near bottom of slope, 1500m., 01.11.1995, *Shaffer-Fehre, Jayasekara 581* (K); Gallandala, 2 km from Udugama, rocky cliff over torrent, in splash-zone, 8.11.1995, *Shaffer-Fehre, Jayasekara, 485* (K). LOCALITY UNKNOWN: Thwaites C.P. 974 (BM.,p.p.; CGE in part:

K;PDA p.p.); *Mrs Chevalier* (BM); *Freeman 270C, 271D, 272E, 273F* (BM); *Barkly* (BM); *Alston 1093* (K).

2 Christella dentata (Forssk.) Brownsey & Jermy, Brit. Fern. Gaz. 10: 338. 1973. Holttum., J. S. Afr.Bot. 40: 143. 1974.

Polypodium dentatum Forssk., Fl. Aegypt. Arab.: 185. 1775. Type: Arabia, *Forssk.* (C)
Nephrodium molle (Sw.) R.Br., Prodr. Fl. N. Holl.: 149. 1810. Hooker, Sp.Fil. 4 : 67. 1862. Bedd., Handb.Ferns Brit. India: 277. 1883, p.p.; Handb. Suppl.: 76. 1892, p.p. *Cyclosorus subpubescens* sensu Holtt., Rev. Fl. Mal. 2: 273, fig. 157. 1955. non *Aspidium subpubescens* Blume sensu Panigrahi & Manton, J. Linn. Soc. (Bot.) 55: 729-743.1958.
Thelypteris taprobanica Panigrahi, Kew Bull. 31: 187. 1976, p.p. Type: Sri Lanka, *G. Wall* (K!).

Illustrations. Sledge, Bull. Br. Mus. nat. Hist. (Bot.), 8 (1): 30. fig. 5D. 1981.

Rhizome short-creeping, 3-4 mm in diam., moderate cover of large brown scales with wide base and narrowing tip 6 (-10) mm × 2mm. Fronds closely spaced; dimorphic, the fertile longer, often with more remote pinnae than the more spreading sterile ones. Stipe of variable length 16-25 (-40 cm), stramineus, deep adaxial groove; lowest 1-2cm of slightly darker base taken up by narrow band-shaped to lanceolate scales, 10 × 1 mm., above often glabrous due to hair loss. Lamina 67 × 22 cm to 58 cm × 25 cm [*Faden 76/531*] elongate lanceolate by virtue of the 1-4 ± gradually reduced, often but not always decrescent, widely spaced basal pinnae. Pinnae with truncate base, sessile to slightly pedicillate (1 mm., *Faden 76/573*), 7.5 -16.5 cm long, 1.1 × 2.2 cm wide, disposed ± horizontally, acuminate tip slightly upturned, or at angle of c. 15° to rachis; apex pinna-like cut 9/10 to costa at base, tip crenate; lobed margin of pinnae with 22 (-28) segments cut from 2/5th to slightly more than ½ to costa; basal acroscopic segments often slightly enlarged, with crenate margin and sometimes branching veins; pinnae abaxially eglandular; indument consisting of soft white hairs variably: 0.5 and 1mm; smaller, straight hairs are an admixture throughout, lamina surface with only short hairs; the larger, slightly curved hairs on all vascular elements; mature lamina surface glabrous in majority of specimens. Rhachis with adaxial groove containing dense acicular hairs, below rounded with soft, white acicular hairs. Costa with hirsute groove above, rounded, prominent and hairy below. Costule hairy above and below. Vein above usually with one to four large hairs; below lowest vein of 6-8 (-10) pairs joins with vein of next segment to give rise to excurrent vein; second acroscopic vein runs parallel to excurrent vein to sinus, second basiscopic vein reaches margin above sinus. Sinus membrane distinct. Sori medial, 1-1.2 mm in diam., 4-6 pairs. Indusia 0.5-0.8 mm in diam. yellow to brown, with full covering of ± dense, white acicular hairs, rarely glabrous.

Spores 0.045 × 0.02 mm., membrane at equator more than ½ but not complete, pale amber with dark branching ridges. (Sledge: "n=72, tetraploid".)

D i s t r. Tropics, subtropics: Africa, Asia, Polynesia south to New Zealand. Introduced: America.

E c o l. Open, lightly shaded, grassy or bushy places, roadside banks, all elevations to 1800m.

S p e c i m e n s   E x a m i n e d. MATALE DISTRICT: Bambaragalla, 750m., 12 Dec., 1976, *Faden 76/531* (K); Rattota, road along Mid Car Tea Estate, cardamom plantation, in shade, 1000m., 25 Oct 1995, *Shaffer-Fehre with Jayasekara & Ekanayake 403* (K); Road and estate, below granite rock, 1000m., 25 Oct 1995, *Shaffer-Fehre with Jayasekara & Ekanayake, 404* (K); road at gentle waterfall over granite in moderate shade, 1100m., 25 Oct 1995, *Shaffer-Fehre with Jayasekara & Ekanayake 408* (K). KANDY DISTRICT: Hantane Range, in forest, July 1844, *Gardner 1105* (CGE); *Gardner 1105* ex herb. J. Smith (BM); Lady Horton's Walk, Kandy, *Robinson 148* (K); Kandy, in jungle, 3 June 1927, *Alston 1093* (K; PDA); Le Vallon, forest, ca 1500m., 9 Feb. 1954, *Sledge 1128* (BM); Kadugannawa, on damp bank in rubber plantation, 12 Dec. 1950, *Ballard 1091* (K); Corbet's Gap, 1200m., 7 Jan.1951, *Sledge 834* (BM); Corbet's Gap, 7 Jan. 1951 *Ballard 1563* (K); two miles east of Panilkanda, edge of jungle near road, 660m., 24 Jan. 1951, *Sledge 924* (BM); Hunnasgiriya, jungle 870m., 16 Jan 1954, *Sledge 978* (BM); Corbet's Gap, 5 April 1954, *Sledge 1402* (BM); near Urugala, on Kandy-Mahiyangana road, 650m., 24 Dec 1976, *Faden 76/562* (K); stream on S.W. side of Wattakelle hill above Wattakelle village N.E. of Kandy, Alt. c. 1200-1500 m., 1 Oct 1993, *Fraser-Jenkins with Bandara 265* (K); Forest 3 ½ km N. of and below Corbet's Gap, on road to Winchfield and Mimure, N.E. of Kandy, Alt. c. 880 m., 14 Oct 1993, *Fraser-Jenkins with Abeysiri & Gunawardana 289* (K); Irrigation channel in woods above village of Pahalena Ella, Knuckles Range N.E. of Kandy, Alt. c. 550 m., 14 Oct 1993, *Fraser-Jenkins with Abeysiri & Gunawardana 300* (K); Denipittiya Pussalawa,at 36/7 km marker, 950m., 31 Oct 1995, *Shaffer-Fehre, Jayasekara & Samarasinghe 433* (scimetar form) (K); Ramboda, Katukitula culvert, sheer rockface, wet by runnels of water, overhung by lavish vegetation, fern in deep shade 860m., 31 Oct 1995, *Shaffer-Fehre with Jayasekara & Samarasinghe 436* (K). BADULLA DISTRICT: Road from Badulla to Passara, from 3M post turn to Nanumukula peak, disturbed road-side bank, 900m., 03 Dec 1995, *Shaffer-Fehre with Jayasekara & Samarasinghe, 618* (scimitar form) (K); Hali Ela to Bandarawela road junction, Urugala Estate, grassy slope in shade of trees but much open light, 700m., 5 Dec 1995, *Shaffer-Fehre with Jayasekara & Samarasinghe 630* (K); Tangamalle Forest Sanctuary near Haputale, cloud forest with clayey soil, 1450m., 05 Dec 1995, *Shaffer-Fehre with Jayasekara & Samarasinghe 644* (K); Beragalla, in soil at base of rock, above gully, 700m., 06 Dec1995, *Shaffer-Fehre with Jayasekara & Samarasinghe 656 + 657* (K).)

KALUTARA DISTRICT : Panadura, roadside in coconut plantation, 2-4m., 14 Feb. 1968, *Comanor 997* (K,PDA). RATNAPURA DISTRICT: road from Rakwana to Deniyaya, at 4 km marker, moist, shady road above tea estate on slope below disturbed primary jungle, 600m., 26 Nov 1995, *Shaffer-Fehre with Jayasekara & Samarasinghe 579* (K). NUWARA ELIYA DISTRICT: Nuwara Eliya, presented 30Jan 1932, *Freeman 267A, 268A, 269B* (BM); Hakgala, 1800m., 23 Dec. 1950, *Sledge 704* (BM); Hakgala Peak 1830m., sub *C. jaculosus*, 23 Dec.1950, *Manton P199* (BM); shady ground by track above Hakgala Gardens, 1830 m., 23 Dec. 1950, *Sledge 704* (BM); Hakgala, 1700-1800m., 14 Nov. 1976, *Faden 76/271* (K); road side rocks and slope, c.1/2 km W. of Nuwara Eliya on Kandy Road. Alt. c. 975m., 25.Oct 1993, *Fraser-Jenkins with Abeysiri 308* (K); road from Rendapola to Ambewala, foot of hill near public water pump, fern in deep shade among bamboo at brook, 1400m., 01.11.1995, *Shaffer-Fehre with Jayasekara & Samarasinghe 452* (K); road just below 4km marker, boulder-strewn slope, in soil under rock overhang, 1500m., 01 Nov 1995 *Shaffer-Fehre with Jayasekara & Samarasinghe 456* (K); Ginigathena 2/6 km., steep, shady bank at road, 10m from torrent, 800m., 17 Nov 1995, *Shaffer-Fehre with Jayasekara & Samarasinghe 524* (scimitar form) (K); same locality at 3/2 km., near Aberdeen waterfalls, fern epilithic on Gneiss boulders, 800m., 17 Nov 1995, *Shaffer-Fehre with Jayasekara & Samarasinghe 525* (K); outskirts of Maskeliya on Hatton road, c. 4m high steep slope over busy road, in deep shade under overhanging bushes and trees, 19 Nov 1995, *Shaffer-Fehre, Jayasekara, Samarasinghe 571* (K).). MONARAGALA DISTRICT: Bakinigahawela on Bibile-Moneragala road, shady stream bank, ca 250m., 25 Dec. 1976, *Faden 76/573* (K). GALLE DISTRICT: Panangula, below coco nut palm., sea level, 8 Nov 1995, *Shaffer-Fehre, Jayasekara 476* (K). HAMBANTOTA DISTRICT: Jungle at Henaratgoda, 14 July 1927, *J. M. de Silva* (PDA). LOCALITY UNKNOWN: *Thwaites C.P.714* (BM., CGE, P); *Thwaites C.P. 974* in part (BM.,CGE, PDA); *Thwaites C.P. 3498* (PDA, one of four sheets).

3. Christella hispidula (Decne.) Holttum., Kew Bull. 31: 312. 1976.

Aspidium hispidulum Decne., Nouv. Ann. Mus. Hist. nat. Paris 3: 346. 1834.
    Type: Timor, *Guichenot (P)*.
Nephrodium hispidulum (Decne.) Baker, Hook. & Baker, Syn. Fil. : 293.
    1867. nomen tantum-
Nephrodium tectum Bedd., Handb. Suppl.: 79. 1892, p.p. quoad King spec.
    ex Perak. *Nephrodium molle* sensu Bedd., Handb. Ferns Brit. India:
    277.1883, p.p., Handb. Suppl.: 76. 1892, p.p., non *Aspidium molle* Sw..

Illustrations. Sledge. W.A. Bull. Brit. Mus. nat. Hist. (Bot.), 8 (1): 30. figs
    5B,C. 1981.

Caudex erect, 1-2.5 cm in diam., dense scales protecting meristem., scales brown, lanceolate with horizontal base ± 6 × 1.5 mm. Fronds tufted. Stipes stramineous, shorter than lamina 18-20 cm., adaxial groove; in dessicates of this softly herbaceous plant many small grooves arise in stipe and rachis; scales at very base, brown, 6 × 1mm., stipe elsewhere softly pubescent with pale hairs ± 1mm. Lamina lanceolate ca 50 cm × 18 cm., 22-26 pairs of sessile, horizontal to weakly ascending pinnae, with truncate base, 4 (-5) lower pinnae decrescent. Pinnae 8 (-10) cm long, 1-1.4 (-2.4)cm wide, in lower third often slightly contracted, basal acroscopic lobes ± ? longer than others, incurved, on lower pinnae with wavy margin; tip of lamina pinna-like, pinnatifid; where decreasing higher pinnae remote (weedy appearance), tip will be pinnatifid but smaller and more remote still (*Faden 76/480*). Pinnae with 22-26 segments, rounded lobes cut ?to costa; tip, ? - ¼ of pinna, narrows fairly abruptly, its margin becomes crenate, then entire; hairs on upper surface of lamina short, stiff, 0.5mm (also with "minute, colourless subsessile gland-like capitate hairs" Sledge). Rachis stramineous, dense cover of soft, slightly curved unicellular hairs 1-1.5 mm long. Costa faintly grooved adaxially, rounded, prominent below, cover of white acicular hairs 0.5-1mm on both sides. Costule with usual hair cover on both sides. Veins have one to few 1mm long hairs on both sides; 6-9 pairs of veins in a segment lowest veins of adjacent segments join to form short, excurrent vein, other veins free, not quite reaching margin. Sinus membrane only rarely distinct. Sori , submarginal, on up to 6 fertile veins; sori on lowest vein always more distinct, close together at base of excurrent vein, being first to arise i.e didymosoral condition in *C. hispidula* (see *C. parasitica*) occurs on same plant with fronds of full soral cover (*Faden 76/653*). Indusia present, persistent, glabrous, one or more hairs rare, brown- yellowish, lobed margin 0.5mm., transparent, thin,pale. Spores monolete, oval 0.03 × 0.007 mm., with few dark, long and short ridges, branching over amber spore body. (Sledge: "n=36", diploid.)

D i s t r. Tropics and subtropics of Africa, Asia, Sri Lanka and America; far less frequent than *C parasitica* or *C. dentata*.

E c o l. On clearings or along jungle roads, in situations with increased light penetration; on clay.

S p e c i m e n s   E x a m i n e d. MATALE DISTRICT: stream gulley in Midland Estate, N.E. of Matale, Alt. c. 900 m., 15. Sept. 1993, *Fraser-Jenkins* with *Jayasekara & Bandara 120* (K). KEGALLE DISTRICT: Allagalla, 600m., 19 Feb. 1954, *Sledge 1151* (BM). KANDY DISTRICT: heights above Kandy, *Walker* in Herb. Hook. (K); Kandy, Lady Horton' s Walk, 600m., 11 Dec. 1950, *Sledge 578* (BM); Laxapana road, road side in shade, 28 Jan. 1954, *T.G. Walker T232* (BM); Laxapana, 900m., 28 Jan. 1954, *Sledge 1060* (BM); Laxapana road, 6 Feb 1954, *Sledge 1100*, 24 March 1954, *Sledge 1356* (BM); Roseneath Valley, Kandy, 630m 3 Feb 1954, *T.G.*

Walker, Sledge 1092 (BM); Kandy, Catchment on bank in secondary jungle, 750m., 4 Feb. *Sledge 1098* (BM). RATNAPURA DISTRICT: hillside above Potupitya, forest patch much degraded, 1500m.,4.Dec.1976, *Faden 76/480* (K); between Kirapatdeniya and Weligepola, swampy places in forest patch above road, 1900-2000m., 30 Dec. 1976, *Faden76/653* (K); path to Morningside bungalow, among luxuriant vegetation above tea estate, in shade, 1010m., 26 Nov 1995, *Shaffer-Fehre with Jayasekara & Samarasinghe 604* (K). LOCALITY UNKNOWN: *Thwaites C.P. 974* in part (CGE, PDA); *Thwaites C.P. 714* (K; PDA, in part); 1899, *Bradford* (K).

4. Christella subpubescens (Blume) Holttum, Webbia 30: 193. 1976.

Aspidium subpubescens Blume, Enum. Pl. Java: 149. 1828. Type: Java, *Blume* (L). *Nephrodium amboinense* sensu Hook., Spec. Fil. 4: 75. 1862; Syn.Fil.: 292. 1867p.p.; Bedd., Suppl. Ferns S. India & Brit. India: 19. 1876; Handb. Suppl. 75. 1892, p.p. non C. Presl; *Nephrodium molle* var. *amboinense* sensu Bedd., Handb. Ferns Brit. India: 278. 1883, excl. syn.; *Nephrodium extensum* var. *minor* Bedd., non *N. amboinense* C. Presl; *Nephrodium molle* var. *major* Bedd., Handb. Suppl. : 76.1892; quoad pl. Sumatr. tantum.

Illustration. Sledge, Bull. Brit. Mus. nat. Hist. (Bot.), 8 (1): 30. fig.5E. 1981.

Rhizome short-creeping, compact. Fronds few, up to 1m tall. Stipe up to 30 cm long, bulky at base, soon narrowing to 0.6 cm -0.2 cm., below scales 1 × 0.1 cm., lanceolate, brown, decreasing in size 0.5 × 0.02 cm above, minutely pubescent to glabrous. Lamina up to 80cm long, pinnate with up to 20(-26) free pairs of mainly alternate, sessile pinnae below, pinnatifid apex; lowermost 1-3 pinnae short and wide, almost auriculate. Pinnae up to 12 × 1.5cm., well spaced, particularly in more slender, fertile fronds, almost their own width free between pairs, more distant towards base, finally remote between auricles; pinnae lanceolate base truncate to cuneate, $^1/_4$ - $^1/_6$ of pinna forms narrow, acuminate tip, completely soriferous in fertile specimens; 18-25 segments per pinna lobed and cut to ¼ or less than ½ to costa; basal acroscopic segment larger, strongly incurved, almost clasping rhachis. Rhachis densely hairy above (hairs curved, white 0.5 mm long) alongside adaxial groove, laterally and below hirsute (0.2 mm). Costa with hirsute groove above, pale rounded, prominent hirsute below, smaller hairs are 'appressed', rather patent. Adaxially costule and veins with few hairs, in specimens seen abaxially costule and excurrent vein hirsute. Veins 4-6 per segment; branching in enlarged basal pinna lobes and auricles. Sinus membrane distinct. Sori medial, c. 1mm in diam., 4-5 pairs per segment, sporangia dense. Indusium glabrous in Sri Lankan specimen, elsewhere often hirsute, sometimes glandular. Spores monolete, reniform with discrete spines, grey on amber body 0.035 × 0.02 mm. (Sledge "n-72, tetraploid".)

D i s t r. NE India, Sri Lanka to S China, Burma, Thailand, Vietnam., Malaysia to New Guinea, Philippines, Fiji. *C. subpubescens* is apparently very rare and maybe overlooked through confusion with other species.

E c o l. Amongst undergrowth near roadside, at moderate altitude.

S p e c i m e n s E x a m i n e d. KANDY DISTRICT: Kadugannawa, amongst undergrowth below Hevea trees near roadside, 300m., 12 Dec. 1950, *Sledge 585* (BM). RATNAPURA DISTRICT : Road from Ratnapura to Palmadulla, on slope above marshy area, disturbed by past open gem mining, 25 Nov 1995, *Shaffer-Fehre with Jayasekara & Samarasinghe 573* (K); Road from Rakwana to Deniyaya, on slope below disturbed primary jungle, 600m., 26 Nov 1995, *Shaffer-Fehre with. Jayasekara & Samarasinghe 580* (K).

5 Christella meeboldii (Rosenst.) Holttum, B.K. Nayar & S. Kaur, Companion Bedd.: 208. 1974; *Dryopteris meeboldii* Rosenst., Rep. Spec. nov Regni veg. 12: 247. 1913.
Type: Southern India, Tellicherry, *Meebold 2133* (WRSL).

Illustration. Sledge Bull. Brit. Mus. nat. Hist. (Bot.), 8 (1): 30. fig. 5F. 1981.

Rhizome short-creeping, apex concealed in long rusty-brown, lanceolate scales 5-8 mm long, to 1mm wide, with slightly hairy margins, auricled base, narrow, acuminate tip. Fronds 30-60cm long, sterile fronds: appear shorter, lamina broader than fertile one, apparently lack lower decreased pinnae (*Ballard 1315*). Stipe (of fertile frond with auricles) 9-12 cm long, (of sterile frond) 24-29cm long, with adaxial groove and scales along basal 1-2cm., much smaller scales occasionally above, stipe elsewhere minutely hirsute to glabrous. Lamina 28-48 cm long, 11-15cm broad herbaceous, hairy, pinnate, c. 14-16 pinna pairs; pinnatifid apex cut at base ¾ to costa, lobes decrease above, margin of tip entire. Fertile frond with six pairs of decreasing pinnae, three of these ± auricles. Pinnae slightly pedicillate (0.5 mm), gently ascending (15-30°) only lower ones decrescent, all with truncate base, largest pinnae 5-8.5 × 1.5 cm; pinnae with 14-19 segments cut ¼ to costa; lobes broadly rounded to truncate with forward pointing tip, appear to be less segmented, basal acroscopic lobe distinctly larger, 1mm longer, broader than other segments. Rhachis with adaxial groove, abaxially prominent, hirsute with both acicular hairs 1mm., and minute 0.1 and 0.2mm long ones. Costa grooved, hirsute above, prominent below with long acicular and minute hairs. Costule above as veins, below the same as costa. Veins hairy above with 1-2 mm long hairs on each vein, below hirsute with long and short acicular hairs and tiny stiff hairs as on other vascular elements and on margin of lobe; veins 5(-6) pairs per segment, the lowest pair joins to give rise to an off-centre excurrent vein i.e acroscopic half of lower segment wider, its vein longer, second vein from this half joins excurrent vein below sinus; second vein of basiscopic next upper segment meets excurrent vein at sinus but does not fuse with it, other veins go free to margin. Sinus membrane rarely apparent. Sori medial to sub-marginal, often only first and second veins fertile, additional sori arise first on basiscopic half of segment leading to 'u'-shape rather than a 'garland'-pattern of sori. Indusia persistent, lobed, rusty-brown, small white hairs nearest cen-tre. Spores monolete, reniform., equatorial membrane narrow, a few, short, mainly unconnected, beaded ridges (Sledge: 'verrucose'), grey on an pale amber body, 0.05 × 0/015mm.

D i s t r. In Southern India; too rare to know exact distribution in Sri Lanka. At higher elevations, 1450m., in the environs of Nuwara Eliya.

E c o l. In open ground along road sides, near water and waterfalls.

S p e c i m e n s E x a m i n e d. NUWARA ELIYA DISTRICT: Badulla road mile 77/3, 29 Dec. 1950, open ground near road, *Ballard 1315* A (2 sheets / K); Kandapola, near Parawella falls, 1450m., 19 Mar. 1954, *Sledge 1327* (BM).

6. Christella papilio (Hope) Holttum, B.K. Nayar & S. Kaur, Companion Beddome: 208. 1974. *Nephrodium papilio* Hope, J. Bombay nat. Hist. Soc. 12: 625, t.12.1899.

Lectotype: India, Darjeeling, 1880, *Levinge* (K).
Nephrodium molle var. *major* Bedd., Handb. Suppl.: 76. 1892, excl.pl. ex
    Sumatra.
Christella papilio var. **repens** Sledge, Bull. Br. Mus. nat. Hist. (Bot.), 8 (1):
37. 1981.
Aspidium extensum sensu Thwaites, Enum. Pl. Zeyl.: 391. 1864, quoad C.P.
    3498, non Blume. Sledge, Bull. Brit. Mus. nat. Hist. (Bot.), 8 (1):
    37-39. 1981.

Rhizome erect (var. papilio) or short-creeping (var. **repens**). Frond to 140 cm long. Stipe stramineous to 20 cm long, up to 5mm in diam., adaxial groove, basal 4 cm darker, curved, with small, acuminate scales 3-5 mm long, base 1-2 mm wide, glabrous above. Lamina to 112 cm long, decrescent, pinnate, c.13 pairs of pinnae (basal one ½ length of others), 5- 6 (-11) pairs of auricles, diminishing in size,below; surfaces ± glabrous, texture papery. Pinnae to 22 cm long 2.3 cm wide, up to 30-40 segments below caudate tip with crenate margins (such margins entire in *C. denticulata*); segments with slightly forward pointing blunt, triangular lobe, not infrequently with tiny mucronate tip cut $^1/_3$ to costa; lobes of strong, pinnatifid tip of lamina cut ½ to costa. Rhachis with continuing adaxial groove, minutely hirsute. Costa grooved, hirsute above, prominent; with white acicular hairs below. Costule ± glabrous above, few acicular hairs below. Veins 1 ½ pair join in excurrent vein, the next run to side of short sinus, veins above free, reaching margin. Sinus membrane distinct. Sori medial 1-1.5 mm in diam., only basal 3-4 veins of segment appear fertile i.e. sori rarely in free lobe. Indusium hirsute, sharply pointed hairs stout, c. 0.1mm long. Spores verrucose, 0.04 × 0.03 mm., grey-black.

D i s t r. Southern India and north-west Himalayas eastwards to Sikkim., Thailand, northern Malaysia and Taiwan. Sri Lanka, in Central - and Uva Provinces.

E c o l. By streams in forests at higher elevations (1500-2200m).

N o t e. Sledge established two varieties of *Christella papilio*; their chromosome numbers were published by Loyal and Manton respectively: *C. papilio* var. *papilio*: erect rhizome, diploid, n=36 and *C. papilio* var. *repens*: creeping rhizome, tetraploid, n=72. Both varieties occur in Sri Lanka but *C. papilio* var. *repens* is here the more frequent of the two. In the PDA collection are 2 specimens with erect rhizomes. They are not accounted for by Sledge (or Sub *Nephrodium amboinense*?). Hope coined the epithet 'papilio' recalling the butterfly-like opposite-auriculate pinnae towards the base of the lamina, before Leveille (1974) established the genus *Christella*. *Christella* replaces *Nephrodium*. Hope's lectotype from north India has an erect and emergent rhizome. In accepting the species by Thwaites and by Wall (collections lacking rhizomes) Hope enlarges the area of distribution for *C. papilio* var. *papilio* by the island of Sri Lanka. (Sledge 1981, abstract.) Beddome had seen the Wall specimens recognizing their creeping rhizomes and was satisfied that they warranted a new species or variety. The PDA specimens: '*C.P. 3498* and *140 Nephrodium amboinense*' have no rhizome; Sledge identifies both as *Christella papilio*.

S p e c i m e n s   E x a m i n e d. (var. *papilio*, with erect rhizome:) sub *Nephrodium amboinense* -on same sheet as *C. zeylanica* (PDA). (var. *repens*, with creeping rhizome:) BADULLA DISTRICT: Tangamalai Sanctuary, Haputale, 1500m., 25 Feb. 1954, *Sledge 1205* (BM); foot path to Nanumukula Peak, in moist forest clearing on stony soil, 1800m., 04 Dec 1995, *Shaffer-Fehre with Jayasekara & Samarasinghe, 624, 625* (K); Tangamalle Forest reserve, near Haputale, on bank beside torrent, in deep shade, 1500m., 05 Dec 1995, *Shaffer-Fehre* with *Jayasekara & Samarasinghe 652* (K). NUWARA ELIYA DISTRICT: Hakgala, by jungle stream., 1650m., 27 Dec.1950, *Sledge 744* (BM); below Hakgala Gardens, jungle, 1650m., 26 Feb. 1954, *Sledge 1212* (BM., holotype!). LOCALITY UNKNOWN: Wall (K). (without rhizome:) *Thwaites C.P. 3498* (BM.,PDA).

7. Christella zeylanica (Fée) Holttum, B.K. Nayar and S. Kaur, Companion Beddome: 208. 1974. Sledge, Bull. Brit. Mus. nat. Hist. (Bot.), 8 (1): 39-40. 1981.

Nephrodium zeylanicum Fée, Mém. Fam. Foug.10:42. 1865. Type: Sri Lanka, *Thwaites C.P. 3391* holotype not seen; BM., CGE, K (errore 3390), PDA, isotypes).

Aspidium extensum sensu Thwaites, Enum. Pl. Zeyl.: 391. 1864 quoad C.P. 3391 non Blume. *Nephrodium extensum* var. *minor* Bedd., Ferns Brit. India: t. 201.1866. Type Sri Lanka, Thwaites (K).

Nephrodium amboinense var. *minor* (Bedd.) Bedd. Ferns S. India & Brit. India: 19. 1876. *Nephrodium molle* var. *amboinense* sensu Bedd., Handb. Ferns Brit. India: 278. 1883, p.p. non C. Presl

Nephrodium amboinense sensu Bedd., Handb. Suppl.: 75. 1892p.p. non C. Presl *Thelypteris srilankensis* Panigrahi in Notes R. bot. Gdn. Edinb. 33:499. 1975. Type: Sri Lanka, *Thwaites CP 3391* (BM!).

Rhizome erect to short-creeping, tip submerged in dense brown scales 2-5 × 1 mm acuminate with straight base, expanded in lower third, with few 2-cellular, glandular ? hairs along margins. Fronds 37-55 cm long, in young plants tufted in mature ones closely spaced in sets of two. Stipes on either side of stele, 6-7 cm long, basal 1-2 cm darker grey or mahogany, stramineous above, glabrous but for top 1 cm below rhachis, stout, curved white hairs ± 0.3 mm long, continue into groove of rhachis. Lamina glabrous, texture papery, with 4-10 (-20) pairs of free, subentire pinnae beneath well-marked terminal pinnatifid pinna; its c. 20 lobes cut from ½- ? to costa, its crenate tip this may be to 13 cm long. Pinnae to 7-9 × 1.2 cm., 17 segments distinguished by crenate lobes, margin of tip entire, lowest two pinnae decreased sequentially by ½ (specimen with up to 6 opposite-auriculate pinnae untypical). Rhachis adaxially grooved hairy, below prominent, rounded, glabrous. Costa just so. Costules, as veins, weakly prominent above and below. Veins in three pairs: 1 ½ pair join in long excurrent vein: first vein joins with that of the adjacent segment, giving rise to an excurrent vein which is met half way along or at sinus by second vein from basiscopic segment, second vein of acroscopic segment goes to margin next to sinus, veins of third pair go to margin. Sinus membrane distinct, sometimes prominent, ending in a little process. Sori 1mm in diam., medial, initiated on lowest vein, in juxtaposition to excurrent vein, appear to occur in pairs, but in segments of apical pinna up to 3-5 in a line. Indusium a thick weft with thinner fringe, rust -brown with white hairs. Amber-coloured glands associated with top of sporangial stalk just below annulus. Spore monolete, reniform., distinct point of attachment, verrucose ('ridged' according to Sledge), 0.03 × 0.015 mm.

D i s t r. Sri Lanka and (teste Holttum) Nicobar Islands.

E c o l. Not known; probably close to water source at moderate elevation.

S p e c i m e n s   E x a m i n e d. KEGALLE DISTRICT: " Kitool Galla", Ambagamuwa, small form of habit, *Wall* before 1873 (BM); Kitulgala, 150m., Sabaragamuwa Province, 1887, *Wall* (P, n.v.). LOCALITY UNKNOWN: *Thwaites C.P.3391* (BM.,CGE, K (errore 3390) lge. specimen; PDA).

HYBRIDS

Manton & Sledge, Phil. Trans. R. Soc. Lond. B, 238: 127-185. 1954. Panigrahi & Manton, J. Linn Soc.(Bot.) 55:729-743. 1958. Holtt., Kew Bull. 31: 295.1976. Sledge, W.A., Kew Bull. 34: 77-81. 1979. Sledge, W.A., Bull. Brit. Mus. nat. Hist. (Bot.) 8(1): 40-41. 1981.

According to Sledge (1981), cytological evidence proving occurrence of hybrids in wild populations of *Christella* was first demonstrated by Manton (1954). The number of such plants found after research in the field gave rise to suspicion that hybridisation was frequent in wild populations. Panigrahi & Manton (1958) studied particularly hybrids in the *C. parasitica* group, experimentally synthesizing them. They found that important distinguishing characters behave as simple dominants in F1 hybrids. Such hybrids may bear such close resemblance to only one of the parent species that recognition, in absence of cytological evidence, becomes difficult. Dominant characters were: creeping rhizome, non decrescent frond, subfoliar glands present. Holttum (1976) advised that more evidence was required to confirm dominance of non-decrescence, because in Thelypteridaceae decrescence is the most frequent mode of growth.

The characters: depth of cut of pinnae, number of anastomosing veins, hair length did not behave as simple dominants or recessives and F1 hybrids were intermediate in these. Difficulties originating from this prompted the advice that in herbaria frequently occurring plants with mixed characters are best regarded as hybrids.

When growing hybrids experimentally, Panigrahi and Manton's results showed that hybrids may be chracterized by a high proportion of abortive spores. Some hybrids generated in that study and maintained at Kew had sound spores in subsequent years. This fact illustrates that good spores do not disprove hybrid origin and can arise on hybrids in successive harvests.

Christella dentata × parasitica. *C. malabariensis* sensu Holttum, Kew Bull. 31: 317.1976, p.p., non *Nephrodium malabariense* Fée.

Frond outline of *Christella dentata*. Lamina short hairy beneath, without acicular hairs, some capitate hairs above. Pinnae, only one pair reduced and deflexed; pinnae with abundant glands below, resembling those of *C. parasitica*. Veins, one pair anastomosing. Indusium hairy.

N o t e. (Relating also to species): Some specimens which may be *C. parasitica* or *C. dentata* appear to form a varietal group, distinct by the top-half of each pinna rising in a very regular fashion, giving an over-all candelabra-like image or that of a scimitar if the individual pinna-pair is considered; this is here referred to as "scimitar form".

S p e c i m e n s   E x a m i n e d. MATALE DISTRICT: Bambaragala, road along Midland Tea Estate, steep slope with partly eroded soil cover, secondary montane forest above, 750m., 25.10.1995, *Shaffer-Fehre with Jayasekara & Ekanayake, 401* (K). KEGALLE DISTRICT: Ginigathena, 600 m., Dec. 1950, *Manton Z 30. Alston 11745* ex Trop. Fern House, Kew (BM). KANDY DISTRICT, Kandy, Lady Horton's Walk, 600 m., 7 Dec. 1950, *Sledge 509* (BM); Lady Horton's Walk, 24 Mar. 1954, *Sledge 1357* (BM). BADULLA DISTRICT, Pusella, depression below iron bridge, near shallow stream with boulders, 300m., 03 Dec 1995, *Shaffer-Fehre with Jayasekara & Samarasinghe 616* (K). NUWARA ELIYA DISTRICT, road to Kandapolle, ditch below cultivated ground, 1750m., 31 Oct .1995, *Shaffer-Fehre with Jayasekara & Samarasinghe 449* (K). GALLE DISTRICT, Panangala, steep herbaceous bank over road, 8 Nov 1995, *Shaffer-Fehre with Jayasekara 481, 482* (K); same district, Karelegama, in ditch along road side, 8Nov 1995, *Shaffer-Fehre with Jayasekara 483* (K).

Since *Christella dentata* and *C. parasitica* are the commonest species of *Christella* in Sri Lanka and India where they often grow together, this is almost certainly the most frequently occurring hybrid. Sledge prepared the above description (here paraphrased to suit our format) from the cultivated plant which, according to him., shows good evidence of each species; he discusses that the voucher also agrees with non-decrescence of frond being a dominant character (Panigrahi & Manton,1958), but disagrees with them in having good spores. Sledge judges that several forms of this hybrid exist and accepts a wider range, these are discussed in a paper in Kew Bulletin, 1979 and in Sledge, 1981.

Christella hispidula × parasitica

Lamina of *Christella parasitica* with only lowest pair of pinnae a little shortened; one pair of veins joining, second pair to margin well above sinus base; pinnae eglandular beneath, thinly hairy on costae, veins, indusia; long, acicular hairs absent. Sporangia mostly abortive; no good spores.

S p e c i m e n s   E x a m i n e d. Lady Horton's Walk, Kandy 600 m., Dec. 1950, *Manton P43*; 1974, *Alston 11742* ex Trop. Fern House Kew (BM).

This plant was a triploid. The frond taken by Alston from cultivated stock agrees well with Panigrahi & Manton's silhouette (loc. cit. fig. 4) of a synthesized hybrid of same parentage, save that lowermost pair of pinnae is somewhat shorter than next pair. Absence of subfoliar glands in the wild hybrid is presumably due to the *Christella parasitica* parent, an eglandular form of the species. Parents and *C. dentata* are frequent on the wooded hillside of Lady Horton's Walk but venation and triploid cytology would appear to rule out *C. dentata* as second parent.

Christella meeboldii × parasitica

Rhizome 'sub-erect to short-creeping' frond decrescent, 3-6 pairs of progressively reduced pinnae; these cut less than ½ way to costa, short hairy on costa; veins above and below, without acicular hairs have abundant glands (of *C. parasitica*) beneath, two pairs of veins joining (as in *C. meeboldii*); indusia glabrous or nearly so; spores ± verrucose or shortly ridged.

The mixed characters of this plant seem only explicable on the assumption of above parentage, and that non-decrescence of the frond is not a dominant character in this hybrid combination.

Typical *C. meeboldii* and *C. parasitica* were both collected here by Ballard.

S p e c i m e n s E x a m i n e d. NUWARA ELIYA DISTRICT: Hangiliella beyond Welimada on Badulla road from Nuwara Eliya, c. 900 m., 29 Dec. 1950, *Manton P. 307*; *Alston 11737* ex trop. Fern House, Kew (BM).

Christella subpubescens × dentata

This hybrid, prepared by Panigrahi in 1976 at Leeds (K), was not mentioned by Sledge (1981). Holttum (1976) appends the note "tetraploid" (the parental ploidy). The soral cover is fully developed, and basal acroscopic lobes of the ± 8 lower pairs of pinnae enlarged (*C. dentata*), lowest four pinnae well formed, gradually decreasing, widely spaced (*C. dentata*), the last three of these being auricles with enlarged, crenate acroscopic basal lobe (*C. dentata*). Pinnae with caudate tip (*C. subpubescens*), segments with truncate-falcate lobes cut ± 1/5 to costa. Two pairs of veins or even one of the third pair, occasionally joining (*C. subpubescens* 1, *C. dentata* 1.5-2). The indument too tends to that of the dominant species, *Christella dentata*, as would be expected from a frequently occuring taxon.

9. AMPELOPTERIS

Kunze, Bot. Ztg. 6 : 114. 1848. Type species: *Ampelopteris elegans* Kunze [= *A. prolifera* (Retz.) Copel.]. Sledge. W.A. Bull. Brit. Mus. nat. Hist. (Bot.), 8 (1): 13-14. 1981.
A monotypic genus distributed throughout the wetter parts of tropical and subtropical regions of the Old World.

Ampelopteris prolifera (Retz.) Copel. Gen. fil.: 143. 1947. (K p.p.!; LD). Manton & Sledge, Phil. Trans. Roy. Soc. London, B, 238: 127-185. 1954. Holtt., Blumea, 19: 25. 1971. Sledge, Bull. Brit. Mus. nat. Hist. (Bot.) 8(1): 1-54. 1981. Holtt. Fl. Males., Ser.II, 1(5): 387. 1982 "1981".

Hemionitis prolifera Retz., Obs. Bot. 6: 36. 1791. Type : Southern India
   Koenig (GOET, large fragm. at BM!, pinnule at K!)
Meniscium proliferum (Retz.) Sw., Syn. fil.: 19. 207. 1806; K. Iwats., Mem.
   Coll. Sci. Kyoto Univ. B, 31 : 196. 1965.
Polypodium proliferum (Retz.) Roxb. ex Wall., num. list. 312. 1829. nom.
   illeg., non Kaulf. 1824, nec Roxb. 1844; Thwaites, Enum. Pl. Zeyl. 439.
   1864 (as to name only, not as to syn. *Meniscium thwaitesii* Hook.).
Goniopteris prolifera (Retz.) C. Presl, Tent. Pterid.: 183.1836; Bedd., Ferns
   South. Ind.: t.172. 1864; Bedd., Handb. Ferns Brit. Ind.: 296, fig.
   153.1883. Trimen, J. Bot. 23: 139. 1885.

Rhizome creeping, 0.4cm in diam.; scales scarce, lanceolate, 1mm long ×
0.5mm wide. Fronds aggregate, lax, dimorphic: paripinnate fronds: of indefinite apical growth (more than 2m), new plant where apex touches soil;
imparipinnate fronds: up to 30cm long; terminal pinna (11cm) larger than
lateral ones (7cm); stipes 16-55cm long, 0.1-0.4 in diam.. Buds present in
axils of most primary pinnae in both forms of frond, giving rise randomly to
tufts of secondary fronds. Lamina paripinnate, elliptic in lower half, pinnate,
with 15-20 pinna pairs of evenly in decreasing lengths, pinnae in top half
abruptly decreasing in size, texture herbaceous to subcoriaceaous; pinnae generally subsessile truncate base apex acute, margins crenate or subentire, obtuse in shortened distal pinnae. Lamina imparipinnate, broadly deltoid, with
5-6-(10) pinnae pairs, lateral pinnae to 15cm long, 1-1.75 cm broad, subsessile
with truncate base and acute apex, margins crenate or subentire, free apical
lobes up to 2mm. Rhachis abaxially rounded, smooth;with unicellular hair,
occasionally forked, dense in adaxial groove; that groove in costa deepening
towards truncate base of pinna, abaxially slender but prominent, costules
adaxially visible as very fine lines. Veins of paripinnate form: 3-4 pairs of
veins form sinuous excurrent vein; of imparipinnate form: up to 10 pairs,
4-6 forming sinuous excurrent vein, free veinlets in lobes reach margin. Hairs
acicular, white, 0.3 - 0.5 mm long on costa and costule of abaxial pinna. Sinus
membrane not apparent. Sori exindusiate, medial, round or ± elongate, smallest at tip of pinnule, largest over veinlet close to costa, here often confluent
with neighbouring sorus; line of sori producing garland effect. Stalks of sporangia bearing hairs with terminal glandular cell.Cytology: n=36

D i s t r. Generally widespread in the wetter tropics and sub-tropics of the
Old World from Africa to New Caledonia, rare in Sri Lanka.

E c o l. In marshland, or amongst boulders by river, prefers open situations
or light shade. All localized records are from the 'dry zone' east of the central
massif, and at low elevations.

N o t e. Sledge (1981) notes that the shorter pinnae of paripinnate frond
appear always to be sterile; the majority of primary fronds are paripinnate,
larger ones fertile; tufts of fronds arising secondarily from axillary buds

imparipinnate in all samples seen; both forms may occur as primary fronds on the same rhizome.

S p e c i m e n s   E x a m i n e d. TRINCOMALEE DISTRICT: Kannia, hot wells, Dec. 1885, *Ferguson s.n. in*1885 (PDA); hot wells close to Trincomalee, *Freeman, 295, 296* (BM); Kannia hot wells near Trincomalee, in marsh, 16 Jan. 1951, *Sledge 879* (BM); Trincomalee, hot wells, in wet ground in immediate vicinity of springs, 16 Jan. 1951, *Ballard 1487* (K). MATALE DISTRICT (formerly Tamankaduwa District): Near Matale boundary, *Nevill s.n.* in 1893 (PDA). POLLONARUWA DISTRICT: Minnery, *Beckett 1131* (PDA). BADULLA DISTRICT: Uma-Oya, in water, *s.coll. s.n.*, in 1881 (PDA). LOCALITY UNKNOWN: *Thwaites C.P. 3916* (PDA).

10. ᛁ ᛌRONEPHRIUM

C. Presl, Epim. Bot.: 258. 1851. Epimeliae botanicae *In* Abh. Kön. Böhm.Ges. Wiss.ser.5, vol. 6. Holttum in Blumea 19: 34. 1971; 20: 105. 1972.

Caudex suberect to long-creeping; fronds simply pinnate with subentire pinnae and terminal pinna-like segment, the basal pinnae not reduced; most pairs of veins anatomosing to form excurrent vein between costules; lower surface of pinnae often pustular when dry, acicular or round-hooked hairs frequent on one or both surfaces; sori indusiate or exindusiate and then spreading along veins, sporangia often bearing hairs or glands near annulus ; spores with a continuous wing and a few cross wings. n=36

About 60 species ; India and Sri Lanka, southern China southwards throughout Malesia; north-eastern Australia and Pacific Islands.

Holttum divided the genus in to three sections. The smallest of these, Sect. *Grypothrix* Holtt., is distinguished by round-hooked hairs on pinnae and sporangia *P. thwaitesii* and *P. triphyllum* belong here. They differ distinctly from *P. articulatum* and *P. gardneri* which are each placed in different sections.

Type species : *Pronephrium lineatum* (Blume) C. Presl [based on *Aspidium lineatum* Blume]

KEY TO THE SPECIES

1 Lamina oblong with more than 8 pairs of free pinnae; sporangia without hairs
  2 Up to 20 pairs of pinnae, their margins shallowly cut ¼ way to costa, 3-5
    pairs of veins form connivent anastomoses; indusia large, glabrous . . . .
  . . . . **1. P. articulatum**
  2 10-12 pairs of pinnae, their margins coarsely crenate, shortly ciliate; 7
    pairs of veins form connivent anstomoses; exindusiate . . . .
  . . . . **2. P. gardneri**
1 Lamina deltoid with less than 6 pairs of free pinnae; sporangia setose with hooked hairs

3 Lamina deltoid with 3-5 pinnae, terminal pinna entire, larger than lateral ones; elongate sori occupy whole length of veins, all fusing with neighbouring sori
· **3. P. triphyllum**
3 Lamina subdeltoid with 2-4 pairs of free pinnae pairs below pinnate or pinnatifid terminal pinna; sori circular, median, fusing with neighbouring sori only in lowest pinnae · **4. P. thwaitesii**

1. Pronephrium articulatum (Houlston & T. Moore) Holttum, Blumea 20: 116. 1972. Sledge, W.A. Bull. Brit. Mus. nat. Hist. (Bot.), 8 (1): 45.1981.

Nephrodium articulatum Houlston & T. Moore, Gdnrs'Mag. Bot. Hort. Flor. Nat. Sci.: 293. 1851. Type: cult. Hort. Bot. Kew ex Ceylon, 1845, *Gardner1104* bis (BM!; CGE; K!).

Nephrodium pennigerum Baker in Hook. & Baker, Syn. Fil. 2 ed.: 292. 1874 p.p.; C.B. Clarke, Trans. Lin. Soc. (Bot.) II,1:532. 1880, quoad plantae Zeylanicae; Bedd., Handb. Ferns Brit. India: 277. 1883 p.p.; id Handb. Suppl.: 73. 1892.

Dryopteris megaphylla sensu Christ, Index Fil.: 277. 1906 p.p. quoad plantae Zeylanicae, non *Aspidium megaphyllum* Mett.

Nephrodium abruptum sensu Hook., Spec. Fil. 4: 77-78, t.241B. 1862 p.p.; Bedd., Ferns S. India: 31, t.86. 1863, non *Aspidium abruptum* Blume. *Aspidium abruptum* sensu Thwaites, Enum. Pl. Zeyl.: 391. 1864, non Blume.

Rhizome short-creeping, (may sometimes appear erect). Stipes to 50cm long, sometimes longer, with scattered broad, thin brown scales especially at base, sparsely hairy throughout. Lamina 50-80-(100) cm., narrowly lanceolate base truncate, pinnate up to 20 pairs of patent pinnae below terminal one; pinnae up to 20 × 3 cm., apex acuminate, base narrowly cuneate in basal pinnae which are not reduced; pinnae sessile, acroscopic base of lamina slightly adnate, increasingly so towards acuminate pinna-like tip; base of pinnae gradually changing throughout lamina from narrowly cuneate at base to truncate towards tip; lobes of margins becoming slightly falcate only in largest pinnae, generally rounded, very even over ¾ length of pinna, top ¼ of pinna attenuating abruptly into acuminate tip with simple margin. Rhachis minutely pubescent throughout, more densely in adaxial groove, together with minute scales (0.75 mm long, 0.3 mm wide). Costa with 10-12 veins, 6-7 pairs fuse with excurrent vein the rest free, reaching margin. Sinus membrane prominent. Sori medial, round (1 mm in diam. at base, 0.5 mm at pinnule tip). Indusia glabrous. Spores monolete, ovate, translucent with few dark brown ridges.

D i s t r. In India, Sri Lanka, Bangladesh, Burma, northern Thailand and western China.

E c o l. In the forests and at road sides. Rare! Central Province between 600-1250m.

N o t e. Where fertile fronds are not entirely mature, sori will arise gradually, starting at top of frond and, in turn, at tip of pinna so that largest pinnae at base might still be without sori when tip of frond is fully covered.

S p e c i m e n s   E x a m i n e d KANDY DISTRICT: Kandy, *Blume 1808* (BM); Kandy, Lady Horton's Walk, 610m., 16 Feb. 1954, *Sledge 1144* (BM); Kandy, Lady Horton's Walk 25. March 1954, *Sledge 1363* , *Trevor Walker T894* (BM). RATNAPURA DISTRICT: *Thwaites C.P. 3271* (BM.,CGE, K, PDA: Haldunmulle, April 1856. NUWARA ELIYA DISTRICT: *Thwaites C.P. 3271* (BM.,CGE, K, PDA: Wattegodde 1856. LOCALITY UNKNOWN: *Gardner*, anno 1845, cult. RBG, K 1851 (BM); *Randall 1868* in herb. Rawson 3243 (BM); 1884, *Wall* (PDA); *Mrs Chevalier 1887* (BM); *Robinson 155* (K).

2. Pronephrium gardneri Holttum, Kew Bull. 26: 81. 1971.
Sledge, Bull. Brit. Mus. nat. Hist. (Bot.), 8 (1): 46-47. 1981. Type: Sri Lanka, *Gardner 1137* (K holo.!)

Goniopteris urophylla sensu Bedd., Ferns S. India : t. 239. 1864, non C. Presl
Polypodium granulosum sensu Thwaites , Enum. Pl. Zeyl.: 394.1864 p.p. non C. Presl
Nephrodium urophyllum sensu Bedd., Ferns S. India Suppl.: 18. 1876, non (C. Presl) Keyserl. Handb. Ferns Brit. India : 274. 1883, p.p.

Illustration. Bedd. Ferns South. India, t. 239. 1863 [based on *Thwaites s.n.* in Herb. Bedd. (K!)]

Rhizome short-creeping. Fronds 70 cm or longer, spacing not established. Stipe 50cm or longer, stramineous 5mm in diam. at base, glabrous, save in groove; scales in basal 3 cm; scales brown, various, discontinuous m sizes: c. 2×0.03 mm., c. $3-5 \times 2$ mm., c.$8-10 \times 0.03$ mm. Lamina narrowly oblong, shortly acuminate at apex, base truncate or broadly cuneate with 10 -12 pairs of pinnae; on lower surface, in vicinity of costule a massing of green-translucent pustules in some specimens. Pinnae narrowly oblong, shortly (2-3 cm) acuminate at apex, base truncate or broadly cuneate, lowest pinnae narrowed toward base; very shallowly-lobed margins, crenate,minutely ciliate. Rachis 5 mm in diam., glabrous, hirsute in adaxial groove. Aerating tissue at base of costae. Costae stramineous, prominent on abaxial side, minutely pilose. Costules brown, 4.5-5.5 mm apart. Veins from 10 to 11 pairs, slightly raised, up to 9 pairs meet excurrent vein which has many apparent discontinuations, free vestige of excurrent vein always points towards margin; two pairs of veins join sinus membrane, three or four free veins reach margin. Sinus membrane prominent. Sori medial, round, exindusiate. Spores monolete, kidney-shaped, translucent with few dark brown ridges 0.03-0.04 mm long, 0.02 mm wide, its translucent wing with minutely erose margin.

D i s t r. Endemic, in forest near Kandy (Udawatakelle).

E c o l. In forests at moderate altitude (c. 600 m).

S p e c i m e n s   E x a m i n e d. KANDY DISTRICT: *Macrae 863* (Superintendent PDA, 1827-1830); Hantane Range, in forests, August 1844, *Gardner 1137* (CGE; K holo! 2 sheets; 3rd sheet stored with type is 'type of tab 239' from Beddome's Fern Herbarium); Oodawella forest, uncommon, terraneous *Wall*, 48/65, before 1873 (3 sheets, 3/3 with rhizome) (BM). LOCALITY UNKNOWN: *Thwaites C.P. 3063* (K).

3. Pronephrium triphyllum (Sw.) Holttum, Blumea 20: 122.1972. Sledge, Bull. Brit. Mus. nat. Hist. (Bot.), 8 (1): 47. 1981.

Meniscium triphyllum Sw. J. Bot. Goettingen 1800 (2): 16. 1801; Hook & Baker Syn. Fil.: 39. 1868. Bedd., Handb. Ferns Brit.India: 397, t. 231. 1883. Type: no locality or collector (S-PA).

Rhizome long-creeping, branching, rarely upright, 2 mm in diam., with shallow grooves, covered in minute, stiff hairs and some narrow lanceolate scales which increase in number towards apex, covering base of young crozier. Scales brown, narrowly lanceolate 4 × 0.3 mm., covered on both sides with two-cellular hairs; radial walls of small rectangular to polygonal cells of similar dimensions throughout. Fronds at irregular intervals of 1.5-4cm fertile frond, has smaller blade and almost always overtops sterile one. Stipe 1/10th to almost longer in fertile frond, grey, flattened and 2 mm wide over ca 1.5 cm of its base, above stramineous, 1 mm in diam., in dried specimens twisted with several grooves, described as trigonal for living plants; moderately hirsute with very short, stiff hairs, bearing few remaining scales of those described above. Lamina subcoriaceous, deltoid, with three to five pinnae. pinnae on fertile fronds more slender than on sterile ones, terminal pinnae are largest ones in all cases : fertile lamina: terminal pinna 10 × 2.4 cm., basal pinnae 4.5 × 1.5; sterile lamina: terminal pinna 13.5 × 3.5 cm., basal pinnae 6 × 2.5cm; most pinnae become abruptly acuminate in top 1/6 of their length, at their base they may be lanceolate, cuneate or truncate with all stages in between. Pinnae alternate or opposite; lateral pinnae shortly pedicillate or sessile, terminal pinnae always supported by short rhachis. Rhachis 1-2cm., densely hirsute with short, stiff hairs, these also cover pedicel where present. Costae and costules abaxially raised, adaxially with faint grooves; abaxially moderately (sterile) or densely (fertile) hirsute, adaxially more sparsely hirsute with minute stiff hairs. A moderate number of small yellow glands towards abaxial base of fertile pinnae. Veins from costules alternate or opposite (middle of pinna); 7-10 pairs join excurrent nerve which forms series of pale lines in dried specimens; the last two or three free pairs of veins reach wavy and partially simple margin, sinus visible on margin, veins in acuminate tip of pinnae frequently anastomose. Sinus membrane not apparent. Sori elongate, fusing with next sorus forming wave-like pattern along longitudinal axis of pinna, exindusiate. Sporangia with round-hooked hairs at face/annulus margin 0.12 mm long. Spores monolete, dark translucent with dark irregularly connecting ridges on face membranous fringe 0.03 mm long, 0.02 mm wide.

D i s t r. India, Sri Lanka, Burma, Thailand, Malesia, China, Japan, Taiwan, Philippines, Australia (Qld.).

E c o l. Damp, shady places at low to moderate elevations.

S p e c i m e n s   E x a m i n e d. MATALE DISTRICT: Matale East, Dorsomundella, *Wall* (K); Puwakpitiya Pass, Matale, Dec. 1860 (BM). KANDY DISTRICT: Hunnasgiriya, April 1857; Awagala Estate, 19.Feb. 1954, *T.G. Walker T474* (BM). GALLE DISTRICT: Hinidoon, *Thwaites C.P. 1293* (CGE, K, PDA) RATNAPURA DISTRICT: Pelawatta, 20 Jan. 1951, edge of pass leading to Pahala Hewessa , *Ballard 1530* (2 fine sheets) (K); Gongalla Hill, 11. March 1954, *Sledge 1261* (BM); Gongala near Bulutota, alt. up to 1070 m., 16 March 1954, *T.G. Walker T758* (BM). LOCALITY UNKNOWN: *Wall* before 1873 Herb (BM); *Beddome* 1885 (BM); *Walker* (K).

4. Pronephrium thwaitesii (Hook.) Holttum, Blumea 20: 122. 1972. Sledge, Bull. Brit. Mus. nat. Hist. (Bot.), 8 (1): 47-48. 1981.

Meniscium thwaitesii Hook., Fil.exot. sub t.83. 1859. Thwaites, Enum. Pl. Zeyl.: 382. 1864. Bedd. Ferns Brit. India: 399,fig. 323. 1883. Type: Sri Lanka, *Thwaites C.P. 3145* (K !).

Rhizome long-creeping, 2.5-3mm in diam., with shallow grooves, minutely pubescent, glabrescent, branching occasionally. Fronds dimorphic: fertile fronds, with smaller, more narrow lamina, overtop sterile ones e.g. 58:34 cm; sterile fronds spaced ± evenly 1-1.5cm apart. Stipe (sterile) 18.5-20cm long, (fertile) 28cm-35cm long, 0.8-1.5mm in diam. with an adaxial groove, pale brown pubescent just below rhachis, base bent and twisting towards meristem., baseal 1 cm., greyish and with few scales above, stramineous, glabrescent. Scales along lowest 2cm of stipe 1.5mm long , 0.2mm wide. Lamina (sterile), 15-25cm long, 13-15cm wide at truncate base, of subdeltoid contour; pinnae mostly at 90° to rhachis 2-0.5cm apart, free below dominant pinnatifid tip of 9-12 fused pinnules (accounting for- 3/5-2/3 length of lamina) lamina decurrent below tip; margin of tip altering from deeply incised, falcate to crenate and entire, acuminate end.; acroscopic lobe at base of pinnule pronounced; base of pinna cuneate to truncate, tip acuminate to blunt. Pinnules defined by crenate margin (2-3 lowest pinnae) or by costules below entire margin; lamina (fertile) 2-3 lowest pinna pairs with crenate margins, distincly larger, 4.5-9cm long, 1-1.6cm wide ascending (c. 60°), narrowly triangular or expanding in the middle; above 6-12 pedicillate to adnate pinna pairs with entire margins, almost horizontal, c. ½-¹/₃ the length and width of basal pinnae below pinnatified pinna-like tip, that accounts for $^1/_3$ - $^1/_2$ of the length of the lamina; Rhachis with adaxial groove occupied by short ascicular hairs, abaxially rounded, prominent. Costae and costules abaxially round (with sometimes almost imperceptable grooves in herbarium specimens). Indument moderate: ascicular hairs on costa and costule of lower surface, almost glabrous above. Short stout hairs on abaxial, rounded costa of fertile frond. Up to four veins may join in excurrent vein. Sinus membrane distinct in most mature pinnae. Sori spreading along veins, merging with adjoining sorus, appear to run in four waves or fewer parallel to costa. Exindusiate. Sporangia with one or two slender, thick-walled, septate hairs, 0.15 mm long, along annulus / face-margin, tip acuminate, slightly curved or top ? bent into shallow crook. Spores monolete, ovate, translucent with few dark brown ridges.

D i s t r. In Southern India (Nilgiri Hills) and Sri Lanka in Forests of Central Province.

E c o l. In forests and at disturbed road sides from 900-1200m.

S p e c i m e n s   E x a m i n e d. LOCALITY UNKNOWN: *Thwaites C.P. 3145* (CGE; K, holotype !; PDA); *Robinson 39a* (K); Beddome collection (K); *C.B. Clarke 10/77* (K); *Wall* reed *11/81* (K); *Skinner* (K). Thomas Moore's fern herbarium (K); Sri Lanka, ex herb. Hance, *Bradford 129* (BM).

11. PSEUDOCYCLOSORUS

Ching, Acta Phytotax. Sin. 8:322. 1963. Emend, Holttum, J. S. Afr. Bot. 40: 137. 1974.

Caudex erect or short creeping; fronds bipinnatifid with abrupt transition at the base to numerous small pinnae and often reduced to small tubercles; aerophores present at base of lower and reduced pinnae; upper surfaces of costa grooved; veins free, usually raised below, the acroscopic basal one passing to the base of the sinus or the two veins converging but not fusing at the sinus base; surfaces glabrous or with acicular hairs, eglandular; sori indusiate, sporangia without hairs or glands but usually with a septate hair on the stalk. n=36 (35 also reported).

About 12 species; three in Africa and nine in tropical and subtropical Asia to Japan and Luzon. (Sledge, 1981)

Type species: *Pseudocyclosorus tylodes* (Kunze) Ching

Pseudocyclosorus tylodes (Kunze) Ching, Acta Phytotax. Sin. 8: 323. 1936 [as 'xylodes']. Spore morphology in Thelypteridaceae I: Pseudocyclosorus, Grimes, Kew Bull. 34(3) 517-520. 1979. Sledge, Bull. Br. Mus. nat. Hist. (Bot.), 8 (1): 25-26. 1981. Manickam & Irudayaraj, Pter. Fl. Western Ghats : 175. 1992.

Aspidium tylodes Kunze, Linnaea 24: 244, 283. 1851 ['xylodes' loc. cit.: 281]; Thwaites, Enum. Pl. Zeyl. : 391. 1864. Type: India, Nilgiris, *Schmid* (B).

Lastraea tylodes (Kunze) T. Moore, Index Fil. :107. 1858.
Lastraea ochthodes var. *tylodes* (Kunze) Bedd., Ferns S. India: t.107. 1863.
Nephrodium prolixum var. *tylodes* (Kunze) Baker, *In* Hook. & Baker, Syn. Fil.:268. 1867. Handb. Ferns Brit. India: 240. 1883. Index Fil., suppl. 3: 102. 1934 [as 'xylodes].
Thelypteris tylodes (Kunze) Ching, Bull. Fan meml Inst. Biol. (Bot.) 6: 296. 1935 [as 'xylodes]. Copel., Fern Fl. Phillip.: 330. 1960 [as 'xylodes'].

Illustrations. Beddome, Ferns South. India, t. 107. 1863 (as *Lastrea ochthodes* var. *tylodes*). Manickam & Irudayaraj, Pter. Fl. Western Ghats, t. 133. 1992.

Rhizome erect. Fronds tufted, base of tuft ± 3cm wide. Stipe up to 50cm long, swollen base deeply furrowed, adaxial groove, glabrous except for ovate brown scales near base, scales 1-1.5mm long, 0.75-1mm wide, tip rounded; pneumatophores can extend to just above swollen base of stipe, 10-12 tuberculate pairs several cm apart, discrete, alternating, ca 1cm apart (see costa).Lamina oblong-lanceolate 40-60 (-100)cm long, 10-20cm wide, with ca 25 pairs of alternate horizontal to ascending pinnae, pinnatifid end portion with denticulate tip; texture subcoriaceous, basal pinnae decrescent; lower pinnae abruptly reduced to tubercles on stipe, ± black when dry. Pinnae free, sessile, adnate towards tip of lamina, pinnae elongate-oblong, acuminate, in large fronds lowest pinnae slightly contracted towards their base; pinnae 1.4-2cm wide with c. 35-40 alternate or opposite lobes two basal lobes larger; lobes elongate-rounded to slightly triangular-falcate sparcely hairy margin recurved,, deeply cleft ¾ to costa, sligtly more than ½ towards tip, margin of acuminate tip crenulate to entire. Rachis prominent, glabrous, pale, ± qadrangular in transverse section of dried specimens, here all sides, the lateral included, collapse and become grooved, internodes more narrow than nodes (i.e.3 mm as compared to 4mm). Costa prominent, round to triangular below, grooved, hairy above, arising on rachis from centre of a dark brown aerophore (compare stipe). Costules and veins prominent on both sides, pale. Veins free, prominent on both surfaces lowest of c. 10 pairs connivent but not fusing at sinus all others go free to margin. Sinus membrane distinct. Sori 0.3-0.5 mm in diam., inframedial, basal pair alone medial, 0.75mm in diam.. Indusium persistent, glabrous, dark reddish-brown when dry, attached to vein by margin closest to costule folding over swelling sorus, horse shoe-shaped. Spores monolete, oval 0.05mm × 0.025mm., brown, spinulose.

D i s t r. Southern India and Sri Lanka.

E c o l. In mountain forests of the Central Province above 1000m.

N o t e. Very large plants may have pinnae up to 30cm long but 10-15cm is a normal size. A small form is occasionally found in which the whole frond scarcely exeeds 30cm and the pinnae are then 4-6cm long. Sledge's specimens (nos. *551 & 571*, BM) from Corbet's Gap are such forms and are very closely matched by Gamble *12122 & 15317* (K) from the Nilgiris and by two sheets from Sri Lanka at Paris. These small forms bear a superficial resemblance to *Trigonospora ciliata* and have sometimes been misidentified as that species, although the presence of abortive pinnae on the stipes and the prominent raised veins are sufficient to distinguish them. (Sledge 1981). Holttum (1974) also discusses the two spellings of our species name: *xylodes* and *tylodes*; he opines that *tylodes* (tylo- Greek: with knobs, lumps or projections) is the spelling which Kunze "almost certainly intended". *P. ochthodes* is not a Sri Lankan fern.

S p e c i m e n s   E x a m i n e d. MATALE DISTRICT: Riverstone Estate, Central Province, 1100m., 19 Jan. 1977, *Faden 77/183* (K); forested stream gulley beside and below road, 1 km W. of top pass between Rattota and Laggala, N.E. of Matale, Alt. c. 1200 m., 15. Sept. 1993, *Fraser-Jenkins* with*Jayasekara & Bandara 130* (K); Rattota, Medawatta road along Midland Tea Estate, at 19/10 km marker, lower secondary montane forest, in coarse sand at foot of lichen-covered rock, 750m., 25 Oct 1995, *Shaffer-Fehre with Jayasekara & Ekanayake, 406* (mixed collection) (K); Rattota, Riverstern, mountain forest above bubbling brook, granite not exposed *418* (K). KANDY DISTRICT: Corbet's Gap, by track, through jungle, 1200-1300m., 9 Dec. 1950, *Sledge 551, 571* (BM); Rajamalle Division of Moray Tea Estate, heights above fishing hut, bare rocks of quarry with water flowing into pool below, c.1600m., 19 Nov 1995, *Shaffer-Fehre with Jayasekara & Samarasinghe 562* -small form- (K); Road to Gartmore Tea Estate, bare rocks with stream., 1600m., 19 Nov 1995, *Shaffer-Fehre with Jayasekara & Samarasinghe 568* (K). BADULLA DISTRICT: Road from Nanumukula via Spring Valley Tea Estate, in rocky side valley smothered with vegetation of angiosperms, 1300m., 04 Dec 1995, *Shaffer-Fehre with Jayasekara & Samarasinghe 628* (K); road from Bandarawela to Haputale, steep slope with marsh above road, 05 Dec 1995, *Shaffer-Fehre with Jayasekara & Samarasinghe 635* (K). CENTRAL PROVINCE: common in the higher forests, terraneous, *Wall 44/53* (BM). NUWARA ELIYA DISTRICT: Nuwara Eliya, shady woods, Sept. 1844, *Gardner 1108* (CGE; K); Ramboda, *Beckett* (BM); Hakgala, stream banks in jungle, 28 Feb. 1906, *J.C. Willis* (PDA); Ramboda, bank by road side, 1575m., 17 Dec. 1950, *Sledge 659* (BM); between Pattipola and Horton Plains, by stream in jungle, 19 Dec. 1950, *Sledge 688* (BM); Horton Plains, by stream in jungle, 2000 m., 20 Dec. 1950, *Sledge 688* (BM); Ramboda, Katukitula culvert, sheer rock face, moist, with water runnels trickling, overhung by lavish vegetation, 860m., 31 Oct 1995, *Shaffer-Fehre with Jayasekara & Samarasinghe 437* (K); road from Kandapola, to Ragala at 12/4km marker, on grassy slope among herbaceous vegetation, 1725m., 31 Oct 1995, *Shaffer-Fehre with Jayasekara & Samarasinghe 451* (K); road from Rendapola to Ambewala, over ditch at bottom of grassy slope with scrub and legumes, 1500m., 01 Nov 1995, *Shaffer-Fehre with Jayasekara & Samarasinghe 454* (K); Horton Plains, Dayagama road, at road side in mixed primary forest disturbed by road, in deep shade, 1900m., 01 Nov 1995, *Shaffer-Fehre with Jayasekara & Samarasinghe 463* (K). LOCALITY UNKNOWN: *Thwaites C.P. 1361* (BM.,CGE,K, PDA); *Freeman 236, 237* (BM); *Robinson 160* (K); *Walker* (K).

12. PSEUDOPHEGOPTERIS

Ching , Acta Phytotax. Sin. 8: 313. 1963.
Holttum, Blumea 17: 12. 1969. Sledge, Bull. Br. Mus. nat. Hist. (Bot.), 8 (1): 5-6. 1981.

Dryopteris subgen. *Phegopteris* C. Chr. Ind. fil. XXI.1905. p.p.
Thelypteris subgen. *Phegopteris* Ching 'group 4' Ching, Bull. Fan. Inst. Biol.VI: 246. 1936.
Type species: *Pseudophegopteris pyrrhorachis* (Kunze) Ching.

Holttum recognizes 20 species, mainly from south-east Asia. These may have different levels of ploidy. Rhizome short- to long-creeping to erect. Scales at base of stipe cordate to lanceolate, only long attenuated tip free of surface hairs. Stipe and rhachis of varying red shades of burnished mahogany, glossy, only apparently glabrous, blade usually bipinnate, in part bipinnate-pinnatifid. Sori mostly round, ex-indusiate. Typically in mountain habitat. (*Macrothelypteris*, with which it was grouped at some time grows at low altidtudes.)

Represented in Sri Lanka by *Pseudophegopteris pyrrhorachis* (Kunze) Ching

Pseudophegopteris pyrrhorhachis (Kunze) Ching , Acta Phytotax. Sin. 8: 315. 1963.

Illustration. Manton & Sledge, Phil. Trans. R. Soc. B.238: 162, fig 13. 1954.

Polypodium pyrrhorhachis Kunze, Linnaea 24: 257. 1851. Type: India, Nilgiris, *Weigle-Schaeffer 6* (? B, n.v.).
Lastrea pyrrhorhachis (Kunze) Copel., Gen. Fil.: 139. 1947 p.p.
Macrothelypteris pyrrhorhachis (Kunze) Pic.Serm. in Webbia 24: 716. 1970.
Polypodium distans D. Don, Prod. Fl. Nepal.: 2. 1825, non Kaulf. 1824.
    Baker in Hook.& Baker, Syn. Fil.:308. 1867, p.p. Type: Nepal, *Wallich* (n.v.).
Phegopteris distans (D. Don.) Mett., Abh. Senckenb. naturforsch. Ges. 3: 16. 1858. Bedd., Handb. Ferns Brit. India: 292.1883.
Nephrodium microstegium Hook., Spec. Fil. 4:119, t.250. 1862. Type: India, Khasya, *J.D. Hook. & Thomson* (K).
Lastrea microstegia (Hook.) Bedd., Ferns Brit. Ind. : t.39. 1865.

Polypodium paludosum sensu Bedd., Ferns South. Ind. : t. 168. 1863, non
  Blume 1829.
Thelypteris paludosa sensu K. Iwats., Acta phytotax. geobot. Kyoto 19: 11.
  1961. Mem. Coll. Sci. Kyoto Univ. B, 31 : 139. 1965 p.p. non
  Polypodium paludosum Blume

Illustrations. Hook. Spec. Fil. 4:119, t. 250. 1862 (as *Nephrodium micro-*
  stegium). Bedd. The ferns of Southern India, t. 168, 1863 (as *Polypodium*
  paludosum). Bedd., The ferns of British India, t. 39, 1865 (as *Lastrea*
  microstegia (Hook.) Bedd. Manickam & Irudayaraj, Pter. Fl. Western
  Ghats, t. 128, 1992.

Rhizome short-creeping to suberect. Fronds to 120cm long, closely spaced
or tufted; texture herbaceous or coriaceous. Stipes 20-60cm long, base
stramineous, grey when mature, reddish-brown above; scales scarce [4] or
moderately frequent [6], scales ovate 1× 0.5 mm to narrowly lanceolate 5mm
× 1mm., dense on crozier 10 × 1.5mm., transparent, cell walls reddish-brown,
thin cover of superficial and sometimes marginal hairs save on acuminate
apex. Lamina narrowly deltoid reducing gradually towards pinnatifid apex,
pinnate to pinnate-pinnatifid, c. 22-(35) pinna pairs; largest pinnae 11 × 2cm
[4] or 15 × 4cm [6], individual pinnae with truncate base, attenuate tip, lower
½ of basiscopic side of pinna noticeably broader than corresponding acroscopic
part; 1or 2 basal pinna pairs remote, reduced to half size; pinnules sessile,
separated by incision ± down to pinna-rhachis, here sometimes connected by
very narrow wing. Costules in largest pinna 5-7mm apart [4]; or pinnules
pinnatifid throughout; sinus 1-2mm above pinna rhachis, costules in largest
pinna 7-9mm apart [6]; pinnule margins crenulate, or slightly serrate along
lowest margin on basiscopic side of pinna; pattern rarely extends to acroscopic
side of pinna. Rhachis reddish-brown, increasingly tomentose towards apex;
costae prominent abaxially, slightly raised adaxially in both forms with coarse
pale brown hairs [4] or only white, acicular hairs [6]; hairs adxially appressed
± dense, parallel to costa / costule stouter than acicular hairs on abaxial side,
the latter less dense, at 90° to costa or costule. Veins obscure above, a little
raised below. Sinus membrane not apparent. Sori one in each of lower 4-6
lobes -preferentially on acroscopic side if not all lobes covered. Indusia membranous, very small, difficult to see.

D i s t r. Southern India and Sri Lanka, northern India from Kashmir to
Assam., west China, and Vietnam (Tonkin).

E c o l. In marshy habitat, along stream banks and road side clearings, at
altitudes of 1500-2250m Distinct representatives of different ploidies (2,4,6)
found in the field. A frequent species at high elevations, growing in the open
or in lightly shaded places. Readily recognized by its ruddy -coloured rhachis
and opposite pinnae with basal pinnules contiguous with or overlapping frond
axis.

N o t es. Sledge (1981) draws attention to a tetraploid and a hexaploid cytotypes which exist in Sri Lanka. The tetraploid, apparently more frequent, differs from hexaploid in its somewhat narrower and more deeply lobed pinnules which, in middle and distal pinnae have more acute apices. "According to observations of the present author, a coriaceous texture of the lamina, often confluent sori of a larger diam. and innumerable sporangia are correlated with tetraploidy; a thinner, more herbaceous texture of the lamina and very small discrete sori seemingly 'immature' (4 to 8 sporangia), are correlated with hexaploidy" Holttum (1981) refers to a diploid form with "quite distinct fronds".

S p e c i m e n s   E x a m i n e d. BADULLA DISTRICT: Blackwood forest, Welimada-Haputala road , by stream., c. 1500m., 18 Nov. 1976, *Faden 76/345* (K); road from Bandarawela to Haputale Katagalla Estate, 1400m., 5 Dec 1995, below tea slopes on 2m high grassy bank of stream near gully, *Shaffer-Fehre with Jayasekara & Samarasinghe, 641* (K). NUWARA ELIYA DISTRICT: Nuwara Eliya, moist open places, Oct.1845, *Gardner 1151* (CGE; BM; K); Nuwara Eliya, 10 May 1906, *Matthew 637* (K); Nuwara Eliya, *Freeman 287,288, 289, 290* (BM); Nuwara Eliya, 21 Jan. 1908, *Bicknell* (P); Adams Peak, 14 Feb 1908, *Mathew, s.n.* (K); Nuwara Eliya, *Bradford* (CGE); Adam's Peak, 1950m., 14 Dec. 1950, *Sledge 621* (prob. diploid) (BM); by track through jungle near Horton Plains, 2100m., 19 Dec. 1950, *Sledge 681*(hexaploid) (BM); Moon Plains, Nuwara Eliya, 1800m., 23 Dec. 1950, *Sledge 712* (tetraploid) (BM); Moon Plains, Nuwara Eliya, 1800 m., 23 Dec 1950, *Ballard 1198* (K); Forest above Le Vallon tea estate, 1350m., 9 Feb 1954, *Sledge 1113* (BM); rocks and cliff by stream and in woods, south side of ridge, c. 1-2km along track from c. 1km below top of Ramboda pass, on west side, heading N.W. towards Maturata, Alt. 1700-1800 m., 25.Oct 1993, *Fraser-Jenkins* with *Abeysiri 354Aa & 355* (K); Horton Plains, 1900m., Dayagama road, above roadside ditch 1 Nov 1995 *Shaffer-Fehre with Jayasekara & Samarasinghe, 462* (K). LOCALITY UNKNOWN: *Thwaites C.P. 1288* (BM; CGE; K; P; PDA); *Walker* (K); *Robinson 190* (K); *Wall* ex herb. Hope (P).

13. SPHAEROSTEPHANOS

J. Sm., Hook., Gen. Fil. : t.24. 1839. Holtt., Blumea 19: 39. 1971.
Sledge, Bull. Br. Mus. nat. Hist. (Bot.), 8 (1):1- 54. 1981.

Caudex erect to-long creeping; fronds decrescent with varying number of much-reduced basal pinnae; pinnae pinnatifid to shallowly lobed; aerophores at base of pinnae often swollen; veins anastomosing, rarely free, surfaces always ± hairy and sessile spherical glands commonly present on lower or both surfaces; sori round, in a few species ± elongate, usually indusiate; indusia often hairy and/or glandular; sporangia bearing usually spherical glands or setae near annulus, stalks with multicellular hairs and swollen end cell; spores light brown, spinulose or bearing many small translucent wings. Cytology n=36.

About 150 species: mainly in New Guinea and the Phillipine Islands with a few species in Madagascar, Mascarene Islands, Southern India, Sri Lanka, Burma, southern China and throughout Malesia eastwards across the Pacific to Tahiti. Outlying stations in Sao Thome, East Africa, New Caledonia and Australia. (Sledge, 1981.)

Type: *Sphaerostephanos polycarpus* (Blume) Copel. [based on *Aspidium polycarpus* Blume].

KEY TO THE SPECIES

1 Rhizome long creeping, texture subcoriaceous to coriaceous sori supramedial, lower pinnae abruptly reduced . . . . **3. S. unitus**
1 Rhizome erect, texture herbaceous to firm herbaceous, sori medial
  2 Transitions between normal and reduced pinnae abrupt, margins cut 1/3 way to costa aerophores at junction with rachis . . . . **2. S. subtruncatus**
  2 Lower pinnae gradually reduced to deltoid auricles, main pinnae closely spaced, margins crenate or serrate, aerophores absent from junction with rachis. . . . . **1. S. arbuscula**

1. Sphaerostephanos arbuscula (Willd.) Holttum J. S.Afr. Bot. 40: 164. 1974. Sledge, Bull. Brit. Mus. nat. Hist. (Bot.), 8 (1): 43-44. 1981.
Aspidium arbuscula Willd. *In* L. Sp. Pl. 4th ed., 5: 233. 1810. Type: Mauritius. Herb. Willd. 19,763 (B). *Nephrodium arbuscula* (Willd.) Desv. *In* Mém. Soc. Linn. Paris 6: 253. 1827; Bedd, Handb. Ferns Brit. India: 276, fig. 142. 1883. *Cyclosorus arbuscula* (Willd.) Ching *In* Bull. Fan meml Inst. Biol. (Bot) 8: 194. 1938. *Thelypteris arbuscula* (Willd.) K. Iwats. In Acta phytotax. geobot. Kyoto 21: 170. 1965.

Caudex erect, up to 22 cm high, 2.8cm wide. Fronds single in short succession form tuft at apex of caudex, 30-120cm long. Stipe 4-10cm long, contracted at base, but increasing to 5mm in diam., with deep adaxial groove in some specimens, dark grey at base, stramineous above, minutely hirsute. Scales in lowest 4cm portion of stipe ± dense, brown, 4-7mm long, 0.5mm wide with c. 1.2mm wide auriculate base and long, slender, acuminate tip. Lamina ovate-fusiform., herbaceous; top half of lamina with approximately 28 pairs of ± minutely pedicillate (to 1mm) alternate, narrowly oblong, acuminate pinnae, often auricled at acroscopic base; pinnae up to 8.5×1cm., margin crenate to serrate with 15-20 lobes, cut to 1/6 or less to costa, in acuminate tip margin entire; lower half of lamina with 10 or more shorter pinnae which are gradually reduced to deltoid auricles; laminar tissue mainly glabrous, vascular portion hairy. Rhachis with adaxial groove, abaxially round, densely covered in c. 0.3mm long, curved hairs. Costa grooved above, rounded below, like costule prominent on both sides, pale above and dark below, covered in short 0.2mm long, curved white hairs which also frame pinna margin, laminar tissue predominantly glabrous, save for small yellow glands, fewer above than on lower side of lamina. Segments with 4- 6 (-7) pairs of veins, the lowest 1-½- (-2) anastomose and form excurrent vein, the next pair run to the long, often distinct sinus membrane. Sori medial, round, often restricted to first or second vein. Indusium glandular, hairy above centre. Sporangia with small yellow glands at face/annulus margin.

D i s t r. Sri Lanka, Southern India, Mascarene Islands, Madagascar.

E c o l. Common on banks of streams and in wet ground in forest in the interior.from 550-1550m.

N o t e. This species and *Sphaerostephanos unitus* are two of the most frequently occurring thelypteroid ferns in the hill country of Sri Lanka. Several other species with anastomosing veins have coloured glands on or between the veins on under surface of the pinnae, but the species of *Sphaerostephanos* are the only ones in which this character is combined with decrescent fronds. (Sledge 1981).

S p e c i m e n s E x a m i n e d. MATALE DISTRICT: Rattota area, road to Riverstern TV installation, mountain forest above bubbling brook, in shade under bushes, 1300m., 25 Oct 1995, *Shaffer-Fehre* with *Jayasekara & Ekanayake, 417* (K). KANDY DISTRICT: Kadugannawa, moist woods, Oct. 1846, *Gardner 1251* (CGE); Kotmale, 1847, *Fortescue* (CGE); Kandy, 1854, *Bradford* (BM., CGE); Corbet's gap, margin of secondary jungle, 1219m., 9 Dec 1950, *Ballard 1070* (PDA); Corbet's gap, 1200-1300 m., 9 Dec 1950, *Sledge 553, 557* (BM); Corbet's Gap 1200 m 7 Jan. 1951, *Sledge 854* (BM); Hunnasgiriya, 18 Jan. 1954, *Schmid 951* (BM); Hunnasgiriya, open ground by stream., 870 m., 16 Jan. 1954, *Sledge 969* (BM); Ambagamuwa, peaty swamp in jungle, 19 Jan. 1954, *T.G. Walker T137* (BM); same locality, marsh in jungle, 570 m., 19 Jan. 1954, *Sledge 993* (BM); Galaha 940m., 22 Jan. 1954, *Schmid 1036* (BM); Hunnasgiriya, near Madugoda, 900m., in tall grassland in damp place *T.G. Walker T114, T115* (BM); west side of Hunnasgiriya Mt., SSE of Elkaduwa, NE of Kandy, Alt. c. 1200-1300, stream gulley in forest above and SE of "Hunas Falls Hotel" 25.08.1993, *Fraser-Jenkins with Jayasekara, Samarasinghe & Abeysiri 42* (K). BADULLA DISTRICT: Badulla, *Freemann 264, 265* (BM); Oodawella, 1870, 1871, *Randall* in herb. Rawson *3220* (BM); on grassy bank, near Badulla, alt. ca 1000m., 29 Dec 1950, *Holttum S.F.N. 39216* (BM); same locality (77/3) & date, *Ballard 1315* (PDA); Badulla, at foot of Nanumukula, steep road side bank, disturbed soil only partly overgrown, 03 Dec 1995, *Shaffer-Fehre* with *Jayasekara & Samarasinghe 617* (K); road from Bandarawela to Haputale, steep grassy slope with patches of marsh, 1100m., 05 Dec 1995, *Shaffer-Fehre* with *Jayasekara & Samarasinghe 636* (K); road from Haputale to

Wallawaya, Naketiya, boulder slope with thin soil, covered in low scrub, 700m., 06 Dec 1995, *Shaffer-Fehre* with *Jayasekara & Samarasinghe 660* (K). CENTRAL PROVINCE: common in forests, ca 900-1500m., terraneous *Geo. Wall 44/179* before 1873 (BM). RATNAPURA DISTRICT: above Pinnawala on Balangoda road, rocky creek in forest, 19 March 1968, *Comanor 1090* (US); Balangoda-Rassagala road, in lowland forest ca 750 m., 16 Nov. 1976, *Faden 76/306* (K); road from Rakwana to Weddagalla, dry overgrown, wind exposed boulder slope in moderate shade of bushes, 700m., 25 Nov 1995, *Shaffer-Fehre with Jayasekara & Samarasinghe, 576* (K); path to Morningside bungalow, c. 200m from disturbed primary jungle, 825m., 26 Nov 1995, *Shaffer-Fehre with Jayasekara & Samarasinghe, 582* (K). NUWARA ELIYA DISTRICT: Rambodde, shady banks, June 1845, *Gardner 1109* (CGE; K); Adam's Peak, 14 Feb.1908, *Matthew* (K); Ramboda Pass 1560m., 17 Dec. 1950, *Sledge 660* (BM); Ramboda Pass, 1560 m., 17 Dec. 1950, *Sledge 660* (BM);); Denipitiya road to Nuwara Eliya at 26/9 km marker, steep slope above road side gully, disturbed, overgrown with grasses and ferns, 600m., 31 Oct 1995, *Shaffer-Fehre with Jayasekara & Samarasinghe, 428* (K); Hora Wanguwa, Rothchild Tea Estate, at 42/7km marker, crevice in rock face above waterfall, 950m., 31 Oct 1995, *Shaffer-Fehre with Jayasekara & Samarasinghe 435* (K); road from Ginigathena to Norton Bridge, in water at road side below steep, shady road side bank, 850m., 17 Nov 1995, *Shaffer-Fehre with Jayasekara & Samarasinghe 520* (K); near same locality overgrown road side bank, *Shaffer-Fehre with Jayasekara & Samarasinghe 521* (K), *522* (K); Rajamalle Division, Moray Tea Estate, path through foothills of Adam's Peak, rocky, loamy gully, shaded, on border of tea lands, 1500m., 18 Nov 1995, *Shaffer-Fehre with Jayasekara & Samarasinghe 529* (K). MATARA DISTRICT: Deniyaya, 550 m., 5 Feb. 1954, *Schmid 1145* (BM). LOCALITY UNKNOWN: *Thwaites C.P. 1359* (BM., CGE, K, PDA); *Walker* (K); *Robinson 150* (K); *Geo. Wall* (BM., PDA); *Ferguson* (PDA, US 816379).

2. Sphaerostephanos subtruncatus (Bory) Holttum, Kew Bull. 26:80. 1971. Sledge, Br. Mus. nat. Hist. (Bot.), 8 (1): 44. 1981.

Polypodium subtruncatum Bory, in Belanger, Voy. Ind. Or. (Bot.) 2:32. 1833. Type: India, Madura, Mts de Denigall, Belanger (P).

Aspidium molle sensu Thwaites Enum. Pl. Zeyl.: 391.1864. non Sw..

Caudex erect, densely scaly at apex; scales ovate-lanceolate 5×2mm. Fronds tufted, to 2m tall. Stipe 40-60cm long, scaly near the base, elsewhere almost glabrous, dark coloured although stramineous in exsiccates, to 8mm in diam., tetragonal, with deep adaxial groove and more shallow lateral grooves, abaxially rounded; short, soft appressed hairs sparsely distributed along groove, glabrous elsewhere, a few large scales in groove, scales 1mm acuminate, to 5mm long; stipe contracted. Lamina lanceolate-elliptic, up to 75-110 (-140) × 30cm., texture herbaceous, simply pinnate, apex acuminate, falcate lobes cut ½ to ? to costa, fused with crenate margin at tip; abruptly reduced at base; ca 10 basal pinna pairs abruptly reduced to a series of auricles, the lowest pinnae are mere protuberances; pinnae at regular intervals in a given lamina: free space between them rarely less than their width but sometimes 1-2 (-3) times that width; up to 30-45 pairs of pinnae 16 × 1.4 cm with 38-40 pinnatifid lobes cut ½ to costa, lobes blunt, slightly falcate, apex gently acuminate, margin entire, lamina abaxially covered in small yellow, spherical glands. Rachis as pale as stipe, densely covered in pale curved, pointed hairs 0.5mm and weaker hairs of 0.2mm (fuzz) in vicinity of highly reduced pinnae. Costa slightly raised and grooved above, distinctly raised and rounded below, densely covered in curved hairs 0.8mm long above, 0.3mm long below. Costules moderately distinct and slightly hairy above, prominent and densely hairy below, inter costular distance to 5mm. Veins up to 10 pairs excurrent vein formed by lowest anastomosing 1-1.5 pairs. Sinus membrane occasionally distinct. Sori median on veins, up to 10 pairs, round, 1mm in diam.. Indusia mostly persistent, hairy at centre. Spores monolete, 0.04 × 0.02 mm., oval with finely spinulose exine.

D i s t r. Seychelles, south-west India, Sri Lanka.

E c o l. In forest terrestrial along partially exposed stream banks, alt 400-900m. Rare.

S p e c i m e n s   E x a m i n e d. RATNAPURA DISTRICT: hillside above Potupitiya, in degraded forest, 450m., 4.Dec. 1976, *Faden 76/481* (K). LOCALITY UNKNOWN: *Thwaites C.P. 714* (2 sheets), sub *Nephrodium molle* (PDA).

3. Sphaerostephanos unitus (L.) Holttum, J. S. Afr. Bot. 40: 165. 1974. Sledge, Bull. Br. Mus. nat. Hist. (Bot.), 8 (1): 44. 1981.

Polypodium unitum L., Syst. Nat. 10th ed., 2: 1326. 1759, excl.syn. Type: no locality (LINN).

Aspidium unitum sensu Thwaites non (L.) Sw., Enum. Pl. Zeyl.: 391. 1864, non Mett.
Cyclosorus unitus (L.) Ching, Bull. Fan. meml Inst. Biol. (Bot.) 8: 192. 1938; Holtt. 1955. nec Mett.
Aspidium cucullatum Blume, Enum. Pl. Jav.: 151. 128. Type: Java (L).
Nephrodium cucullatum (Blume) Bedd., Ferns South. India: t.88. 1863. Baker in Hook. & Baker, Syn. Fil.: 290. 1867. Bedd., Handb. Ferns Brit. Ind.: 270, fig. 138. 1883.

Rhizome long-creeping, branching, to 1cm in diam., densely scaly at apex; scales lanceolate 6 × 1.25mm., apex acuminate, margin entire with few short acicular hairs. Fronds 32- 130cm long, evenly spaced. Stipe (from rhizome to lowermost auricle) 13-25 cm long, 2-5mm in diam. with deep adaxial groove, base of stipe larger, lowermost 2-3 cm densely covered in lanceolate, brown scales 4-6mm long, 1mm wide at base, sparsely hairy on both sides; above scales scattered white hairs c. 5mm long 0.01 mm wide, closely appressed to stipe becoming more dense along rhachis. Lamina elongate-oblong, measurement and contours above highly contracted pinnae 58 × 20 cm., cut off horizontally where tips of 3 or 4 highest pinnae pairs reach same level, they overtop basal 1/5th of pinna-like tip the end of which forms a narrow, appendix-like extension to the frond. Pinnae up to 38 pairs, ascending (± 60°) subopposite or alternate, sessile, in distal part of frond adnate to main rachis, (4)-8(-12) pairs of basal pinnae abruptly reduced to auricle-like lower pinnae 4-1cm apart. Pinnae 20 × 1.2 cm gently acuminate with broadly cuneate base, linear, only in lowest pinnae basal portion slightly contracted towards cuneate base, margin appears dentate due to recuved lamina, cut ? to costa into rounded, acute, cucullate lobes, abaxial surface of pinna with spherical, sessile yellow-orange glands, adaxial surface glabrous. Rhachis adaxially grooved, abaxially rounded, densely covered in strigose hairs. Costa adaxially grooved, with strigose hairs, abaxially rounded, prominent and hairy just as the slightly raised costules and up to 7 (-10) pairs of veins, 3 to 3.5 lowest pairs form excurrent vein which is more prominent and hairy than costule, the fourth pair joins with massive sinus; veins adaxially appear pale due to overlying elongated, translucent cells. Sinus membrane distinct. Sori marginal on veins 1mm in diam.. Indusia hairy, opaque, yellow, 0.5 mm in diam.. Sporangia with one or more orange glands at annulus. Spores monolete, oval to reniform 0.04 × 0.02 mm., dull grey, minutely spinulose.

D i s t r. East Africa, Madagascar, Mascarene Islands, Seychelles, Southern India, Sri Lanka, Assam and Burma to Vietnam., Malesia to New Guinea, Philippines, New Caledonia, Fiji and Samoa.

E c o l. Moist roadside banks and open ground near streams in the west and centre, to 1650 alt.

S p e c i m e n s   E x a m i n e d: KANDY DISTRICT: Kotmale, 1847, *Fortescue* (CGE); common at Kandy, *Mrs Chevalier* (BM). Kandy,1854, *s.coll.* (CGE); Kandy, 1868, 1871, *Randall* in herb. Rawson *3220* (BM); Peradeniya, 24 May 1915, *Petch* (PDA); Hunnasgiria, 900m., 18 Jan. 1954, *Schmid 954* (BM); Pussalamankada, between Kandy and Maturata, open ground above stream., 540m., 18 Jan. 1954, *Sledge 982* (BM); Haragama Pussalamankada (15 miles from Kandy) at 540 m., 16. Feb. 1954, *T.G. Walker, T 125* (BM); irrigation channel in woods above village of Pahalena Ella, Knuckles Range N.E. of Kandy, Alt. c. 550 m., 14.Oct.1993, *Fraser-Jenkins with Abeysiri &*

Gunawardana 299 (K); Rajamawatha (B492) at 4km marker, close to stream in reflected sunlight, 300m., 03 Dec 1995, *M. Shaffer-Fehre with P. Jayasekara & D. Samarasinghe 613* (K). BADULLA DISTRICT: Badulla, *Freeman 263* (BM); Rajamawatha (B492) at 4km marker, close to stream in reflected sunlight, 300m., 03 Dec 1995, *Shaffer-Fehre with Jayasekara & Samarasinghe 613* (K). CENTRAL PROVINCE: very common on banks and roadsides in the Central Province and Ouva, up to 1500m., *Geo. Wall, 44/171* before 1873; Uduwara, abandoned quarry, marshes on several levels, waterlogged, 725m., 05 Dec 1995, *Shaffer-Fehre with Jayasekara & Samarasinghe 631* (K); Beragalla, grassy slope in wet zone, close to intermediate zone, 700m., 06 Dec 1995, *Shaffer-Fehre with Jayasekara & Samarasinghe 658* (K). RATNAPURA DISTRICT: Rajawaka, moist roadside bank, 460 m., 30. Dec. 1976, *Faden 76/660* (K); Ginigathena, at 4/7km marker 800m., 17 Nov 1995, *Shaffer-Fehre with Jayasekara & Samarasinghe 526* (K); road from Rakwana to Weddagalla, on dry, overgrown, wind-exposed boulder slope, in moderate shade of bushes, 700m., 25 Nov 1995, *Shaffer-Fehre withJayasekara & Samarasinghe 575* (K); near Potupitiya, side of brook above water station, 325m., 27 Nov 1995, *Shaffer-Fehre with Jayasekara & Samarasinghe 611* (K). NUWARA ELIYA DISTRICT: between Hakgala and Ambawela, roadside 1650m., Dec. 1950, *Sledge 716, 722* (BM); road from Rendapola to Ambewala, near public water pump, in deep shade under bamboo, above brook, 1400m., 01 Nov 1995, *Shaffer-Fehre with Jayasekara & Samarasinghe 453* (K); Ginigathena, at 4/7km marker 800m., 17 Nov 1995, *Shaffer-Fehre with Jayasekara & Samarasinghe 526* (K). GALLE DISTRICT: Galle, woods, May 1860, *Dubuc* (E); Talanume Range, near entrance to Kanneliya National Forest, below bridge among rocks over stream., 09 Nov 1995, *Shaffer-Fehre with Jayasekara 514* (K). LOCALITY UNKNOWN: *Thwaites C.P. 973* (BM.,CGE, K, PDA); 1819, *Moon* (BM); *Ferguson* (PDA,US 816403).

14. STEGNOGRAMMA

Blume, Enum. Pl. Jav.: 172. 1828. emend. K. Iwats. in Acta phytotax. geobot. Kyoto 19: 112-126 1963.

Caudex ascending or short creeping; stipe and rachis hairy with unicellular or septate hairs; fronds pinnate with subentire or pinnatifid pinnae, basal pinnae not or little reduced , the upper ones coadnate at base; aerophores lacking; veins free or with goniopteroid anastomoses; surfaces of pinnae hairy, lacking spherical glands; sori exindusiate, elongated along veins , sporangia setose; spores finely spinulose or with small wings. About 12 species from Spain, Macaronesia, tropical and southern Africa eastwards across the warmer parts of east Asia to Japan and south to the Philippine Islands and Indonesia; a few in tropical America.

Type species: *Stegnogramma aspidioides* Blume

Stegnogramma pozoi (Lag.) K. Iwats. Acta phytotax. geobot. 19: 124. 1963.

Hemionitis pozoi Lag. sp. nov., Gen. et Sp. 33. 1816. Type: Northern Spain,
del *Pozo* (S-PA).

var. **petiolata** (Ching) Sledge, Bull. Br. Mus. nat. Hist. (Bot.) 8(1): 49. 1981.
Leptogramma petiolata Ching in *Acta phytotax. sin.* 8 : 319. 1963. Type: Sri
Lanka, *G. Wall.* (PE ?).*Grammitis totta* sensu Thwaites, Enum. Pl. Zeyl.
:382. 1864, non C. Presl
Leptogramme totta sensu Bedd., Handb. Ferns Brit. Ind. : 377 p.p. excl. fig.
215. 1883. Holtt. Thelypteridaceae. Fl. Males., Ser.II, 1(5): 331-560.
1982 "1981".

Illustrations: *S. pozoi*, not specif. var. *petiolata*. Bedd., Handb. Ferns Brit.
India, t. 215 p.p. 1883. Manickam & Irudayaraj, Pter. Fl. Western Ghats,
t. 132. 1992.

Rhizome short- or long-creeping, 5-18cm., up to 0.5cm in diam., covered
in opaque, brown scales, 2-5 mm long covered on both sides in 0.2mm short,
stiff, white hairs. Fronds 20 to 60cm long, evenly spaced, ± 0.5cm apart;
texture herbaceous to subcoriaceous, downy. Stipe hirsute, mostly shorter
than lamina. the basal 4cm and less so up to lamina, bear conspicuous scales
acuminately or bluntly triangular, minutely hairy on both sides, 2-4mm long,
1mm wide, glabrescent. Lamina lanceolate to narrowly oblong -lanceolate 4-
15cm wide, pinnate, up to 9 pairs of free pinnae below broadly adnate distal
ones, 1-(3) lowest pinnae pairs might be slightly reduced, peduncles here
longest, 1-1.5mm., steadily shortening until truncate pinna base becomes sessile
or entirely fused to rhachis. Pinnae 2-6 (10) × 1-1.75cm., truncate at base,
usually suddenly narrowing to blunt or subacute apex, rarely attenuate, lobed
less than half-way to costa with lobes blunt and rounded or only crenate
lobate in small fronds, costa ad- and abaxially moderately prominent bearing,
as do costules and veins, a mixture of some long and many short hairs on both
surfaces and often minute hairs on lamina surface; (3)-5(-6) veins in segments
simple, free, rarely with anastomoses; lowest veins of segments rise separately
and are connivent at sinus, they rarely unite; excurrent vein not formed. Sinus
membrane distinct. Sori linear to punctate, medial, exindusiate with setose
sporangia. Cytology : n=36

D i s t r. *Stegnogramma pozoi* s. lat. is recorded from Spain, Madeira &
West Africa to China and Japan; var. *petiolata* from Sri Lanka and Java.

E c o l. At high altitudes in Forests about Nuwara Eliya, above 1750m and
to 2090m (according to collecting data).

N o t e s. Sri Lankan plants differ from south Indian plants in their generally smaller stature, their shorter (3-4cm) pinnae, which are proportionately

broader and more abruptly narrowed into a blunt or subacute apex and in having many pairs of pinnae free from the rachis and shortly stalked. The Sri Lankan form is distinctive in appearance and matches specimens from Java more closely than those from India. Where Sri Lankan specimens have longer pinnae (than Indian form) they still differ from Indian specimens in the larger number of stalked! pinnae.

S p e c i m e n s   E x a m i n e d : NUWARA ELIYA DISTRICT: Nuwara Eliya, in woods, 1800m., Sept. 1844, *Gardner 1071* (BM; CGE; P); in the forests about Nuwera Eliya; terraneous, *Wall 52/3*, before 1873 Nuwara Eliya, April 1899, *Gamble 27567* (K); Moon Plains near Nuwara Eliya, 1800m., 23 Dec. 1950, *Ballard 1201* (K); near Hakgala by small stream in shade, c.1676 m., 23 Dec.1950, *Holttum SFN 39167* (K); by track from Pattipola to Horton Plains, 1800 m., 20 Dec. 1950, *Sledge 667* (BM); Horton Plains, by stream in shady gully, 2040 m., 19 Dec. 1950, *Sledge 687* (BM); Ramboda Pass, c. 1900 m., 12. March 1954 on bank in forest, *T. Walker T828* (BM); Horton Plains 2070 m., March 1954, *Schmid 1397* (BM); Horton Plains, on roadside banks in forest, c.2090 m., 15 Nov. 1976, *Faden 76/285* (K); Ramboda Pass-Maturata track, in shady forest, 1940m., 17 March 1954, *Sledge 1305, 1316* (BM); Hakgala, 1800 m., 20 March 1954, *Sledge 1340* (BM); one km below top of Ramboda Pass, on west side which heads N.W. towards Maturata, Alt. c.1700-1800 m., among rocks and cliff by stream and in woods, south side of ridge. 25.Oct. 1993, *Fraser-Jenkins* with *Abeysiri 353* (K). LOCAL-ITY UNKNOWN: *Thwaites C.P. 1292* (BM.,CGE, K, P); *Walker* (K); *Bradford* ex herb. Hance (BM); May 1906, *Matthew* (K); *Mrs Chevalier* (BM); *Freeman 334A, 335B, 336C* (BM).

15. TRIGONOSPORA

Holttum, Blumea 19:29. 1971.
Sledge, Bull. Br. Mus. Nat. Hist. (Bot.) 8(1): 1-54. 1981.

Caudex erect; basal pinnae not or little shortened and never reduced to auricles on stipe, pinnae usually deeply lobed; veins free the basal acroscopic one running to the base of the sinus between two segments, the basal basioscopic one reaching the edge above the base of the sinus; sori indusiate; spores trilete, minutely papillose. n=36.

Sledge (1981, loc. cit.) comments on the genus: *Trigonospora* is unique amongst thelypteroid ferns in having trilete spores. All species commonly grow among stones in shallow rivers (oya) or close by river beds. The genus is confined to South-East Asia and was estimated by Holttum (in Blumea loc. cit.) to contain about eight species with the probability of some other species awaiting description. Throughout most of its area the species are clearly de-fined, their fronds tend not to be decrescent, the under surface of their pinnae is without hooked hairs and indusia are evident. In Southern India and in Sri

Lanka the degree of variability (of often quite uniform appearance of taxa) reaches its maximum and the species are, in consequence, more difficult to distinguish. Hooker and Thwaites, therefore included plants from these regions in a variable species identified as Blume's Javan plant, *Aspidium calcaratum*. Beddome recognized a number of different variants and treated them as varieties under *Lastrea calcarata* (Blume) T. Moore. Sledge is of the opinion that most specimens can be identified by following his key (see below). *T. calcarata* and *T. ciliata* are more widely distributed than all others; *T. glandulosa* is distinct but quite rare, so too is *T. angustifrons*.

Type species: *Trigonospora ciliata* (Wall ex Benth.) Holttum

KEY TO THE SPECIES

1 Base of pinnae cuneate with no pinna lobes close to rhachis . . . .
. . . . **7. T. zeylanica**
1 Base of pinnae truncate with pinna lobes close to rhachis
  2 Lower surface of pinnae & indusia glandular . . . . **6. T. glandulosa**
  2 Lower surface of pinnae & indusia eglandular
    3 Basal acroscopic lobes free at least in lower pinnae, lobes very oblique, indusia glabrous
      4 Lamina 2-3 times as long as wide, pinnae oblong, hairy at least on costae . . . . **4. T. calcarata**
      4 Lamina 3-6 times as long as wide, pinnae fusiform., quite glabrous .
. . . . **5. T. angustifrons**
    3 Basal acroscopic lobes not free, lobes oblique, indusia hairy or glabrous
      5 Pinnae 4cm or less, cut 1/2 way to costa, tips blunt or acute . . . . . .
. . . . **3. T. ciliata**
      5 Pinnae 4cm or more, cut 2/3 or more to costa, tips acuminate or caudate
        6 Pinnae 4-8cm long with about 10 pairs of lobes . . . .
. . . . **2. T. obtusiloba**
        6 Pinnae 8-13cm long with 15-20 pairs of lobes . . . .
. . . . **1. T. caudipinna**

1. Trigonospora caudipinna (Ching) Sledge, Bull.Br. Mus. Nat. Hist. (Bot.) 8(1): 15. fig. 2A. 1981.

Thelypteris caudipinna Ching, Bull. Fan meml Inst. Biol. (Bot.) 6: 288. 1936.
    Type: Hainan, *Hancock 108* (K !).
Pseudocyclosorus caudipinnus (Ching) Ching, Acta Phytotax. Sin. 8: 324.
    1963.
Lastrea falciloba Bedd., Ferns S. India : 37, t. 105. 1863, non *Nephrodium falcilobum* Hook. Type: Anamallays, 3000, *Beddome* (K).
Lastrea bergiana sensu Bedd., Ferns S. India & Brit. India suppl.:16 t.370.
    1876, non

Polypodium bergianum Schlechter.

Lastrea calc..rata var. *ciliata* Bedd., Handb. Ferns Brit. India: 235, fig. 121. 1883. Type: *Aspidium ciliatum* Wallich 351, Nepal, prob. Katmandu Valley (May 1830)

Illustrations. Sledge, Bull. Br. Mus. nat. Hist. (Bot.) 8 (1): 16, fig. 2 A. 1981. Manickam & Irudayaraj, Pter. Fl. Western Ghats, t. 138. 1992.

Caudex erect 13cm long, 3cm wide, base of caudex obscured by dense mat of roots. Fronds 60-100cm long, forming tuft at tip of caudex. Stipe 40-60cm long, 3-5mm in diam. with adaxial groove; a few broad scales 1.9mm long, 1.5mm wide occasionally along c. 4cm darker basal portion. Lamina 28-60cm long, 16-25cm wide, pinnate-pinnatifid, 16-(20) pairs of free sessile pinnae, fused, pinna-like tip. Pinnae 0.5-2cm wide, up to 12cm long, tail-like, caudate tip 1.5-3cm long, extremity entire, 2mm wide. Pinnule lobes 14-24, slightly falcate, 2-4mm wide, cut ? to costa; basal acroscopic pinnule alone inclined towards rhachis. Rhachis minutely hirsute, adaxial groove persists to tip, abaxially round. Costae minutely hirsute allround, prominent on both sides, more so abaxially, persistent groove adaxially. Costules hirsute, few acicular hairs as well, slightly prominent on both sides. Veins hirsute, very slightly pominent, lowest of seven pairs meet at sinus, others free running to margin. Sinus membrane distinct or even prominent. Sori median, 0.6 mm in diam. Indusia hirsute 0.5mm in diam.. Spores trilete, minutely papillate.

D i s t r. South India, Sri Lanka, Nepal, Sikkim., Assam., Burma and China (Hainan).

E c o l. On moist banks epecially by streams mainly in mountain forests.

N o t e. Although *T. caudipinna* is not strictly dimorphic, its fronds appear to have distinct degrees of fertility. Those beset densely with sori have more narrow pinnules cut almost to costa, gap between pinnules equals their width. On less fertile fronds sori are often found, if at all, only on two lowest veins close to costa, leaving pinnule largely free, incision between pinnules is here much smaller. Lamina of former frond tends to be coriaceous, that of the latter herbaceous.

S p e c i m e n s   E x a m i n e d. MATALE DISTRICT: Forested stream gulley beside and below road, 1 km W. of top pass between Rattota and Laggala, N.E. of Matale, Alt. c. 1200 m., 15. Sept. 1993, *Fraser-Jenkins* with *Jayasekara & Bandara 131* (K); Rattota, Riverstone VHF Tower, montane forest 2-3m high, beside small torrent spilling over boulders1150m., 25 Oct 1995, *Shaffer-Fehre* with *Jayasekara & Ekanayake, 413* (K); Rattota, Riverstone, montane forest above bubbling brook, granite not exposed, in shade, 1300m., 25 Oct 1995, *Shaffer-Fehre* with *Jayasekara & Ekanayake, 416* (K). KEGALLE DISTRICT: Mawanella, 27 Jan.1954, *Schmid 1070* (BM).

KANDY DISTRICT: Hantane, 1868, *Robinson 156* (K); West side of Hunnasgiriya mountain, SSE of Elkaduwa, NE of Kandy, Alt. c. 1200-1300stream gulley in forest above and SE of "Hunas Falls Hotel" *Fraser-Jenkins with Jayasekara, Samarasinghe & Abeysiri, 43,* (K); Manikkapatana foot path to Ratnapura, Gartmore Tea Estate, primary jungle above Gartmore, 1630m., in rocky gully beside stream., 1630m., 19 Nov 1995, *Shaffer-Fehre with Jayasekara & Samarasinghe, 567* (K). BADULLA DISTRICT: Yelumali, Namunukula, 12 March 1907, *Silva* (PDA); Foot path to Nanumukula Peak, forest gully, 1800m., 04 Dec 1995, *Shaffer-Fehre with Jayasekara & Samarasinghe, 626* (K); Tangamalle Forest Sanctuary near Haputale, cloud forest, in wide gully with boulders, among rocks in deep shade 1500m., 05 Dec 1995, *Shaffer-Fehre with Jayasekara & Samarasinghe, 647, 648* (K). RATNAPURA DISTRICT, path to bungalow, Morningside, thickly herbaceous road side below slope of primary jungle, 1000m., 26 Nov 1995, *Shaffer-Fehre with Jayasekara & Samarasinghe, 598* (K). NUWARA ELIYA DISTRICT: Ramboda, shady banks, June 1845, *Gardner 1107* (BM; CGE; K); Ramboda Pass- Maturata track, ca 1900m., 17 March 1954, *Sledge 1303* (BM); Hakgala, edge of forest, 1800m., 20 March 1954, *Sledge 1341* (BM). LOCALITY UNKNOWN: *Thwaites C.P. 1363* (CGE, K, P, PDA); (literature: prob. Adam's Peak) *Walker, Moon,* anno 1819, proliferous form.

2. Trigonospora obtusiloba Sledge, Bull. Br. Mus. Nat. Hist.(Bot.) 8(1):18-20. 1981.
Type: Sri Lanka , *Sledge 814* (BM holo.!) Sinharaja Forest near Hedigala

Illustrations. Sledge, Bull. Br. Mus. nat. Hist. (Bot.) 8 (1): 19, fig. 3. 1981.

Caudex erect up to 3cm long, 1.5cm wide. Fronds tufted, up to 77cm long, fertile fronds taller than broader, sterile ones. Stipe with adaxial groove, stramineous with dark grey base; minutely hirsute with a few acicular hairs or glabrous, very few shield-shaped scales up to 2mm long, 1.5mm wide along stipe; stipes of fertile lamina doubly as long or slightly less so than those of sterile one. Lamina oblong, ovate-oblong or deltoid ovate in outline, commonly with 7-10 but up to 15 pairs of free sessile pinnae, lowest pair deflexed and contracted at base.Pinnae (sterile) (3-)4-6(-9)× 1-2cm., pinnatifid, cut to ? with about 10 pairs of subpatent or slightly oblique, broad and bluntly rounded segments with hairs on margin to 5mm wide; fertile pinnae and segments more narrow; basal acroscopic segments of either pinna a little enlarged, incurved, often underlying the rhachis but sometimes scarcely different from the rest; pinna apex entire, acute or acuminate. Rhachis with adaxial groove, hairy with mixture of short and acicular hairs allround, grooved costa, costule and veins with variable amounts of hair below, surfaces subglabrous. Lowest of five pairs of veins meet in prominent sinus, the others go free to margin. Sinus membrane distinct in mature pinnae. Sori round, medial, none fused.

Indusia usually hairy, sometimes ± glabrous, occasionally with yellow glands (T845). Spores trilete.

D i s t r. Endemic, but see last voucher.

E c o l. By rivers and streams on shaded ground on all elevations up to 2000m.

N o t e. Sledge warns that "large detached fertile fronds of *Trigonospora obtusiloba* are not easily distinguished from small fertile fronds of *T. caudipinna*". Further "*T. obtusiloba* and *T. caudipinna* have not previously been distinguished from one another. Both are usually labelled *T. ciliata* or *T. calcarata* var. *ciliata* in herbaria. *T. obtusiloba* bridges the gap in size between *T. caudipinna* and *T. ciliata*, this doubtlessly contributed to confusion between the two species. In *T.ciliata* the pinnae are less deeply lobed, lobes of sterile pinnae are never so broad and blunt, and tips of pinnae are usually acute, sometimes shortly acuminate, but never caudate". Fertile fronds (even from the same rhizome) exhibit great variability in completion of spore pattern (cf. *Shaffer-Fehre 600*).

S p e c i m e n s   E x a m i n e d. MATALE DISTRICT: Riverstone estate, Rattota-Dikpatana road, mile post 33, along stream., 19 Jan. 1977, *Faden 77/181* (K); Stream gulley in Midland Estate, N.E. of Matale, Alt. c. 900 m., 15. Sept. 1993, *Fraser-Jenkins* with *Jayasekara & Bandara 121* (K); Rattota on road along Mid Car Tea Estate, at 25/9km marker, by gentle water fall over granite boulders1100m., 25 Oct 1995, *Shaffer-Fehre* with *Jayasekara . & Ekanayake*, (mixed collection) *406 and 407, 410* (K). COLOMBO DISTRICT: Kottawa Forest Reserve (Arboretum), on bank along path, 21 Jan. 1951, *Ballard 1536* (K). KANDY DISTRICT, Rajamawatha (B492), at 5km marker, 300m., 03 Dec 1995, *Shaffer-Fehre with Jayasekara & Samarasinghe, 614* (K). BADULLA DISTRICT, Tangamalle Forest Sanctuary near Haputale, cloud forest, 1450m., 05 Dec 1995, *Shaffer-Fehre with Jayasekara & Samarasinghe, 644, 650, 651* (K). RATNAPURA DISTRICT: Pehala Hewessa, on soil in jungle near stream., 20 Jan. 1951, *Ballard 1527* (K); bridge, mile post 6/5, past Rassagala on Balangoda-Rassagala Road, lowland forest near stream., 16 Nov. 1976, *Faden 76/309* (K); path to Morningside bungalow, thickly herbaceous road side below primary jungle, in deep shade on moist slope, 1000m., 26 Nov 1995, *Shaffer-Fehre with Jayasekara & Samarasinghe, 588, 589, 590, 591, 593, 594, 596, 599, 600, 601, 602* (K); Sinharaja Forest near Hedigala 75m., 5.Jan. 1951, Sledge 814 (BM., holotype!); same location & date, *Ballard 1393* (K); Manikkapatana foot path to Ratnapura, Gartmore Tea Estate primary jungle above estate, 1630m., 19 Nov 1995, *Shaffer-Fehre with Jayasekara & Samarasinghe, 564* (large sheet colln.), *565, 566* (K). NUWARA ELIYA DISTRICT: Adams Peak, *Matthew*, 14 Feb. 1908; near Hakgala, by stream in forest, ca 1676m., 27 Dec 1950, *Holttum S.F.N.39192* (K); Ramboda Pass, c. 1900 m., 17 Mar. 1954, *T.G. Walker T*

826, T827, T 830, T 845 (BM); Nuwara Eliya-Kandy road, between mile posts 46/3 and 46/4, shady moist bank near stream., 2 Jan. 1977, *Faden 77/31* (K); Ginigathena very steep bank with strong flowing brook, plants in fissures of rock, 800m., 17 Nov 1995, *Shaffer-Fehre with Jayasekara & Samarasinghe, 626* (K); Rajamalle Division of Moray Tea Estate, path through foothills of Adams Peak, among scrub on partly exposed rock, boggy in patches, 1500m., 18 Nov 1995, *Shaffer-Fehre with Jayasekara & Samarasinghe, 533, 535, 538* (K); Rajamalle Division of Moray Tea Estate, path through foot hills of Adams Peak, beside slope, among stones in shade 1610m., 18 Nov 1995, *Shaffer-Fehre with Jayasekara & Samarasinghe 55* (K); MONARAGALA DISTRICT: Kuda Oya, on rocks in stream., 28 Dec 1950, *Holttum S.F.N.39204* (K); Kuda Oya on Ramboda road, on bank, 1676-1700m., 28 Dec 1950, *Ballard 1291* (K). LOCALITY UNKNOWN: 1219m., *Beddome* (K).

3. Trigonospora ciliata (Wall. ex Benth.) Holttum, Blumea 19: 29. 1971. Sledge, Bull.Br. Mus. Nat. Hist. (Bot.) 8(1): 20-21. 1981.

Aspidium ciliatum Wall. Cat. 351. 1828 nom. nud.
Lastrea ciliata Hook., Hooker's J.Bot. 9: 338. 1857, nom illeg., non *L. ciliata* Liebm. 1849.
Aspidium ciliatum Wall. ex Benth., Fl. Hong Kong: 455. 1861. Type : Hong Kong, *Bowring 25* (K!).

Illustrations. Sledge, Bull. Br. Mus. nat. Hist. (Bot.) 8 (1) : 16, fig. 2. 1981. Manickam & Irudayaraj, Pter. Fl. Western Ghats, t. 139. 1992.

Caudex erect 2cm long, core 0.5cm wide. Fronds tufted, fertile frond much larger. Stipes 2mm in diam.; adaxial groove, stipe of sterile frond minutely hirsute with very few scattered shield-like scales 1.5mm long, 1mm wide; stipe of fertile frond c. twice as long, ± glabrous, without scales (probably loss rather than absence). Lamina with up to 15 pairs of free pinnae, lowest pair often not deflexed, subcoriaceous; sterile lamina ovate-lanceolate, 6.5cm across middle, fertile lamina elongate- oblong ca 6.5cm across middle. Sterile pinnae, 3.5-4cm long, 0.7-1cm wide, with ca 10-12 pairs of oblique, falcate segments cut ½ to costa, basal acroscopic ones enlarged by ca ½ and incurved; fertile pinnae c. 3 cm long, 0.6-0.9cm wide with 7-9 segments cut ½ to costa, tip in both pinnae entire, acuminate in sterile and slightly caudate in fertile ones. Rhachis with adaxial groove, hirsute with a mixture of very short and longer curved acicular hairs. Costa with adaxial groove, hirsute, abaxially prominent, rounded with acicular hairs. Costules and veins ad- and abaxially prominent, hairy. Veins in basal acroscopic lobe forked elsewhere single, all go to margin. In most segments 4 (-5) pairs of veins, lowest pair meet below sinus. Sinus membrane distinct. Sori medial 0.3-0.5mm in diam., appear crowded. Indusium naked or with sparse hairs. Spores trilete.

D i s t r. North-east India, Sri Lanka, northern Malaysia, Thailand, Sumatra, South China. In Sri Lanka "very rare or, possibly, overlooked" (Sledge).

E c o l. Moist ground by track through jungle.

S p e c i m e n s   E x a m i n e d. BADULLA DISTRICT: Haputale, Tangamalle Forest Sanctuary, in cloud forest, 05 Dec 1995, *Shaffer-Fehre with Jayasekara and Samarasinghe 646* (K). KALUTARA DISTRICT: Pas Dun Corle, Aug. 1865, *Thwaites C.P. 992* (P); Road from Pelawatte to Neluwa, disturbed habitat, side of brook tumbling over rocks, in deep shade under low trees and herbs *515* p.p. (K). NUWARA ELIYA DISTRICT: Between Hakgala and Nuwara Eliya, moist ground in jungle 1650m., 1950, *Sledge 710* (BM), *Holttum 39161* (SING). GALLE DISTRICT: Gallandala near Udugama, in splash-zone on rocky cliff over torrent, 8 Nov 1995, *Shaffer-Fehre with Jayasekara, 485* (K).

4. Trigonospora calcarata (Blume) Holttum, Reinwardtia 8: 506. 1974. Sledge, Bull.Br. Mus. Nat. Hist. (Bot.) 8(1):21-23. 1981.

Aspidium calcaratum Blume, Enum. Pl. Jav.: 159. 1828. Type: Java, *Blume* (L.)
Lastrea calcarata (Blume) T. Moore, Ind. Fil.: 87. 1858.
Nephrodium calcaratum (Blume) Hook., Spec. Fil. 4: 93. 1862, p.p.
Thelypteris calcarata (Blume) Ching, Bull. Fan meml Inst. Biol. (Bot.) 6: 288. 1936.

Illustrations. Sledge, Bull. Brit. Mus. nat. Hist. (Bot.) 8 (1): 22, fig. 4B. 1981. Holtt., Fl. Mal. Ser II 1(5): 372, fig. 5f. 1981.

Caudex erect, to 10cm long. Fronds tufted, up to 52cm long (610). Stipe stramineous with greyish base, hairy in youngest plants, soon minutely hirsute or glabrous save in adaxial groove, ca 1.5mm in diam., up to 25cm long in fertile fronds, stipes of sterile fronds shorter, only ¾ the length in a given specimen. Lamina oblong-lanceolate up to 12 cm wide, 12-15 pairs of free pinnae extending horizontally, at a shallow angle, but mostly at an angle of 60°-65° below much smaller adnate pairs with entire margin and pinnatifid apex. Pinnae oblong, sometimes slightly pedicillate, 4-8cm long, 0.6-1.2cm wide with 10-12 (-15), ± falcate pinnules cut $^2/_3$-$^4/_5$ to costa, basal one, even two, pinnule pairs free, extending parallel to rhachis and sometimes underlying it, apex acuminate, rarely caudate, entire. Rhachis with adaxial groove extending to tip, densely hairy when young, later glabrous. Costa adaxially hirsute, with groove, abaxially rounded, prominent. Costules slightly raised on both sides. Veins, 6 pairs, visible only abaxially, acroscopic vein of basal pair runs, straight, to sinus, basiscopic one makes a curve and merges free with margin above next lower sinus, some veins in free basal pinnules fork. Sinus membrane distinct. Sori medial, the lowest acroscopic sorus is below sinus, c. 0.5mm in diam.. Indusia glabrous. Spores trilete.

D i s t r. Sri Lanka, Sumatra and Java.

E c o l. On stony ground by or in river beds in shade at low or moderate elevations.

N o t e. Pinnae of fertile fronds sometimes narrower (0.6-0.7mm), their pinnules dentate rather than falcate; there are many degrees of difference as tips of pinnae with sterile attributes bear a few sori on the top half of their lamina.

S p e c i m e n s   E x a m i n e d. MATALE DISTRICT: Rattota, Riverstone VHF Tower, montane forest 2-3m high, two strata, beside small torrent over granite boulder, 1150m., 25 Oct 1995, *Shaffer-Fehre* with *Jayasekara & Ekanayake, 412* (K). KANDY DISTRICT, road from Madugoda to Weragamtota, on rock in stream 8 Jan. 1954, *T.G. Walker T 42* (BM); Hunasgiriya near Madugoda, on stream bed, c. 900 m., 16 Jan. 1954, *T.G. Walker T106* (BM); east of Madugoda (3-4 miles), amongst stones in stream in jungle, 750m., 8 Jan. 1954, *Sledge 937* (BM); Rajamalle Division, Moray Estate, path through foothills of Adam's Peak, 1500m., scrub on partly exposed rock, boggy in patches, herbaceous vegetation1500m., 18 Nov 1995, *Shaffer-Fehre with Jayasekara & Samarasinghe, 536, 537* (K). BADULLA DISTRICT: Badulla, *Freeman 230 C* 1908-1923, presented 1932; Badulla, foot path to Nanumukula, forest, in rocky gully with permanently running water, 1800m., 04 Dec 1995, *Shaffer-Fehre with Jayasekara & Samarasinghe, 623* (K). KEGALLA DISTRICT: Kitulgala, amongst rocks by riverside [low alt.] 28.Aug. 1927, *Alston 897* (PDA). MATARA DISTRICT: Deniyaya, 550m., 5 Feb 1954 *Schmid 1138* (BM). RATNAPURA Gallebodde, by stream in jungle, 600m., 26 Jan.1954, *Sledge 1042* (BM); foot of Adams Peak, Carney near Ratnapura 240m., 9 March, 1954, *Sledge 1246 A* (BM); path above Morningside towards Depedene, boulder-strewn brook along shady path through disturbed primary jungle, 26. Nov.1995, *Shaffer-Fehre* with *Jayasekara 608, 609, 610* (K). NUWARA ELIYA DISTRICT: Moist woods at the foot of Adam's Peak , March 1846, *Gardner 1250*. LOCALITY UN-KNOWN: *Thwaites 1363* (PDA partim); *Thwaites 3273* (BM; CGE,K, PDA); *Fosberg, 7 & 8* Dec. 1977, 2500-2600 m., in remnant woods of shallow ravine, from under rocks along stream *Fosberg 57265 & Fosberg 57330* (BM).

5. Trigonospora angustifrons Sledge, Bull.Br. Mus. Nat. Hist. (Bot.) 8(1): 23. 1981.
Type : Sri Lanka, at foot of Adams Peak, *Sledge 1246* (BM holo!)

Important literature. Sledge, Bull.Bᵣ. Mus. Nat. Hist. (Bot.) 8(1): 1-54. 1981.

Illustrations. Sledge, Bull. Br. Mus. Nat. Hist. (Bot.) 8 (1): 22, fig. 4 A. 1981.

Caudex erect 3.5cm high, 1.5cm wide. Fronds tufted at apex, 20-40 (-45cm) long, some, but not all fertile fronds longer than sterile fronds. Stipe

7-10cm long, 1.5mm in diam., dorsal groove, stramineous. Lamina linear-lanceolate, 4-6cm wide, 20-24 cm long, c. 15 pairs of free ascending pinnae (ca 45°), lowest pairs contracted, not deflexed, surfaces quite glabrous above and below, margins not ciliate. Pinnae fusiform., up to 4cm long, 5-8mm wide with 8 pairs or fewer of oblique, falcate segments cut half way (rarely more) to costa in broadest middle region, narrowed at base with lowest pair of segments quite free, acroscopic segment often crenate and close to rhachis ; distal pinnae serrate or crenate, apex of pinnae acuminate, entire. Rhachis glabrous. Costa adaxially with groove, here only occasionally a few hairs. Costules and veins in faint relief below; basal acroscopic vein goes to sinus. Sinus membrane not apparent. Sori round, c. 0.75mm in diam., in mature pinna one per segment,otherwise 5-6 pairs confined to 5-7mm of narrow, entire pinna tip. Indusia glabrous.

D i s t r. Endemic (one specimen at K from the South Andamans is extremely similar to the species but is hairy and the lowest pinnules are not completely free).

E c o l. By streams and rivers at low to medium altitudes about Adam's Peak.

N o t e A well-marked species related to *Trigonospora calcarata*, but distinguished by its narrow fronds, often six times as long as broad and by the whole plant, save the dorsal groove of the stipe and rhachis, being quite glabrous. The pinnae resemble those of *T. zeylanica* in size, their fusiform contour has a strongly cuneate bases, but in *T. angustifrons* a pair of free, erect basal segments adjacent to the rhachis is always present. Apparently rare and localized (Sledge 1981).

S p e c i m e n s   E x a m i n e d. RATNAPURA DISTRICT: at foot of Adam's Peak near Carney, by river, 240m., 9 March 1954, *Sledge 1246* (BM holo!); Dotoluoya -tributary of Bambarakotuwa river, in devastated forest, comon along the waterline and the river, 19 June 1971, *Meijer, 866* (PDA); Doluluwa forest, beyond Rassagala, east of Ratnapura, moist lowland forest, edge of river, 750m., 16 Nov. 1976, *Faden 76/316* (K); Adam's Peak, *Moon* (BM). LOCALITY UNKNOWN: *F.D'A Vincent* (CGE).

6. Trigonospora glandulosa Sledge, Bull.Br. Mus. Nat. Hist. (Bot.) 8(1): 23. 1981.

Illustrations. Sledge, Bull. Br. Mus. nat. Hist. (Bot.) 8 (1): 22, fig. 4 C, D. 1981.

Caudex erect. Fronds tufted, up to 40cm long. Stipes with adaxial groove, densely grey pubescent throughout with a mixture of short crisped hairs and some longer, patent, ascicular hairs 1mm long, stipes of fertile fronds not conspicuously longer than those of sterile ones. Lamina 15-30cm long, 6-8cm wide. narrowly oblong or elliptic, tapering above, c.10-12, less commonly up to 16 pairs of free pinnae, lowest pair not deflexed. Pinnae 3-5cm long, 0.5-1cm wide, deeply pinnatifid with up to 10 pairs of oblique, oblong blunt falcate segments 2mm wide, basal pair quite free and sometimes shortly stalked, often lobed near base, slightly enlarged, close to and often underlying rhachis, sometimes second segment also quite free to base, rest of pinna progressively less deply cut into oblique, blunt, falcate lobes; apex of pinnae blunt or acute, uppermost pinnae and those adnate to apical portion becoming quite entire. Rhachis with continuing adaxial groove, densely hirsute. Costas and costules with scattered acicular hairs above and below, margins of segments fringed with rather stiff hairs, under surface with copious subsessile, pale yellow glands, upper surface with similar glands when young, becoming smooth with age; indusium broad, thin, covered with glands, with or without intermixed setose hairs. Veins moderately raised below, 3-5 pairs per segment. Sinus membrane not apparent, very rarely distinct. Sori medial, arise first on acroscopic half of pinnule. Indusia densely covered in stalked glands, occasionally with intermixed setose hairs.

D i s t r. Endemic.

E c o l. *T. glandulosa* typically grows among rocks in running water, it prefers deep shade.

N o t e. The glands in dry specimens are white and extremely hard to see; magnification of × 40 required; best start with indusia, in dry specimens they have a depression at the centre where glands are more protected from being rubbed off.

S p e c i m e n s   E x a m i n e d: KALUTARA DISTRICT: road from Pelawatta to Neluwa, disturbed habitat, at side of brook tumbling over rocks, in deep shade of trees and herbs 515 (K). RATNAPURA DISTRICT: 8 miles north-east of Ratnapura, in rocky ground by stream in shady ravine, 150m., 4 Jan.1951 *Sledge 808* (BM., holo. !; US); same locality and date, *Ballard 1383* (K); path to Morningside, thickly herbaceous road side below primary jungle, in shade, 1000m., 26 Nov 1995, *Shaffer-Fehre with Jayasekara & Samarasinghe, 585* (K). LOCALITY UNKNOWN: One specimen communicated by *Alston 11746*, from plants propagated at K (1Aug. 1951), origin Sri Lanka, *Sledge 411*, determined by Sledge as *T. glandulosa*.

7. Trigonospora zeylanica (Ching) Sledge, Bull.Br. Mus. Nat. Hist. (Bot.) 8(1): 24-25. 1981. Sledge, Bull.Br. Mus. Nat. Hist. (Bot.) 8(1): 1-54. 1981.

Thelypteris zeylanica Ching, Bull. Fan meml Inst. Biol. (Bot.)6: 287. 1936.Type: Sri Lanka, *Thwaites C.P.3050* (K!).

Nephrodium falcilobum var. *β* Hook., Sp. Fil. 4: 108. 1862. Type: Sri Lanka, *Thwaites C.P. 3050* (K!)

Lastrea calcarata Bedd., Ferns S. India: 82, t.246. 1864, nom illeg., non *L. calcarata* (Blume) Moore 1858. Type: Sri Lanka, *Thwaites C.P.3050* (K!).

Aspidium calcaratum var. *β* (Hook.) Thwaites, Enum., Pl. Zeyl.: 391. 1864. Bedd., Handb. Ferns Brit. India: 237. 1883.

Lastrea calcarata var. *moonii* Trimen in Syst. Cat. Fl. Pl. Ferns Ceylon ex Journ. Sri Lanka Branch Roy. Asiat. Soc.: 114. 1885, nom. nud.

Dryopteris calcarata var. *moonii* Trimen in Willis, Rev. Cat. Fl. Pl. Ferns Ceylon, Peradeniya Manuals 2:116. 1911, nom. nud.

Illustrations. Bedd., Ferns South. India: 82, t. 246. 1863 (as *Lastrea calcarata*). Sledge, Bull. Br. Mus. nat. Hist. (Bot.) 8 (1): 16, fig. 2C. 1981.

Caudex erect often low but up to 3cm high, 2.5cm wide. Fronds tufted at apex, crowded, very variable 11-30 (-40)cm long, fertile fronds with stipes not much longer than sterile ones. Stipe with adaxial groove ca 1mm in diam., stramineous to brown, 1/5 of length at base 1.5-2mm in diam., grey. Lamina 11-30cm long, 3-6 (-9)cm wide lanceolate to fusiform herbaceous to subcoriaceous, simply pinnate, with 9-18 pairs of free, often pedicillate pinnae. Pinnae linear-oblong to oval-oblong or fusiform., 1.8-6.5cm long, (0.5-) 0.7 (-1)cm wide, blunt tip and narrow, cuneate base, no basal pinna segments adjacent to rhachis; pinnae sessile or with pedicels 1mm long , (pedicels increase in length from 1mm at base to 4mm at middle of lamina in large specimens), towards top sessile, with few adnate pinnae below pinnatifid tip; pinna margins almost dentate, between 3-10 pointed segments cut less than ? or up to ½ to costa, tip acuminate, entire; in very small specimens pinnae obovate, margin entire or with 1-3 notches; pinnae largely glabrous but margin especially towards tips of segments with very few straight or round-curved hairs ca 0.2mm long. Rhachis with adaxial groove, hirsute with ± strongly curved hairs 0.5mm long and occasional straight, once- branched hairs 0.8 mm long. Costa adaxially minutely hirsute, with groove, costules smooth, slightly raised as are smooth costa and costules below. Veins 3-4 pairs, basal acroscopic vein goes to sinus, basiscopic vein mostly above sinus approach margin. Sinus membrane not apparent, only rarely distinct in larger specimens. Sori medial round, dark brown, 1mm in diam. Indusium glabrous.

D i s t r. Endemic.

E c o l. By streams or among boulders in streams in deep shade of forest. At home at lower elevations, as confirmed by the most recent gatherings. (50-150m elevation.)

S p e c i m e n s   E x a m i n e d. KALUTARA DISTRICT: Road from Pelawatte to Neluwa, disturbed habitat, side of brook tumbling over rocks, in deep shade under low trees and herbs, 10 Nov 1995, *Shaffer-Fehre* with *Jayasekara 515* (K). RATNAPURA DISTRICT: Sinharaja Forest near Hedigala, stream side, 75m., *Sledge 812* (BM); Pahale Hewissa, by stream., 30m., 31 March 1954, *Sledge 1370* (BM; K; US); same locality, 20 Jan. 1951, *Ballard 1522* (K). GALLE DISTRICT: Kanneliya Forest near Hiniduma, low alt., 7 May 1973, *Kostermans 24736* (US); same locality, 7 Dec. 1976, *Faden 76/504* (K). RATNAPURA & GALLE DISTRICT: *Thwaites C.P. 3050* (BM., CGE, K,PDA: Hinidoon, Dec. 1853; Sinharajah Forest, April 1855). MATARA DISTRICT: Above Enselwatte Estate above Deniyaya, shaded place by rivulet, June 1969, *Kostermans 23658* (US); at foot of Mt. Hiniduma, close to stream with boulders, jungle 8.Nov. 1995, *Shaffer-Fehre* with *Jayasekara 473* (K); *Gallandala 484* (K); Kanelliya National Forest, margin of jungle stream among boulders in water, in shade of Dipterocarp forest, 8.Nov. 1995, *Shaffer-Fehre* with *Jayasekara 506* (K). LOCALITY UNKNOWN: *Thwaites C.P. 992* (BM., CGE, K,PDA: Saffragam., Aug. 1821, *(Moon)*; *Gardner 1257* (BM).

16. THELYPTERIS

Schmidel in Keller, Icon. Pl.: 45, t. 11, 13 1763.

Rhizome slender, long creeping, in marshy ground; fronds pinnate with deeply pinnatifid pinnae, basal ones not or little reduced; veins free, reaching the margins, costae grooved on upper surface, small, flat, thin scales present on lower surface of costae (also filamentous small ones) but sessile spherical glands absent; sori indusiate, short capitate hairs sometimes present on sporangia near annulus; spores spinulose. n=35.

Four species distributed from north temperate Europe through the tropics of the Old World to New Zealand; one species in Sri Lanka.

Type species: *Thelypteris palustris* Schott.

Thelypteris confluens (Thunb.) T. Morton. Contr.U.S. natn. Herb. 38: 71. 1967.

Pteris confluens Thunb., Prodr. Pl. Cap.: 171. 1800. Type: South Africa, *Thunberg* (UPS). Schelpe, Fl. Zamb. Pterid.: 190 tab. 55E. 1970. Sledge, Bull. Br. Mus. nat. Hist. (Bot.), 8 (1): 12. 1981. Tigerschiold, Nord. J. Bot. 9 (6) 1990.

Aspidium thelypteris var. *squamigerum* Schlechtendahl, Adumbr. Fil. Prom. Bon. Spei: 23, t. 11. 1825. Type: South Africa, Cape Peninsula, *Schlechtendahl* (HAL).

Lastrea thelypteris var. *squamigera* (Schltdl). Bedd., Suppl. Handb. Ferns Brit. India: 54. 1892.

Thelypteris palustris var. *squamigera* (Schltdl.) Weath., Contr. Gray Herb. Harv. II, 73: 40. 1924.

Thelypteris squamigera (Schltdl.) Ching, Bull. Fan meml Inst. Biol. (Bot.) 6: 329. 1936 (as squamulosa).

Lastrea fairbankii Bedd., Ferns Brit. India 254 .1867. Handb. Ferns Brit. India 240. 1883. Type: Southern India, Pulney Mts, *Beddome* (K !).

Illustrations: Bedd., The ferns of British India, t. 254. 1867 (as *Lastrea fairbankii* (a pencilled note by C.B. Clarke disputes the drawing of being *L. fairbankii*)). Manickam & Irudayaraj, Pter. Fl. Western Ghats, t. 137. 1992.

Rhizome creeping, black, 4mm in diam. Fronds 35-55 cm long, tufted, fertile frond overtops; stipes variously longer than lamina, longest in fertile frond, 1mm in diam., black base 2cm., above this pale to dark brown. Lamina ovate-lanceolate, 11-15 cm long, 3-4.5cm wide, pinnate-pinnatifid; adaxially glabrous; pinna-pairs well spaced, distant 1-(3) lowest pair(s) of pinnae reduced by half, remote; pinnae 1.5-2.5cm long, 0.25-0.5cm broad, can alternate, most frequently almost opposite; lowest pinnules occasionally free, other -11 alternate to opposite pinnule pairs increasingly less deeply dissected towards fused tip; basiscopic pinnules slightly broader than acroscopic ones, revolute margins make acute- to obtuse pinnules appear triangular when dry. Rachis brown with very few distinct, non-clathrate, cordate scales with dentate margin; costa ,adaxially forms deep groove towards rachis, prominent abaxially, scales here similar to but half the size of those on stipe and rachis. Veins very apparent, go to margin. Sori medial; sporangia numerous. Indusium fimbriate. Spores spinulose. Cytology: n=35

D i s t r. Africa south of the equator, Ethiopia and Sudan, Madagascar; Southern India, Sri Lanka, Sumatra, Mountains of New Guinea, and New Zealand (North Island).

S p e c i m e n s   E x a m i n e d. BADULLA DISTRICT: Swamp near Bandarawela, Sept. 1890: "in great abundance", *s.coll.* (PDA). Close to survey camp, 3km W of Bandarawela, 500m N of rifle range. Alt. 1270m., in open marshy ground by well. 1982. *Tigerschioeld 82047* (S). In Dec. 1995, *Shaffer-Fehre with, Jayasekara could not locate the site*, the area is now used for market gardens.

VITTARIACEAE
(by Monika Shaffer-Fehre*)

(C. Presl) Ching, Sunyatsenia 5(4): 232. 1940.

Vittarieae C. Presl, Tent. pterid. 164. 1836. (as a tribe : "Vittariaceae")

Vittariaceae Link, Fil. sp. 116. 1841. (as a suborder: "Vittariaceae")

Antrophyaceae Link, Fil. sp. 140. 1841. (as a suborder: "Antrophyaceae")

Antrophyaceae Ching, Sunyatsenia 5(4): 231. 1940. (as a family)

Epiphytic, sometimes epilithic, medium-sized to very small ferns with fronds characteristic of a given genus. Represented in Sri Lanka by three of its six (or nine) genera. This description takes account of features displayed by Sri Lankan taxa. Rhizome creeping or suberect. Scales clothing rhizome and lower part of stipe clathrate, often with metallic sheen, seemingly denticulate due to marginally protruding radial cell walls. Leaves simple, rarely forked or cleft, stipe often short, ill-defined or virtually absent. Venation, when present, distinct in transmitted light, specific for genus. Epidermis with characteristic mechanical 'spicular' cells. Sporangia in simple or netted soral lines, protected solely by indusium or recurved laminar margin, or naked in grooves of leathery laminar tissue or on anastomosing veins of lamina. Soral trichomes nearly always present, unicellular or with modified terminal cell (paraphyses), uniseriate or branched. Spores hyaline, trilete or monolete. Gametophyte ribbon-shaped, nearly always bearing filamentous, marginal gemmae. Variability in the family apparently not based on hybridization.
Type: *Vittaria* Sm., Mem. acad. Turin. 5: 413, t. 9. 1793.

Bedd., Ferns of S. India, 1863/64. Bedd., Ferns of S. India, 1873. Bedd. The ferns of Brit. India , 1865/70. Bedd. reprint: The ferns of Brit. India, 1973. Bedd. Supplement to the ferns of S. India and ferns of Brit. India, 1876. Bedd., Handb. to the ferns of Brit. India, Ceylon and the Malay Peninsula, 1883. Reprint with supplement, 1892. Bedd. Handb. to the ferns of Brit. India, Ceylon and the Malay Peninsula, 1892.

---

* Royal Botanic Gardens, Kew, U.K.

Benedict, The genera of the fern tribe Vittarieae. Bull. Torrey Bot. Club 38: 153-190. 1911.

R.D. Dixit & N.C. Nair, Studies in Vittariaceae - I. The genus *Antrophyum* Kaulf. in the Indian subcontinent. J. Indian Bot. Soc. 53: 277-287. 1974. Fée, Mém. Fam. Foug. III, Vit. & Pleur. 1851-1852. Fée, Mém. Fam. Foug. IV, Antroph. 1851-1852.

Illustrations: K.U. Kramer *In* Kubitzki (ed.) Fam. Gen. Vasc. Pl. 1: 272. 1990.

V.S Manickam and V. Irudayaraj Pter. Fl. of the Western Ghats . pl. 83, 84, 85

KEY TO THE GENERA

1 Lamina elongate-lanceolate to ribbon-shaped, sorus marginal, spores monolete
.... **1.Vittaria**
1 Lamina of different shape, sorus positioned elsewhere, spores trilete
2 Lamina filiform., sorus restricted to rhachis, indusiate ....
.... **3. Monogramma**
2 Lamina variously lanceolate, sorus over network of veins covering abaxial surface of lamina, exindusiate .... **2. Antrophyum**

1. VITTARIA

Sm., Mem. Acad. Turin. 5: 413, pl.9, f.5. 1793. Bir, Res. Bull. Punjab. Univ. n.s. 13: 15-24. 1962. K.U. Kramer *In* Kubitzki (ed.) Fam. Gen. Vasc. Pl. 1: 272. 1990. Type species: *Vittaria lineata* (L.) Sm. (*Pteris lineata* L.)

Rhizome short creeping or suberect, densely covered by dark, clathrate scales; radial walls can be thick to massive at centre of scale, very thin along margin, scales often have long, seemingly dentate tip as long as entire blade. Fronds simple and narrow to extremely narrow, ribbon shaped (shoe-lace fern). Stipe indistinct in some species. Venation visible and informative if seen with transmitted light. Sori marginal. Paraphyses in an array of simple, branched or bushy structures. Spores monolete, oval to reniform., pale yellow, transparent without exine-ornamentation.

D i s t r. Santo Domingo and throughout the American Tropics, up to 50 species in warm-temperate regions. In Sri Lanka represented by three species: *V. elongata*, *V. microlepis*, *V. scolopendrina*

E c o l. Epiphytes on bark of trees or lithophytes, in forest shade, not frequent; rare in disturbed habitats

KEY TO THE SPECIES

1 Lamina up to 2.7cm wide, veins numerous, sharply ascending (herring bone-pattern) . . . . **2. V. scolopendrina**
1 Lamina narrow, not more than 10mm wide; veins few, no herring-bone pattern
2 Stipe 0.5 - 1 cm long; midrib prominent abaxially . . . . . . **1. V. elongata**
2 Stipe absent; midrib visible but not prominent abaxially . . . .
. . . . **3. V. microlepis**

1. Vittaria elongata Sw., Syn. fil.: 109, 302. 1806.

Bedd., Ferns S. India: 7, pl. 21. 1863. Thwaites, Enum. pl. zeyl. : 438. 1864.
  Bedd., Handb. Ferns Brit. India: 404, f. 238.1883. B.K. Nayar & S.
  Kaur, 1974. Companion to Beddome's Handbook to the Ferns of Brit-
  ish India. S. Chandra & S. Kaur, S., 1987. A nomenclatural guide to
  R.H. Beddome's ferns of South India and ferns of British India.
Type: country, locality, "*Rottler s.n.*" (herbarium S? n.v.)
Pteris graminifolia Roxb. ex Griff., Calc. Journ. Sci. 4: 502, t. 33. 1844.

Illustrations: Bedd., The Ferns of South. India.: 6, t. 21. 1863. Manickam &
  Irudayaraj., Pter. Fl. of the Western Ghats 107, t. 84. 1992

Rhizome short creeping 3 mm in diam., branching, densely covered in dark, clathrate scales, 1.3 mm long with uniformly thickened radial walls, forming narrow tip above 0.45 mm wide base, radial walls densely pitted towards cell interior. Frond pendulous, linear-lanceolate, dark green. Stipe 0.5-1.5 cm long, round 1mm in diam., scales 3-5 mm long, 0.1mm wide at base, at tip only inner and lateral radial walls of cells apparent (resembling barbed wire). Lamina up to 52.5 cm long, 4 –(7) mm wide, linear-lanceolate, entire, narrowing towards base and acuminate apex, texture coriaceous. Rhachis adaxially indistinct, abaxially raised; veins little distinct but visible in strong transmitted light. Sorus marginal, takes up top part of blade excluding acuminate portion, positioned horizontally between rim of recurved adaxial margin and abaxial indusium., sorus dark brown, 0.5-1 mm wide, includes sporangia, paleae*and branched paraphyses with a bell-shaped terminal cell, 0.12 mm long 0.6mm wide. (*The "paleae" turn out to be sporangia stalks after dehiscence.)

D i s t r. In the Old World tropics and subtropics: Africa, India, Sri Lanka, S. China, SE Asia to Australia. A fern of a wide range of altitudes, from 610 - 2100 m.

E c o l. Epiphytic in jungle and at disturbed sites i.e. secondary jungle.

N o t e. The fronds are more or less straight.

Nevill 05.1885 records that the fern is used as a charm against bears.

Specimens Examined: KURUNEGALA DISTRICT: Doluwakanda Hill, 180m., forest, common, 5 Jan. 1977, *Faden 77/51*(K). MATALE DISTRICT: *Brodie* ("new to Sri Lanka") ex Herb. Jenman, *pres. 1934* (K); On trunk of jungle tree, 1219m., 9 Dec. 1950, *Ballard 1061* (2 sheets, K). Pallegama, 210m., 2 Mar. 1954, *Sledge 1222* (K); north-east of Matale, c.1km west from top of pass between Rattota and Laggala, forested stream-gulley beside and below road, 15. Sept. 1993, *Fraser-Jenkins* with *Jayasekara* and *Bandara* Field No. *141* (2 sheets) (K); Matale, Brodie (K). KANDY DISTRICT: Epiphytic in secondary jungle, Corbet's gap, 1341m., 9 Dec. 1950, *Sledge 563* (K). AMPARAI DISTRICT: rare, in cave on rock Kaduru Kumbura Gala, some 18 miles SW of Panama, May 1885, *Nevill* ( ). RATNAPURA DISTRICT: mile post 7 on Balangoda - Rassagala road (c.2 miles past Rassagala); area called locally "7 mile Rassagala"; remnant forest patch in gully with large boulders, above road, 610m., pendant low epiphyte, common; 17 Nov. 1976, *Faden 76/332* (K); Delgoda, on Kalawana-Weddegala road, above quarry, forest patch, moss-covered, shaded rocks, locally common 04 Dec 1976, *Faden 76/479* (K). NUWARA ELIYA DISTRICT: Hakgala Bot. Garden, Hakgala Peak, 1829 m.,epiphytic on wet jungle trees amongst moss; very large plants. 16 Dec. 1950, *Ballard 1116* (K); Hakgala Mountain, forest, locally common, pendant epiphyte, 1830-2100m., 1 Jan. 1977, *Faden 77/11* (K). LOCALITY UNKNOWN: *C.P. 3806, Thwaites , 1804, 1865*, separate sheets (K). *Mrs. Walker* (K). Sri Lanka, *33a* S.W.K. "*Vittaria rigida* Klfss*" (K). *V. rigida, Wall* (K).

2. Vittaria scolopendrina (Bory) Thwaites, Enum. Pl. Zeyl. : 381. 1864.

Pteris scolopendrina Bory, Voy. 2: 323. 1804.

Synonyms cited by Thwaites:
Taeniopsis scolopendrina (Bory) Sm., J. Bot. 4: 67. 1842
Vittaria zeylanica Fée, Mém. Fam. Foug.: 15, pl.1, f.3. 1851-52. Type: Sri
    Lanka: "habitat in insula zeylanica" *Walker 210* (RB)

Illustration: Bedd., Ferns South. India: 18, 72, t. 212. 1863.

Rhizome short creeping, 6-8 mm in diam., dark, brown-black, covered densely in, clathrate scales. Scales 5.2 - 9 mm long , 0.2-0.5 mm wide with auriculate base, narrow blade above often extending into a long acicular tip; radial walls in central auriculate portion more thin-walled than thick, clathrate radial walls, of remainder of scale; radial walls of moderate, uniform width, faintly pitted, pits open towards cell centre; horizontal radial walls along margin resemble small spines. Fronds evenly spaced 0.5-1.5 cm apart. Stipe (0.3)-0.5-(1) cm long, c.0.3 cm wide, oval, almost black at base, glabrous. Lamina 53-57 cm long, between 1.7 -2.7 cm wide, texture coriaceous; decurrent on rhachis at base, full width attained at a level of about 16cm., from here band-shape maintained , for top 10 cm narrowing to an acute apex. Rhachis adaxially grooved, abaxially prominent, lateral veins parallel, ascending steeply, well seen only in transmitted light. Sorus linear, marginal, dark brown, velvety, revolute adaxial surface overtops abaxial indusium. Paraphyses 0.5mm long, paraphyses with obconical head, 0.053 mm long, 0.04 mm wide, amber-coloured. Paleae (sporangial stalks) 0.3 mm long. Spores monolete, bean-shaped 0.05mm long 0.02mm wide, pale yellow.

D i s t r. Mascarene Islands, India, Sri Lanka, Malaya and Polynesia. Altitude 60-1252 m.

E c o l. Epiphytic and epilithic on rocks in wet forest environment, on isolated hill tops in 'dry zone'.

S p e c i m e n s   E x a m i n e d: KURUNEGALA DISTRICT: Doluwakanda Hill, 180m., forest, locally common, 5 Jan. 1977, *Faden 77/54* (K). RATNAPURA DISTRICT: on rocks in forest, Induru Ganga, Gilimale nr. Ratnapura, 60m., 9 Mar. 1954, *Sledge 1252* (K). NUWARA ELIYA DISTRICT: (?), on rocks, *Macrae* (K); Gilimale Palapatwela , *v. Fridau* anno 1853 (Schmarda's journey 1853-57, GZU H 75) (GZU). LOCALITY UNKNOWN (C.P.1304) [Thwaites type of t. 212 ,752 (?) Bedd's Fern Herb.](K). *Gardner 1304* (K). *Wall* (K). Herb. *Gamble, 1925* (K). Coll. *Gardner* (T. Moore's Fern Herbarium., 1885) (K).

2. Vittaria microlepis Hieron., Hedwigia 57: 202. 1916.
Type: Sri Lanka, without locality, *Gosset s.n;* without locality, *Walker 119.* Further Syntype apparently *C.P. 1113* (*V. lineata* Sw.)

Illustration: Manickam & Irudayaraj, Pter. Fl. of the Western Ghats :107, t. 83. 1992.

Rhizome short-creeping or suberect , 3-4 mm in diam., densely covered in glossy, black, clathrate scales; scales narrow, 4 mm long, 0.5 mm wide (or a wider base -0.7 mm - with ¾ of the scale abruptly narrowing above), cells at centre with thick, dark brown radial walls, radial walls at margin very thin, apparently spiny due to membranous outer periclinal wall collapsing inward (resemblance to barbed wire). Fronds in close succession or tufted, linear lanceolate, pendulous, up to 38 cm long, 3 mm wide. Stipe absent. Lamina decurrent on rhachis, wings extend down to rhizome, apex acute, texture thin. Rhachis in faint relief abaxially. Veins only visible by transmitted light. Sorus dark brown, positioned entirely on abaxial margin of lamina, extending in most fronds from above 7cm at base to 2 - 0.5 cm., clear of tip. Paraphyses turbinate, 0.18 mm long ´ 0.04 mm wide, pale amber in colour. Paleae (sporangial stalks) present. Spores monolete, ovate to reniform., 0.05 mm long × 0.03 mm wide.

D i s t r. Southern India and Sri Lanka. Over a wide range of altitudes 150 m to 1900 m.

E c o l. These epiphytes prefer shade; in disturbed locations exist on road side trees, or in rock crevices.

N o t e: The species is rare, but might have been overlooked. Herbarium specimens share a common characteristic: the top-half of all fronds of a given plant bend in the same direction, looking like a brush in use.

S p e c i m e n s   E x a m i n e d: KALUTARA DISTRICT: Epiphyte on road side tree, 4 miles south of Moragala, 30m., 20 Jan. 1951, *Sledge 883* (K); Megahatenne (2 miles south of Moragalla), on tree trunk by road side, 20. Jan. 1951, *Ballard 1519* (K). RATNAPURA DISTRICT: Near Ratnapura, in rock crevices near road side, 4 Jan. 1951, *Ballard 1376* (K); Epiphytic, near rocky road side bank in jungle near Ratnapura, c.150m., 4 Jan. 1951, *Sledge 810* (K); Carney near Ratnapura, 9 Mar. 1954, *Sledge 1249* (K). NUWARA ELIYA DISTRICT: Nuwara Eliya 1900m., April 1899, *Gamble 27572* (K). East of Nuwara Eliya, forest slope above Hakgala Botanic Gardens on north-east side of Hakgala Mt., 23. Sept. 1993, *Fraser-Jenkins* with *Jayasekara, Samarasinghe* and *Abeysiri 204* (Field no., 2 sheets) (K). LOCALITY UNKNOWN: *V. zosterifolia* var. *flaccida* Mett. pro parte, *Walker 119* (B, type); Gosset p.p. (B, type); *V. microlepis Robinson 33a* (K).

Vittaria microlepis var. **thwaitesii** Hieron., Hedwigia 57: 203. 1916. *V. lineata* sensu Thwaites, Enum. Pl. Zeyl. 381. 1864. Bedd., Ferns Brit. India 407 p.p., as to species from Sri Lanka and S. India only, not as to *Taeniopsis lineata* sensu Bedd., Ferns S. India pl. 54. 1863!

2. ANTROPHYUM

Kaulf. Enum. filic.: 197. 1824. Type species: *Antrophyum reticulatum* (G. Forst.) Kaulf. Enum. filic. 198. 1824. Selected by Benedict from Upolu. (see below)

Solenopteris Wall., Herb.1823 R.D. Dixit and Nair, J. Indian bot. Soc. 53: 277-287. 1974; K.U. Kramer *In* Kubitzki (ed.) Fam. Gen. Vasc. Pl. 1: 274, 275. 1990.

Genus once regarded as family (Antrophyaceae), now divided into three genera: *Antrophyum* s. str., OW, *Polytaenium* and *Scoliosorus* NW.

The generic description refers to Sri Lankan taxa alone: Small to medium sized epiphytic or epilithic ferns with short-creeping or suberect rhizome. Fronds tufted, simple, thin or leathery, shape variable: lanceolate, oblanceolate, linear, spathulate or suborbicular, ± clearly distinguished stipe. Rhachis lost above lower third of frond. Sori, immersed over lateral veins in peripheral grooves of elongated, polygonal (4,5,6) aereoles, appear as 'soral lines' because narrow horizontal border of aereole often not bridged by sorus. Paraphyses septate-filiform or club-shaped, species specific. Spores trilete.

D i s t r. A pantropical genus of about 50 species.Representation in Sri Lanka: Two species (Sledge, 1982): *A. plantagineum., A reticulatum.*

E c o l. Epiphytic or epilithic, 'on moist mossy tree trunks' along stream banks in forests; varying altitudes:

KEY TO THE SPECIES

1 Stipe short, almost indistinguishable from decurrent lamina; paraphyses without support from series of cells, end cell filiform . . . .
. . . . **2. A. reticulatum**
1 Stipe distinct from lamina; paraphyses with branched cellular support, end cell club- to pear-shaped . . . . **1. A. plantagineum**

1. Antrophyum plantagineum (Cav.) Kaulf. Enum filic.: 197-198. 1824.
Index. fil.61. 1906. Bedd., Handb. Ferns Brit. India 403. 1883.
Type: *Hemionitis plantaginea.* Cav. prael. *643.* 1801. locality Marianas, Philippines e Guahan retulit *Chamisso*

Hemionitis plantaginea Cav., Descr. 260. 1802.
Antrophyum reticulatum sensu Bedd., Ferns S. India t.52. 1871.
Literature: R.D. Dixit and Nair, J. Indian bot. Soc. 53: 277-287. 1974. K.U. Kramer *In* Kubitzki (ed.), Fam. Gen. Vasc. Pl. 1: 274, 275. 1990.

Illustration: Beddome, Ferns of South. India t. 52. 1863. R.D. Dixit and Nair, J. Indian bot. Soc. 53: 277-287. figs. 39-45. 1974.

Rhizome short-creeping to suberect ca 3mm in diam., densely covered in clathrate, black scales.Scales 4-5mm long 0,5-0.7mm wide with pads of thin-walled cells across blade of scale, frequently ending in acuminate tip resembling barbed wire (see family description). Fronds tufted, 24 -(to 39)cm long, 5.5cm wide, entire, mid to dark green. Stipe distinct, 5-11cm long, 2 mm wide at base, increasingly wider towards lamina, clathrate scales in 2cm high zone at base. Lamina obtuse-acuminate, tip occasionally replaced by blunt and notched end; texture coriaceous, glabrous. Rhachis distinct, frequently dark dividing into three main veins on entry to lamina. Veins adaxially form raised network of polygonal aereoles 8-16 times as long as wide, these spread towards margin; support soral grooves abaxially; sterile margin outside commissure 2-3mm wide. Sorus linear, dark brown, overlies network of veins, mature first at margin inside commisure, here up to 1.2mm wide, progressively filling soral grooves, 0.5mm wide, towards angustate base of lamina, several grooves often end at same level, giving horizontal cut-off to fertile zone. Paraphyses on cellular support with lateral branches, terminal cells dark rust-brown, pear-shaped 0.1-0.15 × 0.08-0.09 mm. Spores trilete 0.05 mm., transparent, pale yellow.

D i s t r. Predominantly at high elevations 1200-1800 m in the Central Province.

E c o l. Low epiphytes or lithophytes, along shaded stream banks, wet jungle trees and on trees in the protective environment of gullies.

S p e c i m e n s   E x a m i n e d. KANDY DISTRICT: stream gulley in forest above and S.E. of "Hunas Falls Hotel", west side of Hunasgiriya mountain, S.S.E. of Elkaduwa, N.E. of Kandy, Alt. 1200-1300 m., 25 Aug. 1993, *Fraser-Jenkins* with *Jayasekara, Samarasinghe & Abeysiri, Field No. 34* (K); Medamahanuwara, *v. Fridau* anno 1853 (GZU). BADULLA DISTRICT: Glenanore Estate, Blackwood Division, Blackwood Forest, at mile post 15 on Welimada-Haputale road, forest along stream., 1525 m., low epiphyte or lithophyte. 18 Nov. 1976. *Faden, 76/344* (K); Forest near Bandarawela, 15. Sept. 1995, *Jayasekara with Suranjan 438* (K).Central province, 1219m., *Brekelt (?) 216* (K). NUWARA ELIYA DISTRICT: Adams Peak, 14 Feb. 1908, *Matthew* (K); Hakgala 1829m., epiphytic in jungle, 16 Dec. 1950 *Sledge 640* (K); Hakgala Peak, 1829 m 16 Dec. 1950, on wet jungle trees with mosses *Ballard 1124* (K); Hakgala Peak, 1829 m 16 Dec. 1950, on wet jungle trees with mosses *Ballard 1124* (K); near Hakgala, c.1676 m on trees in forest. 23 Dec. 1950, *Holttum S.P.N. 39175*(K); Hakgala Natural Reserve, 2000 m., epiphytic on moss-covered tree trunks, 19 May 1971, *Jayasuriya 172* (K); Hakgala Botanic Gardens and slope of Hakgala Mt. 1720-1820 m., on mossy tree trunks, 14 Nov. 1976 *Faden 76/269* (K). MONARAGALA DISTRICT: Kuda Oya on Ramboda Road 1676-1760 m., 28 Dec.1950, on tree trunk in gully, *Ballard 1283* (K). LOCALITY UNKNOWN: *Thwaites C.P. 3290; Clarke, C.P. 3290.* "Ceylon -42 - I. R." Coll. Col. *Robinson*, C.B. (K).

2. Antrophyum reticulatum (G. Forst.) Kaulf., Enum. filic. 198. 1824; Bedd., Handb. Ferns Brit. India. Suppl. 102. 1892; C. Chr., Index filic. 61. 1906. R.D. Dixit and Nair, J. Indian bot. Soc. 53: 277-287. 1974. K.U. Kramer *In* Kubitzki (ed.), Fam. Gen. Vasc. Pl. 1: 274, 275. 1990.

Hemionitis reticulata G. Forst., Prod. 79. 1786.
Antrophyum reticulatum sensu Bedd., Handb. ferns Brit. India. 401-403. 1883.

Illustrations: Bedd., Handb. Ferns Brit. India. 401, fig. 235. 1883.
R.D. Dixit and Nair, J. Indian bot. Soc. 53: 277-287, figs 15-18. 1974.

Rhizome short-creeping ca 8mm in diam., almost invisible among tight ball of roots, densely covered in almost black, clathrate scales. Scales 3-4mm long, 0.5mm wide, almost black, clathrate radial walls of almost uniform width, margin of scale only apparently spiny (cf. family description). Fronds tufted, elongate-lanceolate, up to 48cm long, 3,8 cm wide, entire, widening with increasing maturity, mid- to dark-green. Stipe indistinct, flat, ca 5.5 cm long if extending to lowest reticulation of lamina. Lamina lanceolate, margin slightly wavy, decurrent on stipe down to base, texture coriaceous, glabrous. Rhachis flat, from an early stage as dark as stipe, very distinct in lower third of lamina, then, within a ca 2cm wide sterile zone, abruptly decreasing in width, changing to narrow soral groove and becoming lost in elongate-reticulate network of venation. Veins form reticulate net of polygonal aereoles ± 16 times as long as wide, predominantly parallel to longitudinal axis of lamina, network raised on adaxial surface; in absence of soral lines, comissure inside sterile marginal zone veins buried in deep laminar tissue. Sori cover abaxial surface; soral grooves, ca 0.3mm wide, trace reticulate venation; linear sori ca 1mm wide when mature, dark brown. Paraphyses without stalk, filiform., sinuous or twisted 2-6 times with blunt or pointed (but never acuminate) tip 0.3mm × 0.02mm- 0.45 × 0.5mm., pale honey-brown. Spores trilete 0.05 mm.

D i s t r. From Galle to Central Province in a wide range of altitudes from 305-1829m.

E c o l. Epiphyte on forest trees.

S p e c i m e n s   E x a m i n e d. KURUNEGALA DISTRICT: *A. reticulatum* (Forst.) Kaulf., Doluwakanda Hill, epihyte in forest, occasional, 5 Jan. 1977, *Faden 77/50* (K). KALUTARA DISTRICT: Pelwatta on tree trunk by path leading to Pahala Hewessa, 20 Jan.1951, *Ballard 1532* (K). NUWARA ELIYA DISTRICT: Nuwara Eliya 1524-1829m., var. b. Blume, *Gardner 1229* (K); Adams Peak, 914m (lge spec.!) (K). GALLE DISTRICT: as *A. callifolium* Blume: Himdoom., *v. Fridau* anno 1853 (GZU); as *A. coriaceum* (Don.) Wall., Narawelle(?) near Galle, on trees, *Champion* (K); Sri Lanka, Haycock Mt., 305m., 22 Jan. 1951, *Sledge 915* (K). MATARA DISTRICT: Masmulla Forest Reserve, 26 Sept. 1996, *Jayasekara with Samarasinghe 457* (K); same date, place and collectors *466* (K). LOCALITY UNKNOWN: *A. reticulatum* Kaulf., *Gardner 1173* (K); *Gardner 1305* (K); *Gardner 1307* (K); *Gardner 1308* (K); Ceylon - 41 - I.R., coll. *Robinson, C.B.* (K); var. *d.* Blume, *Gardner 1228* (K); coll. Major *Skinner*, anno 1932 (K).

3. MONOGRAMMA

Comm. ex Schkuhr, 24. Kl. Linn. Pfl.-Syst. 1:82, 3: pl. 87. 1809.
K.U. Kramer *In* Kubitzki (ed.), Fam. Gen. Vasc. Pl. 1: 274 + 276. 1990.

Type species: *M. graminea* (Poir.,) Schkuhr, E. African Islands
Vaginularia Fée, Congres scientif. de France, 10. session, I: 178. 1843.
(-Exposit des genres, etc., p.97, G.38.) Mem. Fam. Foug. III, 1851-1852.
Pleurofossa *Nakai , 1936*

The genera *Monogramma* and *Vaginularia* were united by Kramer and described as: small to minute epiphytes with creeping rhizome, fronds remote to aggregate, simple, linear, herbaceous, rhachis ensuring entire vascular supply; soral line over rachis sunken in leaf tissue or protected by flange of leaf tissue borne on costa or 'on a lateral vein'; a few 'obscure lateral veins' are suspected by the present author, of being misinterpreted spicular cells. Paraphyses assume shape of sinuate to twisted soral trichomes without further cellular support. Spores trilete. Kramer states: "On the basis or absence of lateral veins this genus has been divided in two. The presence of a flange protecting the soral line formed the basis for another segregate, *Pleurofossa*. These characters may serve for species distinction, but not for separating genera."

Monogramma paradoxa (Fée) Bedd., Suppl. ferns Brit. India: 24. 1876.

Pleurogramme paradoxa Fée, Mém. Fam. Foug. III: 38, pl. 4, f.4. 1851-52

Type: habitat in Oualan= Kusaie, Polynesia (5.19N 162.59E) *Martens 267* (LE)
Monogramma junghuhnii Hooker *In* Bedd., Ferns South. India : 71, t. 210*. 1863.
Monogramma junghuhnii sensu Bedd., Bedd. Ferns S India : 71,t. 210* non Hooker ?.1863.
Vaginularia junghuhnii Mett. Fil. Hort. Lips.: 25, t.27, fs 24-28. 1856.
Monogramma junghuhnii (Mett.) Hook., Sp. Fil. 5: 123 p.p. 1864. Bedd. Handb. Ferns Brit. India: 375, fig. 214*. 1883. (as *Monogramme paradoxa* Fée)
Two syntypes: locality: (?) but both Junghuhn and Zollinger collected in Java
a) *Junghuhn s.n.* in herbarium Kunze (B?) (LZ destroyed),
b) *Zollinger 1890* (B?) or Bogor ?, Leiden ?
* same illustrations (reduced in 1883)

D i s t r. in the warmer parts of the OW from Madagascar and Sri Lanka to Melanesia.

Illustration Bedd., The ferns of Southern India Brit. t. 210 (as *Monogramme junghuhnii*)

Rhizome wide-creeping, branching, 1-1.5mm in diam., black; densely clothed in in clathrate, iridescent scales one cell thick, of variable shape:2-3 cells wide with long extended linear tip or broader, 5-7 cells wide with shorter tip. Marginal cells with very thin outer periclinal; these thin walls collapse on cell below, remaining radial walls causing illusion of marginal spines; all other radial walls with thick secondary or, in central cells, massive tertiary thickening. Fronds simple, grass-like, closely aggregated, 2-6cm long, green. Stipe 0.5-2mm long, transition to blade imperceptible. Lamina partly decurrent on stipes, 0.5-1.5mm wide, parenchyma with translucent spicular cells,

0.2-0.7mm long, that can be confused with vascular tissue, here inner profile of radial wall smooth or in larger cells pitted, cells fusiform or shallowly sigmoid at blade margin which they support and define.Nerve single, divides dichotomously to enclose receptacular tissue as a fusiform island, resulting ridges bordering sorus 0.3-2cm long. Indusium single, almost linear, originates from left ridge, clasping over sorus at base then narrowing in shallow steps; dichotomous strands do not fuse but end each in a narrow lip at same level below tip of blade; gap to tip bearing stomata. Paraphyses 0.2-0.4 mm long 0.02-0.07mm wide, septate, undifferentiated trichomes, slightly widening towards large blunt end cell. Spores trilete-globose, 0.03mm.

D i s t r. At moderate elevations at about 900m in Central Province

E c o l. Epiphytes on bark of trees and lithophytes.

S p e c i m e n s   E x a m i n e d. as *M. paradoxa* (Fee) Bedd. KEGALLA DISTRICT: Lonach near Ginigathena at Norton Bridge, 914m., rocks by stream., 13 Dec. 1950, *Sledge 603* (K). KANDY DISTRICT: Laxapana road, 28. Jan. 1954, *T.G. Walker, T228* (BM); *Vaginularia paradoxa, Thwaites 1281* (K); *V. paradoxa* (Fée) Mett., Rassawa (?) anno 1848, *Gardner, C.P. 1281*(K); *M. graminea, Gardner 1281* (Thomas Moore's Fern Herbarium., (K); *M. graminea*, Sri Lanka, *Gardner 1281* (K); *M. junghuhnii* (Hook.) Beddome's Herb.; *C.P.1281* (used for t. 210 F S I) (K). LOCALITY UN-KNOWN: *M. junghuhnii* Hook., 1836 I. R., coll. *Col. Robinson* (K); *V. junghuhnii* Mett., coll. *Skinner* (presented 1932, K); *V. junghuhnii* Mett., *Gardner 1281* (illustrated: Mettenius t. 27, f. 24-27) (K); *V. junghuhnii, Clarke C.P. 1281* (K).

WOODSIACEAE

(by C.R. Fraser-Jenkins*, *Diplazium* by M.J. Zink**)

(Diels) Herter, Rev. Südamer. Bot. 9: 14. 1949. Type genus. *Woodsia* R.Br.
Athyriaceae Alston, Taxon 5: 25. 1959, nom. superfl.
Hypodematiaceae Ching, Act. Phytotax. Sin. 13: 96. 1975.
Misapplied names. *Dryopteridaceae* Ching, *Aspidiaceae* Mett. ex Frank, nom.
   illeg.

Plants usually terrestrial, sometimes lithophytic, with small to very large
sized, usually typically fern-like fronds. *Rhizome* varying from thin (rarely
thick) and wide-creeping to massive and erect, or even trunk-like, the creeping
rhizomes bearing occasional, often thin stipe-bases along their length, the
upright ones densely surrounded (encased) by old stipe-bases which often
serve as storage-organs, ovate to lanceolate scales mostly limited to apices.
Stipes terete, usually grooved adaxially, often swollen at the base, surrounding
two, usually strap-shaped vascular strands which unite apically into a
U-shaped strand, in large species with small additional strands on each side,
bearing scattered or dense, narrowly lanceolate to ovate scales, at least at the
very base; *rhachis* terete, usually adaxially grooved, the groove may or may
not be continuous with grooves from the pinna-costae; *lamina* rarely simple,
usually compound and often highly dissect, up to four times pinnate, pinnules
anadromous, or becoming catadromous, veins free, or sometimes anastomosing,
surfaces glabrous, or frequently bearing fibrils, hairs or glands. *Sori* ± medial,
sessile, highly variable from ± orbicular or reniform to, more usually, elongated
and curved, J-shaped, U-shaped or straight; the indusium in species with
rounded or reniform sori attached at one side and covering the sorus (in the
area), or inferior, those with elongated sori with the indusium attached along one
side and sometimes highly fugaceous, few taxa are exindusiate. *Chromosome
base number* (in the area) x = 40, 41. Sometimes both in one genus.

A small family of about 10 genera, (overestimated, especially in China),
but *Athyrium* and *Diplazium* both contain about 650 species world-wide,
numbers are uncertain in the largest genus, *Diplazium*, occurring world-wide,

* c/o A.M. Paul, Botany Department, The Natural History Museum, Cromwell Rd.London
   SW7 5BD
** Plant Ecology and Systematics, Biology Department, TU Kaiserslautern, P.O. Box 3049,
   D-67653 Kaiserslautern, Germany

especially in the palaeo-tropical lowland and montane areas, but extending to the far north, less numerous in the further southern hemisphere. Four genera occur in Sri Lanka.

In the present and modern sense, excluding *Peranema*, the name *Woodsiaceae* should be used for this family and not *Athyriaceae* under the principle of priority and in accordance with Pichi Sermolli's (1977, but not 1970) choice. Ching (1978) oversplit the family unnecessarily, into three (see synonyms above), others (e.g. Kato & Kramer 1990) included the clearly distinct *Onocleaceae* Pichi Sermolli within it. Alston (1956), Pichi Sermolli (1977) and the present scheme adopt a middle position. Until the 1930s or 1940s the major part of the family, those with elongated sori, was unnaturally grouped with the *Aspleniaceae* due to the trend of Hookerian overemphasis of soral characteristics alone (see Sledge 1973a for details), consequently most *Athyrium* and *Diplazium* species were placed under *Asplenium* in the older literature. Species with ± orbicular sori, (*Hypodematium.*, *Gymnocarpium*) were usually placed in *Polypodium.*, *Phagopteris*, *Aspidium* or *Dryopteris*.

Two ± natural groups are recognised within the family, the subfamilies *Woodsioideae* (type: *Woodsia*), a monotypic group, with ring-shaped inferior indusia, and *Athyrioideae* (type: *Athyrium*) which is the larger group and the only one present in Sri Lanka. Ching's (1978) treatment of a family, *Hypodematiaceae*, might suggest that a third subfamily should exist, the *Hypodematoideae* Nayar, but *Hypodematium* can be rather easily accommodated within the *Athyrioideae* from its frond-morphology and somewhat open-sided, reniform sori and indusia; its well-known dorsiventral rhizome would appear to be only a secondary development arising from its creeping, lithophytic habit.

KEY TO THE GENERA

1 Sori orbicular or reniform
  2 Lamina, axes and indusia bearing numerous unicellular acicular hairs, rhizome creeping, bearing fronds along the top surface only . . . .
  . . . . **1. Hypodematium**
  2 Lamina, axes and indusia without hairs, or with multicellular, non-acicular hairs, rhizome ascendent, bearing an apical crown of fronds . . . .
  . . . . **2. Deparia**
1 Sori elongated, J-shaped, or occasionally U-shaped
  3 Frond-axes and lamina hairy . . . . **2. Deparia**
  3 Frond-axes and lamina non-hairy, though sometimes with minute unicellular glands
    4 Fronds simple . . . . **4. Diplazium**
    4 Fronds pinnate to tripinnate
      5 Veins anastomosing . . . . **4. Diplazium**
      5 Veins free

6 Stipe-bases and lower stipes thick (more than 4 mm in diam.) . . .
. . . . **4. Diplazium**
6 Stipe-bases and lower stipes thin (less than 3 mm in diam.)
7 Pinnae simple, very shallowly lobed, with large rectangular lobes
. . . . **4. Diplazium**
7 Pinnae compound, or if ± simple, deeply pinnatifidly lobed, with
small ± rounded, pointed, or cuneate-based lobes . . . .
. . . . **3. Athyrium**

1. HYPODEMATIUM

Kunze, Flora 16: 690. 1833.

References. B.K. Nayar & N. Bajpai 1970. A reinvestigation of the morphology of *Hypodematium crenatum*., Amer. Fern J. 60: 107–108 et tt.; D.S. Loyal 1972. Morphology of *Hypodematium crenatum* (Forssk.) Kuhn: comments on a recent paper, Amer. Fern J. 62: 88–92 et tt.; R.C. Ching 1975. Two new fern families, Act. Phytotax. Sin. 13: 96–98; R.E.G. Pichi Sermolli 1977. Tentamen Pteridophytorum genera in taxonomium ordinam redigendi, Webbia 31(2): 313–512.
Type species: *H. onustum* Kunze (= *H. crenatum* (Forssk.) Kuhn).

Medium-sized (up to c. 70 cm tall) lithophytic ferns. *Rhizome* thick, long creeping, densely covered in scales, dorsiventral anatomy, i.e. the usually discrete frond-traces arise from the upper surface and the roots below. *Stipe* swollen at the base, narrow, hard and stiff, terete, scales only at the base, containing two vascular-strands, grooved adaxially; *rhachis* similar to upper stipe, the groove open to and thus continuous with those joining it from the pinna-costae; *lamina* three to four times pinnate, deltate, lowest basiscopic pinnules elongated, lowest pinnules in first pinna anadromous, veins free, lamina bearing scattered hairs ad- and abaxially, just as *costules* and subsequent axes are densely covered in unicellular acicular hairs on both sides; *ultimate segments* stipitate or narrow-based, adnate, asymmetrical and slightly decurrent at their bases above; apices lobed, ± bluntly pointed. *Sori* medially on predominantly acroscopic veinlets of lobes, reniform., surrounded by a reniform to hippocrepiform indusium., sides ± down - curved over sorus, attached at basiscopic side in notch of the sorus, bearing long, locular hairs and lifting on maturity. *Spores* monolete, perisporate. *Chromosome base-number* x = 40 and/or 41.

A small genus of 6–7 species (over-estimated in China), of Sino-Himalayan and Sino-Japanese distribution, but one species more widespread and extending eastwards to SE Asia and westwards to Africa and Macaronesia.

The genus had long been placed in the *Dryopteridaceae* (or *Aspidiaceae*) on account of its reniform indusia and has even been sunk within *Dryopteris*.

Sledge (1972) placed it with the tectarioid ferns, i.e. in the *Dryopteridaceae* subfamily *Tectarioideae*. Because it seemed rather difficult to ally with other genera. Ching (1975) created a new family for it. But the two vascular strands in the stipe, open-based costal grooves and usually somewhat widely notched, reniform sori, with the indusium attached basally in the notch ally it to the athyrioid ferns, as aptly pointed out long ago by Hope (1901: 744). It is not clear whether there are really two chromosome base-numbers in the genus and at least the number of 40 needs confirmation, but both numbers occur in the *Woodsiaceae* (and *Dryopteridaceae*) and some other athyrioid genera contain species with either number.

Hypodematium crenatum (Forssk.) Kuhn in von Decken, Reisen Ost-Afr., Bot. 3(3): 37. 1879.

Polypodium crenatum Forssk., Flor. Aeg.-Arab.: 185. 1775. Type. from Yemen.
Aspidium odoratum Bory ex Willd., Linn. Spec. Plant., ed. 4, 5: 286. 1810, non Mett. 1858.
Hypodematium onustum Kunze, Flora 1833(2): 690. 1833.
Hypodematium ruppellianum Kunze, Farnkr. Schkuhr. Farnkr., Suppl. 1: 41 et t. 1840.
Aspidium chrysolepis Fée, Mém. Fam. Foug. 8, Desc. Foug. Exot.: 107. 1857.
Lastrea eriocarpa (Decne.) Bedd., Ferns S. Ind.: 33–34, t. 95. 1863.
Dryopteris crenata (Forssk.) O. Kuntze, Rev. Gen. Plant. 2: 811. 1891.
Hypodematium eriocarpum (Decne.) Ching, Flor. Tsingling. 2: 130. 1974.
Hypodematium hirsutum (D. Don) Ching, Id. Fern J. (1–2): 49. 1985 [„1984'].

Rhizome thick, branching only occasionally, fully prostrate, often wedged tightly into crevices and extending on rock-surface beyond them., roots arising from bottom surface, bearing fronds towards apices of the rhizome which is completely clothed with a dense mass of golden-brown to dark reddish, thin, lanceolate scales with long, attenuate apices, and which bears a series of yellow-grey, straw-like old frond-bases ± at intervals (occasionally close-packed) along its length from the top surface. Fronds usually 1–3(–4), erect or sometimes horizontal, seldom pendent, up to c. 70 cm tall. Stipe of medium thickness, stiff, grooved adaxially, ± same length as, or slightly shorter than lamina, the very base widened bearing a prominent and dense tuft of scales similar to rhizomal ones, glabrous above. Lamina varying from bipinnate and tripinnatifid in smaller plants to tripinnate and quadripinnatifid, deltate, as wide as, or often wider than long, (base up to c. 30 cm wide), softly herbaceous, pale-green, densely covered with stiff, upright, acicular, whitish hairs all over axes and both surfaces, particularly below, mainly on veinlets; pinnae, up to c. 10(–12) pairs, crowded or overlapping, deltate to elongated triangular, lowest ones up to c. 15 cm long, 12 cm wide, with up to c. 1 cm long stalks (shorter in smaller fronds), mostly symmetrical, only lowest pair basiscopically developed, being up to 0.5 times as long as acroscopic ones (lowest basiscopic one being the longest in frond), the lower ones ± subopposite, bipinnate or tripinnatifid, with up to c. 9 pairs of pinnules. Rhachis glabrous. Costae and costules bearing a covering of dense, unicellular, acicular hairs; pinnules ± symmetrical, very shortly stalked or first narrow, above more widely adnate, elongated triangular, ± adjoining, c. 2 cm up to c. 0.8 cm long, 0.9 – 0.4 cm wide, apices ± acute or ± bluntly pointed, pinnate, up to c. 5 pairs of pinnulets. Sori c. 6–12 per pinnulet, borne throughout the lamina and often present even in very small sporelings, medial on veins, or often nearer the midrib, reniform. Indusium ± thick, white (but appearing grey due to the black sporangia below), reniform., often with a rather open basiscopic notch and here attached to lamina, tall and surrounding the sorus, densely clothed with ± long, stiff, acicular hairs, lifting somewhat and turning light-brown on ripening, but persistent for some time until partly dropping off on old fronds. Spores very small, uniform., perispore of small, short knobs. Cytotype diploid, sexual, n = 41 (with 41 bivalents at meiosis).

D i s t r. Occurs at the interface between the dry and wet zones in north-central Sri Lanka. Extremely rare (known from only one locality apart from a possibly adventive plant). A widespread Sino-Himalayan species whose distribution extends westward to Africa and Macaronesia and eastward to southeast Asia, Sri Lanka, to S.W. and C. China, Taiwan, Thailand and Malesia. Reports from Japan, Polynesia, E. China and perhaps Taiwan refer to other, related species.

E c o l. Occurs in crevices of limestone cliffs at c. 700 m. in Sri Lanka.

N o t es. Fraser-Jenkins (1993) described subsp. *loyalii* Fraser-Jenk. to represent the Himalayan (and S. Indian) tetraploid plant, first discovered by Loyal (in Mehra & Loyal 1965), which has smaller segments, a more dissect frond and fewer and considerably shorter laminar hairs. Loyal (in Loyal, Paik & Tiwana 1977) also found a sterile triploid hybrid between the two. It is possible that the tetraploid could be treated as a species in its own right. The Arabian and Cape Verde plants, as well as the Sri Lankan plant, appear similar to the west Himalayan diploid plant.

subsp. **crenatum**

As described above. The only taxon present in Sri Lanka.

S p e c i m e n s   E x a m i n e d. KANDY DISTRICT: Nitre Cave, *Naylor-Beckett 820* p.p. (in *Thwaites C.P. 3888* p.p.) (PDA); Lagalla [probably the same locality], *Naylor-Beckett 820* p.p. (in *Thwaites C.P. 3888* p.p.) (PDA); at entrance to and beside Nitre Cave (Wawula Gallena [= Bat Cave], c. 1.5 miles above and S.E. of Pahalena Ella, E. of Mimure, N.N.E. of Corbet's Gap (Attala Meluwa), N. of Hunasgyria, N.E. of Kandy, 700 m., *Fraser-Jenkins Field no. 302*, with *Abeysiri, Abeysiri* and *Gunawardena*, 14 Oct.

1993 (K, US, PDA, NMW). NUWARA ELIYA DISTRICT: Nuwara Eliya, 1050 m., *Schmid 1335*, Feb. 1954 (BM) [probably adventive]. LOCALITY UNKNOWN: *Wall s.n.* (K).

2. DEPARIA

Hook. & Grev., Icon. Fil.: t. 154. 1829.
References. W.A. Sledge 1962. The Athyroid ferns of Ceylon, Bull. Br. Mus. nat. Hist. (Bot.), 2: 277–323; R.C. Ching 1964. On some confused genera of the family Athyriaceae, Act. Phytotax. Sin. 9: 41–84; M. Kato 1977. Classification of *Athyrium* and allied genera of Japan, Bot. Mag. Tokyo 90: 23–40; M. Kato 1984. A Taxonomic Study of the Athyrioid Fern Genus *Deparia* with Main Reference to the Pacific Species, J. Fac. Sci. Univ. Tokyo, sect. 3, 13: 375–429.
Type. *D. prolifera* (Kaulf.) Hook. & Grev.

Lunathyrium Koidz., Act. Phytotax. Geobot. 1: 30. 1932.
Dryoathyrium Ching, Bull. Fan Mem. Inst. Biol. 11: 79. 1941.
Parathyrium Holttum., Kew Bull. 13: 448. 1958, nom. superfl.
Athyriopsis Ching, Act. Phytotax. Sin. 9: 63. 1964.

Misapplied names. *Athyrium* Roth, *Diplazium* Sw., *Cornopteris* Nakai.

Medium up to c. 160 cm tall, terrestrial ferns. Rhizome varying from thin and long-creeping with fronds arising at intervals, to thick and upright with an apical crown of fronds. Stipe swollen at the base, grooved adaxially, containing two vascular strands, base thick or thin, ± succulent, bearing ± scattered scales, at least below, and usually bearing multicellular, columnar hairs above and some intermediate, hair-tipped scales. Lamina varying from being deeply pinnatifid to tripinnate and quadripinnatifid, linear-lanceolate to deltate-lanceolate, veins free, lamina bearing multicellular, columnar hairs and small, scattered, fibrillose scales on the veinlets. Rhachis similar to upper stipe, with the groove separate and not open to those reaching it from the pinna-costae, hairy and scaly. Costae, costules and subsequent axes bearing similar hairs and scales on both sides; ultimate segments often ± unlobed, usually symmetrical. Sori orbicular to elongated (and curved or straight), medial on veinlets or running along them., not reaching the segment-margin; indusium naked, thin or fugaceous, often with fimbriate edges, varying from subveniform (dryopteroid) to elongated (athyrioid). Spores monolete, perisporate. Chromosome base-number x = 40.

A genus of about 40 species (over-estimated in China), with many of the species forming complexes of closely similar, related species or subspecies. The species are mainly of Sino-Himalayan and Sino-Japanese distribution, extending to Africa, but with smaller centres of evolution in S.E. Asia, Hawaii, Madagascar and one species in N. America.

Its species have been misplaced into *Athyrium* and especially *Diplazium,* from both of which they are readily distinguishable by their columnar hairs. The species-complexes still require further elucidation. M. Kato (1984) and others have recognised four sections, two of which occur in Sri Lanka, but the sections are very closely related and perhaps not of very much taxonomic consequence. All four, however, were treated as genera by Ching (1941, 1964 and 1978). Because of its large, tripinnate fronds and round sori, *D. boryana* (Willd.) M. Kato has persistently been treated, in south Asia, at least, in *Dryoathyrium* as a separate genus, however, there are several other species within *Dryoathyrium* that have less dissect fronds and oval or even U-shaped or J-shaped sori, clearly linking it to the rest of the genus. *Dryoathyrium* is therefore not recognised here, in accordance with Kato's treatment.

KEY TO THE SPECIES

1 Fronds simple .... **1. D. lancea**
1 Fronds pinnatifid to tripinnate-quadripinnatifid
  2 Fronds linear-lanceolate and pinnatifid, with short lobes, or just bearing pinnate at the very base .... **2. D. zeylanica**
  2 Fronds widely lanceolate to deltate, pinnate-bipinnatifid to tripinnate-quadripinnatifid
    3 Fronds pinnate-bipinnatifid to bipinnate, sori elongated
      4 Rhizome long-creeping subterraneously, lower pinnae not usually the longest .... **3. D. petersenii**
      4 Rhizome erect, lower pinnae usually the longest ....
      .... **4. D. polyrhizos**
    3 Fronds tripinnate to quadripinnatifid, sori orbicular ... **5. D. boryana**

Sect. Athyriopsis

(Ching) M. Kato, J. Fac. Sci. Univ. Tokyo, sect. 3, 13: 401. 1984.

1. Deparia lancea (Thunb. ex Murray) Fraser-Jenk., New Species Syndrome in Indian Pteridology and the Ferns of Nepal 101. 1997. Sato et al., J. Pl. Res. 113(1110): 162. 2000 (as (Thunb.) R.Sato, comb. superfl. et illeg.).

Asplenium lanceum Thunb. ex Murray in L., Syst. Veg., ed. 14: 933. 1784. Id., Fl. Jap. 333. 1784. Thwaites, Enum. Pl. Zeyl. 385. 1864, non *Diplazium lanceum* Bory.
Type: Japan, *Thunberg s.n.* (UPS).
A.subsinuatum Hook. & Grev., Ic. Fil. 1: pl. 27. 1827. Type: Nepal, 1820, *Wallich s.n.* (K)

Diplazium lanceum (Thunb. ex Murray) C. Presl, Tent. Pterid. 113. 1836. Bedd., Handb. Ferns Brit. India 174. 1883, non Bory 1833.

Diplazium subsinuatum (Hook. & Grev.) Tagawa, Col. Ill. Jap. Pterid. 203. 1959. Sledge, Bull. Br. Mus., nat. Hist. (Bot.) 2(11): 295. 1962. id., Bot. J. Linn. Soc. 84: 17. 1982.

Rhizome long-creeping, scaly; scales narrowly subtriangular, up to 3 mm long, concolorous. Stipes distant, slender, 10-15 cm long, dark-stramineous, scaly at base; scales linear, long-acuminate, entire, almost black. Lamina simple, narrowly lanceolate with subentire to crenate margin, 15-30 ′2-3 cm., attenuate towards base and apex, glabrous, subcoriaceous; costa prominent beneath, slightly grooved above; veins pinnate, the acroscopic branch of each group usually fertile; sori linear, distant, elongate from near the costa to near the margin, mostly simple, sometimes diplazioid.

D i s t r. Widespread in E. Asia and the Himalayas, Sri Lanka, Philippines.

E c o l. Terrestrial in forest up to 1000 m altitude.

N o t e. Hitherto usually placed in *Diplazium* as *Diplazium subsinuatum* but included here following Fraser-Jenkins (1997) and based on most recent research including molecular techniques (Sato et al. 2000).

S p e c i m e n s   E x a m i n e d. MATALE DISTRICT: Matale East, 600-900 m., *s.coll.* in *C.P. 1335* (PDA); Matale East and Saffragam., 610-914 m.., *Wall 38/200* in Herb. Barkly (BM). KANDY DISTRICT: Murata, Jan 1848, *Gardner s.n.* in C.P. 1335 (PDA). RATNAPURA DISTRICT: Sabaragamuwa Prov., Kitulgala, 260 m., 28 Jan 1954, *Schmid 1082* (BM., G); Bambarabotuwa Rorest Reserve, in shade of rain forest near stream., 650 m., 19 July 1971, *Jayasuriya 245* (PDA). NUWARA ELIYA DISTRICT: Wattegoda, Feb 1854, *s.coll.* in *C.P. 1335* (PDA); *Gardner* in *C.P. 1335* (K); *Thwaites C.P. 1335* (BM., E, K). WITHOUT LOCALITY: *ditto* in Herb. Barkly (BM); *ditto* in Herb. Bedd., in Herb. Hook. (all K); Central Prov., *Naylor Beckett 36* (BM); *Robinson s.n.* (K).

2. Deparia zeylanica (Hook.) M. Kato., J. Fac. Sci. Univ. Tokyo, sect. 3, 13: 406–407. 1984.

Asplenium zeylanicum Hook., Sp. Fil. 3: 237. 1860. Type. From Sri Lanka: "Kotmalee Oya, on banks", *Gardner 1249* (K).
Diplazium zeylanicum (Hook.) T. Moore, Ind. Fil: 340. 1862.
Diplazium pinnatifidum Feé, Mém. Fam. Foug. 10.: 24 et t, 1865, non Kunze. 1834.
Athyrium zeylanicum (Hook.) Milde, Bot. Zeit. 28: 354. 1870.
Lunathyrium zeylanicum (Hook.) Edie, Ferns Hong Kong: 42. 1978.
Triblemma zeylanica (Hook.) Ching, Act. Phytotax. Sin. 16: 24. 1978.

Rhizome thin, long-creeping subterraneously, occasionally branching, bearing scattered ovate-lanceolate, mid-brown scales, predominantly near the apices. Fronds at intervals, usually 1–6, erect at base, arching, up to c. 40 cm. Stipe ± thin, fragile; grooved adaxially, stramineous, c. ? the length of lamina, base widened, bearing scattered, mid- to dark-brown, linear-lanceolate, entire scales throughout its length and a few sparse, white, columnar hairs. Lamina mostly pinnatifid, becoming pinnate towards base, linear-lanceolate, widest (c. 4 cm) towards middle, softly herbaceous, mid-green, bearing a few scattered fibrils and columnar hairs on rhachis and pinna-costae, bearing up to c. 12 pairs of separate lobes or pinnae, which become cut and fused throughout the top ? or more of lamina, forming a long, shallowly lobed apical segment; pinnae or laminar lobes becoming distant, ± narrowly based and slightly auriculate acroscopically at base of lamina, widely attached above, then fusing with one another increasingly widely from rhachis, towards tip, up to c. 2 cm long, apices obtusely rounded, without teeth or with a few weak serrations. Rhachis similar to upper stipe. Sori c. 8–14 per pinna or lobe, borne in the upper ? of lamina, lowest acroscopic one sometimes diplazioid (i.e. double, with two sori lying back to back), linear, extending along veins just short off costa and segment-margin. Indusium thin, white, linear, attached along one side and lying flat over sorus, slightly toothed or erose, lifting and shrivelling on ripening. Spores small, uniform., perispore of ± dense papillae which sometimes join into groups. Cytotype unknown.

D i s t r. Lower-mid levels in central Sri Lanka, Nepal. Very rare and apparently not collected recently. Possibly to be considered of S.E. Asian origin or affinity.

E c o l. Hardly known but stated (by Gardner) to occur on banks at c. 1200 m.

N o t es. *D. lobatocrenata* (Tag.) M. Kato from Japan is very close indeed to *D. zeylanica*, though with somewhat dimorphic fronds, but its high chromosome number (pentaploid apomort) hardly corresponds with the small spores of *D. zeylanica*, though several species of *Deparia* are known to have more than one cytotype.

S p e c i m e n s E x a m i n e d. RATNAPURA DISTRICT: Adam's Peak Sanctuary, Bogawantalawa – Boralanda, rocky stream area, wet, 1400 m., 19 March 1968, *Comanor 1072* (PDA). NUWARA ELIYA DISTRICT: Kotmale Oya, on banks, *Gardner 1249* (K); *Gardner 1249* (in *Thwaites C.P. 3101*) (PDA); Ambagamuwa, Nov 1854, Gardner *s.n.* in *C.P. 3101* (PDA); Ambagamuwa and Kotmalee, *Wall 38/203* in Herb. Barkly (BM). WITHOUT LOCALITY: *Thwaites C.P. 3101* (BM., BO, E, K); *Wall s.n.* (K, E); *Hutchison s.n.* (E); *Anderson s.n.*, 1899 (E); *Ferguson s.n.* (US 815523); *Robinson C 76* (K).

3. Deparia petersenii (Kunze) M. Kato, Bot. Mag. Tokyo 90: 37. 1977.

Asplenium petersenii Kunze, Anal. Pterid.: 24. 1837. Type. From China (Kwangtung) (C).
Diplazium lasiopteris Kunze, Linnaea 17: 568. 1843.
Allantodia deflexa Kunze, Bot. Zeit. 1848: 191. 1848.
Diplazium decussatum J. Smith ex Moore & Houlston, Gard. Mag. Bot. 3: 231. 1851.
Asplenium thwaitesii A.Br. ex Mett., Abhandl. Senckenb. naturf. Ges. 3: 227. 1859.
Type. Sri Lanka.
Athyriopsis petersenii (Kunze) Ching, Act. Phytotax. Sin. 9: 66. 1964.
Athyriopsis thwaitesii (A.Br. ex Mett.) Ching, Act. Phytotax. Sin. 9: 66. 1964.

Misapplied name. *Deparia japonica* (Thunb.) M. Kato.

Rhizome thin, long-creeping subterraneously, occasionally branching, bearing scattered, pale, lanceolate scales at its apices Fronds usually 1–6, erect at the base, arching, up to c. 60 cm tall, borne at intervals, or occasionally crowded. Stipe ± fragile and succulent, widened at base grooved adaxially, pale-green, c. ? the length of lamina, bearing ± dense, pale, lanceolate, entire scales predominantly towards the base and numerous white, columnar hairs throughout. Lamina pinnate and varying from shallowly lobed to bipinnatifid, ± triangular-lanceolate, widest (up to c. 16 cm), just above base, but rarely at base, softly herbaceous and fragile, pale-green (yellow-green when exposed), bearing minute fibrillose scales, mainly on abaxial veins, mixed with ± dense, prominent, white columnar hairs, bearing up to c. 14 pairs of pinnae, these joined at bases in upper lamina forming a deeply, but progressively more shallowly lobed, almost abruptly caudate-attenuate apex to the lamina; pinnae distant, lanceolate, or elongated triangular-lanceolate, up to c. 2 cm wide, ± sessile, shortly stalked at base, mostly symmetrical, though the lowest ones may be wider at the middle of their basiscopic sides, inserted at 90°, ± deflexed at base, the lower ones ± opposite, their apices usually abruptly caudate-attenuate and varying from ± unlobed to being gradually less lobed to shortly below the apex, deeply pinnatifid with up to c. 10 pairs of lobes; pinna-lobes symmetrical, joined at their bases, ± rectangular or with round-cornered or widely obtuse apices, becoming round-pointed in luxuriant plants, up to c. 6 mm wide and 1 cm long, usually less, those shortly above middle of pinna in basal pinnae often the longest, inserted ± at 90°, or slightly sloping, scale from almost entire, to bearing a few insignificant, ± blunt teeth to noticeably serrate-lobed in luxuriant plants. Rhachis similar to upper stipe; costae and costules bearing scattered, minute, fibrillose scales and ± dense columnar hairs mainly abaxially. Sori c. 6–10 per pinna-lobe, borne in the upper ? or less of lamina, lowest acroscopic one often diplazioid (i.e. double, with two sori lying back to back), linear, ± straight, or curved, sometimes becoming J-shaped, extending along veins (except for hooked tip, when present). Indusium thin, translucent, white, linear, attached along one side and lying flat over the sorus, visible when young, shrivelled, pale-brown when ripe. Spores ± small, uniform., perispore of dense, blunt papillae. Cytotype tetraploid, sexual, n = 80 (80 bivalents at meiosis) and hexaploid, sexual, n = 120 (120 bivalents at meiosis), only the latter recorded from Sri Lanka. Some other cytological reports, including some of the east-Himalayan ones of Bir (1961) may refer to *D. japonica*, due to taxonomic confusion, but there still appear to be two cytotypes within *D. petersenii* itself.

D i s t r. Mid- to high-level regions of central Sri Lanka. Common. A widespread south-east Asian element which has spread east along the Himalayas; Sri Lanka, S. India, throughout the length of the Himalaya from N. Pakistan and N.W. India through Nepal, Sikkim and Bhutan to Arunachal Pradesh, S. to Khasia, Burma, Thailand, Laos, Cambodia, Vietnam., across S.W. and south China, Taiwan, S. Japan, Malesia; as different subspecies in the Solomon Islands, New Caledonia, Fiji, Samoa, Rarotonga, Tahiti, Norfolk Island, N.E. Australia, New Zealand (introduced). Naturalised in Hawaii, the Azores, N. America and Brazil. Much confused throughout its range with the more temperate *D. japonica* (see 'notes').

E c o l. Open forests, forest-edges, plantations and road-sides, frequently in secondary habitats, from c. 1200–2100 m. in Sri Lanka.

N o t es. Variable by degree of lamina development, particularly length and obtuseness/acuteness of pinna-lobes; geographical variants have been treated by Kato (1984) as three subspecies and two varieties. Species closely related, cytological and phytogeographic revision of genus, therefore, remains necessary. *D. petersenii*, though evidently a good species, has been much confused with the closely similar *D. japonica*. *D. japonica* has more sloping, generally more acute and often longer-toothed pinna-lobes, often more attenuate pinna-apices, taller, more convex indusia and is hexaploid (the Indian *D. japonica* is diploid), it is also less tropical in distribution than *D. petersenii*, though both occur frequently in the Himalaya.

Sledge (1962) originally treated *D. petersenii* under the name, *Diplazium ± lasiopteris*, but later found that to be synonymous (Sledge 1977); he also said that *D. thwaitesii*, under which name the Sri Lankan plants had generally been known, to him to be identical. But Kato (1984: 426) pointed out slight morphological differences in the Sri Lankan plants which might correspond with the apparently hexaploid chromosome-number found in Sri Lanka, while most other reports are of tetraploids; the Sri Lankan plants are not *D. japonica*, however.

subsp. petersenii

As described above. The careful and detailed conclusion of Kato (1984) was that the Sri Lankan plants apparently belong to this subspecies, though most appear very slightly distinct in cytotype and in having more densely scaly axes.

S p e c i m e n s   E x a m i n e d. MATALE DISTRICT: forested stream-gulley beside and below road, c. 1 km W. of top of pass between Rattota and Laggala, N.E. of Matale, *Fraser-Jenkins Field No.132*, with *Jayasekara & Bandara* (PDA, K, US). KANDY DISTRICT: forested stream gulley, S.E. of Gombaniya mountain, above Nilloomally, N.E. of Kallebokka, N. of Wattegama, W. Knuckles Range, N.E. of Kandy, c. 130 cm., *Fraser-Jenkins Field no. 91*, with *Jayasekara, Samarasinghe, Bandara & Bandara* (PDA, K, US); in secondary jungle, Corbet's Gap, 1320 m., *Sledge567* (BM); forest c. 2 km. N. of and below Corbet's Gap (Attala Meluwa) on road to Mimure, N. of Hunasgyria, Knuckles Range, N.E. of Kandy, *Fraser-Jenkins Field no. 280*, with *Abeysiri, Abeysiri & Gunawardena* (PDA, K, US); Hunnasgyria, *Thwaites C.P. 1343* p.p. (PDA); Hantane, *Thwaites C.P., 1343 p.p.* (PDA); Nilambe, Gonavy Estate, 1891, *Jeffries s.n.*, (PDA); above Le Vallon Estate, in jungle, 1,500 m., *Sledge 1127* (BM); Murata, *Gardner s.n.*, June 1848 (in *Thwaites C.P. 1343 p.p.*) (PDA). BADULLA DISTRICT: Badulla, *Freeman 190D* (BM). NUWARA ELIYA DISTRICT: Ramboda, *Gardner s.n..*, Oxct. 1853 (in *Thwaites C.P. 1343* p.p.) (PDA); Maturata, *Freeman 187A* (BM); Nuwara Eliya, *Freeman 188B and 189C* (BM); Horton Plains, 2040 m–2100 m., *Sledge 682* and *785* (BM); Pedrotalagalla, c. 6000 ft., 26 Dec 1950, Holttum S.F.N. 39188 (BM., SING); Ramboda Pass, c. 1900 m., 17 March 1954, T.G.Walker T841 (BM); Nuwara Eliya, *Wall 38/227* in Herb. Barkly (BM). LOCALITY UNKNOWN: *Thwaites C.P. 1343* p.p. (K, BM., E); *Wall s.n.* (K, E); *Naylor-Beckett 35* (BM., E); *Hutchison s.n.* (E); *Ferguson s.n.* (US 815518); *Hancock 21* (US 1277327); *Robinson 78* (K).

4. Deparia polyrhizos (Baker) Seriz., J. Jap. Bot. 54(6): 180. 1979.

Asplenium polyrhizon Baker, in Hook. & Baker, Syn. Fil., ed. 2: 490. 1874.
    Type. From Sri Lanka: *"Thwaites C.P. 3951"* p.p. (K).
Diplazium polyrhizon (Baker) Sledge, Bull. Br. Mus. nat. Hist. (Bot.) 2(11): 298. 1962.

Misapplied name. *Diplazium decussatum* J. Smith ex Houlston & T. Moore.

Rhizome thick, upright, ± unbranched, surrounded by old stipe-bases, bearing pale-brown, lanceolate scales at its apex and a crown of fronds. Fronds up to c. 6 per rhizome, arching, up to c. 60 cm. tall. Stipe ± thin, fragile, grooved adaxially, pale-green (old, encasing frondbases persisting) as long as, or slightly longer than lamina, base widened, bearing scattered, pale-brown, narrowly lanceolate, accuminate, entire scales, mainly at base, and scattered, white, columnar hairs throughout. Lamina pinnate and deeply bipinnatifid, ± deltate or elongated-triangular; widest at the base (up to c. 20 cm), softly herbaceous and fragile, pale- to mid-green, bearing scattered, white, columnar hairs predominantly on abaxial veins, bearing up to c. 10 pairs of pinnae which become joined at their bases and gradually decrease in size towards the lobed frond-apex; *pinnae* ± distant, or crowded towards tip, elongated triangular-lanceolate, up to c. 1.5 cm wide, ± sessile, lowest ones may be shortly stalked, symmetrical, inserted at right-angles, lower ones usually slightly deflexed or arched backward, basal ones ± opposite, their apices pinnatifid, gradually decreasing, slightly attenuate, ± deeply pinnatifidly lobed (c. ? cut), with up to c. 12–13 pairs of lobes. Rhachis similar to upper stipe; costae and costules bearing scattered columnar hairs, predominantly abaxially; *pinna-lobes* symmetrical, joined by their bases, the lowest opposite-pair on each pinna often shorter than the next, rounded-rectangular with ± widely obtuse apices, up to c. 5 mm wide and 7.5 mm long, ± of equal length along pinna until tapering to apex, inserted at slightly less than 90°, ± entire, or, more usually, crenate-dentate around apices. Sori c. 4–8 per pinna-lobe, borne in the upper ? of the lamina, lowest acroscopic one often diplazioid, linear, ± straight, or slightly curved, extending along veins. Indusium thin, white, linear, attached along one side and lying over the sorus, somewhat convex, laciniate, shrivelling back and becoming pale-brown on ripening. Spores ± small, uniform., perispore of dense, narrow, blunt papillae. Cytotype tetraploid, sexual, n = 80 (80 bivalents at meiosis).

D i s t r. Higher level regions of central Sri Lanka. Very rare. A Sri Lankan endemic whose nearest relatives are in S.W. China and Burma. Possibly to be considered of S.E. Asian affinity, but with no close relatives there today, otherwise of Sino-Himalayan affinity. Sri Lanka.

E c o l. Damp luxuriant forests, at approximately 2000 m.

N o t e. As pointed out by Sledge (1962), Beddome's (1876: at 292 and 1883) reports of it from south India and with a creeping rhizome was in error for *D. petersenii*. Its upright rhizome and wider frond-base distinguish it readily from the latter.

S p e c i m e n s   E x a m i n e d. NUWARA ELIYA DISTRICT: Kandapola, near Nuwara Eliya, swampy part of the forest, 1800–2100 m., *Thwaites C.P. 3951* (PDA, K); ditto, *Sledge 1320*, with *T.G. Walker* (BM); forested stream below and N. of road, c. 3 km. E. of Farr Inn on road towards Ohiya, Horton Plains, S. of Nuwara Eliya, c. 1950 m., *Fraser-Jenkins Field no. 427*, with *Abeysiri* (PDA, K, US). LOCALITY UNKNOWN: *Wall s.n.* (E); *Beddome s.n.* (K, BM), *Hutchison s.n.* (E); *Robinson s.n..*, 1870 (K).

Sect. Dryoathyrium

(Ching) M. Kato, J. Fac. Sci. Univ. Tokyo, sec t. 3., 13: 383. 1984.

5. Deparia boryana (Willd.) M. Kato, Bot. Mag. Tokyo 90: 36. 1977.

Aspidium boryanum Willd., Linn. Spec. Plant., ed. 4, 5: 295. 1810. Type. from Réunion (isotype P).

Aspidium edentulum Kunze, Bot. Zeit. 4: 474. 1846.

Nephrodium divisum Hook., Sp. Fil. 4: 133. 1862.

Polpodium subtripinnatum C.B. Clarke, Trans. Linn. Soc., Lond., 2 Bot., 1: 528. 1880.

Phegopteris kingii Bedd., Suppl. Ferns Brit. Ind.: 84. 1892.

Dryopteris kingii (Bedd.) C. Chr., Ind. Fil. 3: 273. 1905.

Dryoathyrium boryanum (Willd.) Ching, Bull. Fan Mém. Inst. Biol. 11: 81. 1941.

?*Dryoathyrium edentulum* (Kunze) Ching, Bull. Fan Mém. Inst. Biol. 11: 81. 1941.

Lunathyrium boryanum (Willd.) H. Ohba, Sci. Rep. Yokosuka City Mus. 11: 53. 1965.

Rhizome thick, decumbent, bearing mid-brown, linear-lanceolate scales at its apex. Fronds usually 2–4 in a crown, arching, up to 2 m tall. Stipe thick, (encasing frond bases persisting), ± rigid, grooved adaxially, dark brown, as long as lamina, base widened, bearing scattered, mid-brown, narrowly lanceolate, entire scales near base, which become small and fibrillose above. Rhachis similar to upper stipe, bearing a few small, fibrillose scales. Lamina bipinnate and tripinnatifid to tripinnate and quadripinnatifid, ovate-deltate, slightly longer than wide, widest (up to c. 80 cm) shortly above base, softly herbaceous, mid- to dark-green, bearing ± scattered, white, columnar hairs mainly on adaxial veins; pinnae (up to c. 14 pairs) ± crowded, elongate triangular-lanceolate, basal one up to c. 40 cm long,15 cm wide, the lower ones with stalks up to c. 1.5 cm long, mostly ± symmetrical, basiscopic pinnules in lowest 1 or 2 pairs longer than acroscopic ones inserted at ± 90°, but lowest pair slightly backward deflexed, to nearly bipinnate, with up to c. 15 pairs of pinnules; *costae* and costules surrounded by scattered, fibrillose, pale-brown scales and scattered, white, columnar hairs, mainly adaxially; pinnules symmetrical, sessile, narrowly attached at their bases, widely attached near pinna tip, elongated triangular-lanceolate, separated, up to c. 10 × 2.5 cm., those towards the pinna-apex joined by a characteristic very narrow, decurrent, basiscopic-basal wing of lamina along costa, stretching between all except the basal few pairs of pinnules, apices acute, deeply pinnatifid, pinnate only at base in large fronds, with up to c. 12 pairs of pinna-lobes or pinnulets ± rounded rectangular, widely attached to the costule and mostly joined at their base by a narrow wing of lamina, ± crowded, or distant in the lower parts of large fronds, symmetrical, their apices rounded, bearing lateral and apical crenate dentations, veinlets not quite reaching the margins. Sori c. 6–12 per pinnule-lobe, ± throughout lamina, medially on veins, small, orbicular. Indusium very thin, fugacious, white, circular, with a basiscopic ± closed sinus (subreniform), not quite entire, or slightly toothed, lying flat over the top of the sorus until completely shrivelling on ripening. Spores small, uniform., perispore of dense, thick, blunt papillose structures. Cytotype diploid, sexual, n = 40 (40 bivalents at meiosis).

D i s t r. Mid- to higher level regions of central Sri Lanka. Scattered. A widespread south-east Asian element which has spread along the Himalayas and west to Réunion and east to Oceania. Sri Lanka, S. India, throughout most of the Himalayan outer ranges from Jammu through N.W. India and Nepal to Sikkim., Bhutan, Armachal Pradesh and south to Khasia, Burma, Thailand, Malesia.

E c o l. Luxuriant forests, usually on slopes above streams, also in secondary habitats such as on slopes above roads, when well shaded by trees, from c. 900–180 m.

N o t es. Often confused, from a distance, for a large *Diplazium* species, but instantly recognisable by its circular sori. The more compound species of *Cornopteris* are often overlooked by being confused with *D. boryana*, but are without hairs, have smaller segments and minute papillae in the open groove at the junction of the pinna-costae and rhachis, as well as having an *Athyrium*-like tooth formed from the edge of the groove at the junctions of each costule with the pinna-costa.

Many authors have kept *D. boryana* in a separate genus, *Dryoathyrium.*, but this is unwarranted when the other species in the same section are considered and is best treated as a rank below that of genus.

S p e c i m e n s   E x a m i n e d. MATALE DISTRICT: forested stream-gulley beside and below road, c. 1 km W. of top of pass between Rattota and Laggala, N.E. of Matale, 1200 m., *Fraser-Jenkins Field no. 133*, with *Jayasekara & Bandara* (K, PDA, US). KANDY DISTRICT: Corbet's Gap, by stream in shady forest, 1100 m., *Sledge 1030* (BM); [Raxawa], *Thwaites C.P. 3097* p.p. (K, BM., E, PDA). BADULLA DISTRICT: [Haputelle], *Thwaites C.P. 3097* p.p. (K, BM., E, PDA). NUWARA ELIYA DISTRICT: Hakgala, by stream in forest, 1650 m., *Sledge 1213* (BM). LOCALITY UNKNOWN: *Hutchison s.n.* (K); *Ferguson s.n.* (US 816399); *Robinson s.n.* (K).

3. ATHYRIUM

Roth, Tent. Flor. Germ. 3: 58. 1799.

References. W.A. Sledge. 1956. The nomenclature and taxonomy of *Athyrium nigripes* (Blume) Moore, *A. shenopteris* (Kunze) Moore and *A. praetermissum., sp. nov.*, Ann. Mag. Nag. Hist. 9: 453–464; W.A. Sledge 1962. The Athyrioid ferns of Ceylon, Bull. Br. Mus. nat. Hist. (Bot.), 2: 277–

323; R.C. Ching 1964. On some confused genera of the family Athyriaceae. Act. Phytotax. Sin. 9: 41–84 M. Kato 1977. Classification of *Athyrium* and allied genera of Japan, Bot. Mag. Tokyo 90 23–40. C.R. Fraser-Jenkins 1996. A revision of: Himalayan athyrioid ferns. Dehra Dun , 2001.
Type species: *Athyrium filix-femina* (L.) Roth

Anisocampium C. Presl, Epim. Bot.: 52. 1849.
Solenopteris Zenker ex Kunze, Linnaea 24: 267. 1851.
Pseudothyrium Newman, Phytol. 4: 370. 1851.
Microchlaena Ching, Bull. Fan Mém. Inst. Biol. Bot., 8: 322. 1938, non
      Wall. ex Wight & Arnott.
Pseudocystopteris Ching, Act. Phytotax. Sin. 9: 76. 1964.
Cystanthyrium Ching, Act. Phytotax. Sin. 11: 23 et t. 1966.
Kuniwatsukia Pich.Serm., Webbia 28: 455. 1973.

Medium c. 1 m tall, terrestrial ferns. Rhizome ± thick, ascendent or upright (in the area), with an apical crown of fronds. Stipe stramineous or pale green, swollen at the very base, ± thin, fragile, bearing scales at its base, not hairs, containing two vascular strands, grooved adaxially. Rhachis similar to the upper stipe, with the groove open to and continuous with those entering it from the pinna-costae. Lamina varying from simply pinnate to tripinnate, lanceolate to deltate, veins free, or, rarely, anastomosing, glabrous, or bearing a few scattered, fibrillose scales and sometimes minute unicellular glands, or minute, unicellular pubescence (but not in the area), particularly on the costae, the abaxial surface of the costae frequently bearing a spine-like tooth formed from its lateral ridge at the point of insertion of the costules. Spine-like setae on either side of their midrib above, usually somewhat small, lobed, toothed. Sori varying from orbicular to hippocrepiform., to elongated and either straight or curved into a U or J shape borne medially on or along the veinlets, not reaching the segment-margin. Indusiate (in the area),or sometimes exindusiate. Indusium thin, fugacious, entire or fimbriate, varying from sub-reniform to elongated, following the shape of the sorus. Spores bilateral, monolete, bean-shaped, perisporate. Chromosome base number x=40.

A large genus of about 220 species, but greatly overestimated in China. The greatest number of species are Sino-Himalayan or Sino-Japanese, but with another large centre of evolution in S. E. Asia and rather few species reaching Oceania, Africa, Europe and the new world.

The species are usually well-marked though there are several groups of close-knit species; many are confusingly variable in response to environmental and developmental factors. Too many genera, subgenera and sections have been reognized in China, but six sections have been recognized from mainland Asia by Fraser-Jenkins (1996), of which four occur in Sri Lanka, though it is probable that one or two more other sections exist in S.E. Asia and further afield.

KEY TO THE SPECIES

1 Veins anastomosing fronds imparipinnate, simply pinnate and the pinnae shallowly lobed (to less than half their depth on each side) . . . .
. . . . **1. A. cumingianum**
1 Veins free, fronds paripinnate, bipinnate or more, or if simply pinnate the pinnae deeply or pinnatifidly lobed
  2 Lamina tapering markedly to its base, pinnae deeply lobed but not becoming ± pinnate towards their bases . . . . **2. A. hohenackerianum**
  2 Lamina not or only slightly narrower at the base, pinnae ± pinnate or more at their bases
    3 Lamina lanceolate
      4 Segments bearing short or insignificant teeth
        5 Pinnules or pinna lobes not bearing setae above
          6 Lamina thick, bipinnate (except in immature plants), pinnules ± narrowly lanceolate . . . . **3. A. puncticaule**
          6 Lamina thin, pinnate-pinnatifid, pinna-lobes ± ovate to ovate-lanceolate . . . . **4. A. anisopterum**
        5 Pinnules or pinna-lobes not bearing setae above . . **5. A. setiferum**
      4 Segments bearing long, prominent teeth . . . . **6. A. solenopteris**
    3 Lamina widely, ± triangular-lanceolate or deltate
      7 Segments with long, prominent teeth . . . . **6. A. solenopteris**
      7 Segments ± untoothed or toothed, but if so teeth are not markedly long or prominent
        8 Segments lobed, toothed, lamina apex gradually decreasing . . . . . .
        . . . . **7. A. praetermissum**
        8 Segments unlobed, ± untoothed, lamina-apex suddenly and abruptly decreasing . . . . **8. A. wardii**

Sect. Nipponica

Ching & Hsieh in Hsieh, Bull. Bot. Res. 6(4): 131. 1986.

1. Athyrium cumingianum (C. Presl) Ching in C. Chr, Index Fil., Suppl. 3:40. 1934.

Anisocampium cumingianum C. Presl, Abhandl. K. Böhm.Ges. Wiss., 5, 6: 419. 1851.
Type from the Philippines (PRC)
Goniopteris aristata Fée, Mém. Fam. Foug. 5, Gen. Fil.:253. 1852.
Aspidium otaria Kunze et Mett., Abhandl. Senckenb. naturf. Ges. 2: 318. 1858.
Pleocnemia aristata (Fée) Bedd., Ferns S. India: 28, t.83. 1863.
Dryopteris otaria (Kunze ex Mett.) Kuntze., Rev. Gen. Plant. 2: 813. 1891.

Rhizome ± thin, long-creeping subterraneously, occasionally branching, clothed throughout with lanceolate, pale brown scales and bearing fronds at intervals. Fronds usually 2-4 per rhizome, erect at the base, arching, up to c. 70 cm tall; stipe ± thin, fragile, grooved adaxially, stramineus, same length or longer than lamina,lowest part of base widened, bearing few linear-lanceolate, attenuated scales, otherwise naked. Rhachis naked. Lamina simply pinnate, ± imparipinnate (terminating abruptly in a segment similar to one of its pinnae), widely triangular-lanceolate or sub-rectangular, widest at the base (to c. 20 cm wide) slightly succulent herbaceous, dark green (drying dark grey), naked, bearing up to c. 7 pairs of pinnae and a long, pinna-like apical segment; pinnae narrowly lanceolate, distant, widely cuneate at their stalked bases, up to c. 10 cm long, widest shortly below their middle, with attenuate apices, shallowly multi-lobed at the sides; pinna-lobes usually up to ¼ (occasionally up to ?)the depth of the pinna on each side, small, rounded or becoming pointed, length slightly irreguler, toothed with many small, acute teeth. Veins free except for the basal pair in each lobe which fuse with those from the adjacent lobe and run to the margin. Sori c. 4-6 per lobe (throughout width of abaxial pinna, slightly away from costa), Borne throughout most of lamina except very base, small, orbicular to ovate, medially on veins. Indusia very thin, white, small and fugaceous, circular to reniform with a ± wide, basiscopic sinus, sometimes becoming slightly widened out, hardly covering whole sorus, margins laciniate, rapidly shrivelling and disappearing on ripening. Spores small, regular, with wide, ± clear wings of perispore and a network of ridges. Cytotype unknown.

D i s t r. At low altitudes at north-eastern edge of central highlands of Sri Lanka. Very rare. A South-East Asian element. Sri Lanka, South India, S.W. China, Thailand, Laos, Philippines.

E c o l. In damp places in open low-altitude forests, from c. 100-600m alt. in Sri Lanka.

N o t es. Has been confused with the rather similar but larger, narrower pinna'd and free-veined Sino-Himalayan *A. cuspidatum* (Mett.) M. Kato [Syn. *Kuniwatsukia cuspidata* (Mett.) Pich.Serm.] which has been reported from Sri Lanka, however, Gardner's original specimen(K!), though correct *A. cuspidatum.*, probably originated from elsewhere (see Clarke 1880). It is also similar to *A.sheareri* (Baker) Ching from China, Korea and Japan.

Because of its partly anastomosing veins, combined with a nearly circular sorus and simple pinnae, it has long been placed in a separate genus *Anisocampium.,* along with *A. sheareri*, but the vein anstomosis is similar to the situation with *Diplazium esculentum* (Retz.) Sw. (formerly in *Anisogonium*) in relation to other *Diplazium* species and is not of generic significance. Both species are otherwise obvious members of their genera with close relatives with all free veins.

S p e c i m e n s   E x a m i n e d. KANDY DISTRICT: Near Weragamtota, shady forest, Alt. 120 m., *Sledge 953* (BM); Shady forest by Mahaweli River, Minipe, 150m., *Sledge 945* (BM). LOCALITY UNKNOWN: *Gardner s.n.* in Thwaites *C.P. 1299* (PDA, K, BM); *Wall s.n.* in Herb. Hance 20619 (BM); *Hooker & Thomson 28* (BM).

Sect. Athyrium

2. Athyrium hohenackerianum (Kunze) T. Moore, Index Fil.: xlix. 1857.

Allantodia hohenackeriana Kunze, Farnkr. Schkuhr Suppl. 2: 63, t. 126. 1850. Type. From S. India (B).

Rhizome thick, short, ascendent, densely clothed with small, linear, pale-brown scales at apices. Fronds up to c. 10 in apical tuft, ± erect, up to c. 35 cm tall; stipe thin, fragile, short, (encasing old stipe-bases, branched and tufted, persisting), up to c. ¼ the length of lamina, bearing ± dense (especially at base), narrowly lanceolate, pale-brown, entire scales; rhachis similar to upper stipe, bearing scattered scales; lamina pinnate and deeply pinnatifid or almost bipinnate, narrowly lanceolate, widest (up to c. 5 cm) slightly above middle, tapering below, thinly herbaceous, mid-green, ± glabrous, without costular setae above; pinnae, up to c. 20 pairs, ± widely spaced, ± parallel-sided or gradually tapered from base, mid ones up to c. 2.5 cm long, 1 cm wide, shortly stalked, apices ± obtuse, nearly symmetrical or with basal lobes on basiscopic side slightly shorter and slightly more oblique than those on acroscopic side, inserted ± at 90° to rhachis, ± deeply pinnatifidly lobed, or just becoming pinnate at bases; pinnules rectangular with obtusely rounded apices, their bases attached to next ones or narrowed in lower lobes of each pinna, sparsely lobed, laterally and apically with acute teeth. Sori c. 1-6 per pinna-lobe, borne ± throughout lamina, medially along veins, small, elongated, often curved, or hook-shaped; indusium small, thin, white, entire, lying over top of sorus until shrivelling somewhat and becoming brown on ripening. Spores small, uniform., pale-brown, narrow perispore of clear wings. Cytotype diploid, sexual, n = 40 (40 bivalents at meiosis).

D i s t r. Lower altitudes on North side of central mountains of Sri Lanka. Uncommon. A Sri Lanka S. and C. Indian (Deccan) endemic. Sri Lanka, W.C. India (from Bombay and Sind southwards), S.India.

E c o l. Semi-open or shaded, rocky banks and road sides, from c. 400-700m in Sri Lanka.

N o t es. Often confused with the similarly small, C. and S.Indian(also foothills of the W.C. Himalaya) *A.falcatum* Bedd. which can be distinguished by its markedly acroscopically auriculate pinnae and the lobes usually without teeth.

Sledge (1962) mistakenly altered the spelling of the epithet to "hohenackeranum."

S p e c i m e n s   E x a m i n e d. MATALE DISTRICT: Matale East, *Thwaites C.P. 3867* (K, CGE, PDA); Matale East, between Pallegama and Etanwela, moist open bank, 630m., *Sledge 1236* (BM); Lagalla, 600m., *Naylor Beckett 769* (K, BM); Managalla, 600m., 4 Feb. 1891, *Hancock s.n* (K). KANDY DISTRICT: Weragamtota, 450m., *Sledge 952* (BM). LOCALITY UNKNOWN: *Hancock 4* (US 1277181).

Sect. Polystichoides

Ching & Hsieh, Bull. Bot. Res. Harbin 6(4): 130. 1986.

3. Athyrium puncticaule (Blume) T. Moore, Ind. Fil.: 186. 1860.

Aspidium puncticaule Blume, Enum. Plant. Javae: 159. 1828.
Type. From Java (L).
Aspidium macrocarpon Blume, Enum. Plant. Javae: 162.1828.
Athyrium macrocarpon (Blume) Bedd., Ferns S. Ind.: t. 152-153. 1863, non
    Fée. 1852.
Athyrium lanceum (Kunze) T. Moore, Ind.Fil.: 185. 1860.
Asplenium fallax Mett., Ueber Einig. Farngatt. vi, Asplenium: 194. 1859.
Athyrium fallax (Mett.) Milde, Fil. Eur.: 49. 1867.

Rhizome thick, long, ascendent, usually unbranched, densely clothed with linear-lanceolate, pale- or slightly reddish-brown scales at the apex. Fronds up to c. 10 in apical tuft, arching, medium-sized to c. 60 cm tall; stipe ± thin, succulent, (thickened old stipe-bases persisting), ± long, up to c. ¾ the length of lamina, bearing scattered, almost hair-like, entire, pale scales throughout and a dense tuft of linear, pale-brown, entire scales at base. Rhachis similar to upper stipe. Lamina bipinnate, widely lanceolate to ovate, widest, up to 25 cm., shortly below middle, lowest one or two pinna-pairs reduced in length, fragile and slightly succulent-herbaceous, mid-green (but often drying black), ± glabrous, without costular setae above; pinnae of c.18 pinna-pairs ± contiguous, elongated narrowly triangular, with auriculate acroscopic base and a slightly reduced and dimidiate basiscopic one, mid pinnae up to c. 15 cm long, 4 cm wide, shortly stalked, acute apices, asymmetrical, pinnules on acroscopic side being slightly longer and more developed than the more obliquely inserted ones on basiscopic side, inserted at ± 90°, pinnate at acroscopic bases deeply pinnatifidly lobed elsewhere; pinnules obtusely- or rectangular-ovate with rounded apices, their bases mostly joined, basiscopically decurrent, narrowed in lowest acroscopic pinnule, almost unlobed but for lateral shallow lobes in basal pinnules, and deeper lobes in basal acroscopic pinnule of each pinna, lobes obtusely rounded, with few short lateral and apical teeth, or ± without teeth. Sori c. 3-10 per pinna-lobe, confined to upper ? of lamina, medial on veins, large, horseshoe- or kidney-shaped, occasionally only curved-elongate, or hook-shaped. Indusium large, thin, white, with lacerate margins, lying widely over top of sorus until shrivelling somewhat, becoming brown on ripening. Spores very large, uniform., dark, narrow perispore of elongated folds. Cytotype (in Sri Lanka) hexaploid, sexual, n = c. 120 (120 bivalents at meiosis).

D i s t r. Higher altitudes in the central mountains of Sri Lanka. Fairly common. A S.E. Asian element. The furthest E. Himalayan region (Meghalaya: Khasi hills), S. India, Thailand, S. China, Malaya (Cameron Highlands), Sumatra, Java and Borneo.

E c o l. Shaded banks in forest, or shaded, wet rocks by waterfalls, from c. 1500-2100m in Sri Lanka.

N o t es. A member of a difficult complex of closely related species, which have not previously been properly separated. The identity of the Sri Lankan material as being *A. puncticaule* is not certain as the plants have slightly wider, less caudately apexed fronds and more deeply dissect pinnae than elsewhere in its range, though they are otherwise very close. The true S.E. Asian *A. puncticaule* has, as yet, not been investigated cytologically. If they are not conspecific, the Sri Lankan plant is most probably a separate and endemic species.

The nomenclature of this plant was confused by Sledge (1962), who at first followed Christensen (1905) in calling it *A. macrocarpon*, later (Sledge 1982) mistakenly named it as *A. lanceum.*, both being synonyms of *A. puncticaule*. But he mistakenly applied the name *A. puncticaule* to the next species, *A. anisopterum.*, as the type of the former (L!) is a narrow, less developed specimen, superficially similar to the latter.

S p e c i m e n s   E x a m i n e d. NUWARA ELIYA DISTRICT: Nuwara Eliya, woods, *Gardner 1112* (BM) and Jan. 1847 in *Thwaites C.P. 1372* (PDA); Nuwara Eliya, *Freeman 174B, 175C* (BM); *Freeman 174B, 175C, Chevalier s.n.* (BM); Kikilimane, near Nuwara Eliya, 2040 m., *Sledge 1342* (BM); Ramboda, Jan.1854, *Gardner s.n* in *Thwaites C.P. 1372* (PDA); Ramboda Pass, by track from summit of pass to Maturata, 1920m., *Sledge 1311, 1312* (BM); Pidurutalagala, 2025 m., *Sledge 728* (BM); stream-gully above new road to radio-station, upper part of south side of Mt. Pidurutalagala, N. of Nuwara Eliya, c. 2150 m., , 26 Oct. 1993, *Fraser-Jenkins Field nos. 399-401* with *Abeysiri* (K, PDA, US, NMW); Horton Plains, 2100 m., *Sledge P. 257* (BM); Adam's Peak,1950m., *Sledge 616* (BM); Nuwara Eliya, up to 7000 ft., *Wall 38/194* in Herb. Barkly (BM). LOCALITY UNKNOWN: *Thwaites C.P. 1372* (BM., K, E, US); *Gardner 1064* (K); *Gardner 1103* (CGE); *Hooker & Thompson 207*, p.p. (BM); *Mrs. Walker s.n.* (K); *Robinson 71* (K); *Wall s.n.* (K, E); *Hutchison s.n.* (E).

4. Athyrium anisopterum Christ, Bull. Herb. Boiss. 6: 962. 1898. Type. From S.W. China.

Dryopteris thysanocarpa Hayata, Icon. Plant. Formos. 4: 160, t. 100. 1914.

Rhizome thick, short, ascendent, usually branched and tufted, densely clothed with small, linear, pale-brown scales at apices. Fronds up to c. 6 in an apical tuft, ± arching, medium-small, to c. 40 cm tall, stipe thin, fragile, succulent, (encasing old stipe-bases persisting), ± short, up to c. ½ the length of lamina, occasionally longer, bearing a few, scattered, almost hair-like, entire, pale scales throughout and ± dense, narrowly lanceolate, pale scales at base; rhachis similar to upper stipe; lamina bipinnatifid, bipinnate at bases of lower-mid pinnae, lanceolate, widest, up to c. 6 cm., shortly above base, lowest pair of pinnae slightly shorter than the next, fragile, slightly succulent-herbaceous, pale- to mid-green, ± glabrous, without costular setae above; pinnae, up to c.15 pairs, slightly separate, elongated, narrowly triangular, but with an auriculate acroscopic base and a slightly reduced, dimidiate basiscopic one; mid pinnae up to c. 3 cm long, 1 cm wide, shortly stalked, apices acute, asymmetrical, pinnules on acroscopic side slightly longer, more developed than smaller, more obliquely inserted ones on basiscopic side, inserted ± at 90°, sometimes just becoming pinnate at acroscopic bases, more usually deeply pinnatifidly lobed; pinna-lobes ovate to ovate lanceolate, with slightly pointed or obtusely rounded apices, bases joined, basiscopically decurrent, narrowed at base in lowest acroscopic pinnule, here shallow lateral lobes, otherwise unlobed, lobes obtusely rounded, with a few short lateral and apical teeth. Sori c. 3-10 per pinna-lobe, confined to upper two-thirds of lamina, medial, horseshoe- or kidney-shaped, occasionally only curved-elongate, or hook-shaped, large. Indusium large, thin, white, with lacerate margins, lying widely over top of sorus until shrivelling somewhat and becoming brown on ripening. Spores large, uniform., ± dark; perispore narrow, undulate. Cytotype (in Sri Lanka) tetraploid, sexual, n = c. 80 (80 bivalents at meiosis).

D i s t r. Higher altitudes in the central mountains of Sri Lanka. Fairly common. A Sino-Himalayan element. Sri Lanka, S. India, throughout the Indo-Himalaya, S.W. China, Taiwan, Burma, N. Thailand and Malesia.

E c o l. Shaded, wet, rocky banks in forests, often by waterfalls, from c. 1500-2100 m in Sri Lanka.

N o t es. A member of the *A. puncticaule* and *A. foliolosum* group. Two taxa were formerly included under the name *A. anisopterum* in the Himalayan region, both fairly common, true *A. anisopterum* and the smaller, narrower, more toothed, tetraploid sexual *A. micropterum* Fraser-Jenk. (= *A. macrocarpon* var. *atkinsonii* (Hook. & Baker) Tardieu, see Fraser-Jenkins 1997 and 1988, in prep.). The identity of the Sri Lankan and probably also S. Indian and S.E. Asian plants as being *A. anisopterum* is not certain and these plants are slightly intermediate towards the diploid sexual *A. foliolosum* in their taller, slightly thicker fronds. Unfortunately the cytotype of both the Himalayan and S. Indian plants, though reported as diploid, still requires confirmation. If the Himalayan and Sri Lankan plants are cytologically distinct in addition to their slightly different morphology, the Sri Lankan plant most probably represents a distinct, S.E. Asian species. Sledge (1982) mistakenly altered his original nomenclature for this plant to *A. puncticaule*, which is not the same species, as he thought.

S p e c i m e n s   E x a m i n e d. KANDY DISTRICT: Hantane, *Thwaites C.P. 1372,* p.p. (PDA); Wattekelly Hill, 1500 m., *Naylor Beckett s.n.* (E).NUWARA ELIYA DISTRICT: road-bank just below top of Top Pass on west side, upper Ramboda Pass, W. of Nuwara Eliya, c. 1950 m., 23 Sept. 1993, *Fraser-Jenkins Field no. 228 with Jayasekara, Samarasinghe and Abeysiri* (K, PDA, US, NMW); on S. side of ridge, c. 1-2km along track from c. 1 km below top of Ramboda Pass (top part of Top Pass on W. side) which heads N.W. towards Maturata, W. of Nuwara Eliya, in woods, c. 1700-1800 m., 25 Oct. 1993, *Fraser-Jenkins Field nos. 335-338,* with Abeysiri (K, PDA, US, NMW); Ramboda, *Thwaites C.P. 1372,* p.p. (PDA); Mt. Pidurutalagala, 1920 m., *Ballard 1187* (K); between Pattipola and Horton Plains, in jungle, 1950 m., *Sledge 671* (BM). LOCALITY UNKNOWN: *Thwaites s.n.* (BM); *Gardner s.n.* in Herb. Smith p.p. (BM); 1880, *Mrs. Walker* (K); *Gardner 1112* (K).

Sect. Strigoathyrium

Ching & Hsieh, Bull. Bot. Res. Harbin 6(4): 133. 1986.

5. Athyrium setiferum C. Chr., Ind. Fil. 1: 146. 1905. Type. From Nepal.

Asplenium gracile D. Don, Prodr. Flor. Nepal.: 8. 1825, non *Athyrium* Fourn.
Athyrium tenuifrons var. *tenellum* Wall. apud T. Moore ex R. Sim., Priced Cat. Ferns 6: 22. 1859.
Asplenium tenellum (Wall. apud T. Moore ex R. Sim) Hope, J. Bombay Nat. Hist. Soc. 12: 529-531. 1899, non Roxb. in Griff.
Athyrium tozanense (Hayata) Hayata, Icon. Plant. Formos. 4: 235. 1914.
Athyrium supraspinescens C. Chr., Contrib. U.S. Nat. Herb. 26(6): 297, t. 19. 1926.
Athyrium yui Ching, Bull. Fan Mém. Inst. Biol., Bot., 10:6. 1940.

Rhizome ± thick, erect, quite tall, usually unbranched, densely clothed with lanceolate, pale-brown scales at apex. Fronds up to c. 6, in apical tuft, ± erect, medium size, c. 40 cm tall. Stipe thin, fragile, (encasing, thickened old stipe-bases persisting), length similar to that of lamina, at base with a tuft of short lanceolate, entire, pale-brown scales and in its lower half scattered, small, ovate, pale scales. Rhachis similar to upper stipe. Lamina bipinnate, lanceolate to ovate-lanceolate, widest, up to c. 7 cm., about, or shortly below middle, lowest one or two pinna-pairs usually slightly shorter than next, fragile, delicate, mid- to dark-green, glabrous, with prominent, long, delicate costular setae above pinnule-midribs and at pinna-apices; pinnae, up to c. 12 pairs, usually quite well separated, elongated triangular, up to c. 3.5 cm long, 1 cm wide, shortly stalked, ± obtuse or rather abruptly acute, faintly narrowly acute at their apices, ± symmetrical; pinnules small, ovate, with ± narrowly obtuse apices, sessile, narrowed to their bases, but lowest ones very shortly stalked, slightly falcate, with decurrent basiscopic bases, on acroscopic side pinnule-apices rounded-rectangular with few, small teeth, basal acroscopic pinnule often larger than basiscopic ones and larger than subsequent pinnule, varying from almost unlobed to shallowly, or occasionally somewhat deeply lobed inserted slightly obliquely on costa, pinnate. Sori c. ? per pinnule, confined to upper ? of lamina, sub-medial, small, ± straight, larger ones hook-shaped. Indusium small, narrow, thin, white, margins entire to slightly toothed, lying over top of sorus until shrivelling considerably, turning brown on ripening. Spores very small, uniform, pale-brown, smooth, perispore apparently absent. Cytotype diploid, sexual, n = 40 (40 bivalents at meiosis).

D i s t r. Upper-mid to higher altitudes in the central mountains of Sri Lanka. Fairly common. A Sino-Himalayan element. Sri Lanka, S. India, throughout the Himalaya from Kulu eastwards, though uncommon in the west, Meghalaya etc., S.W. China, Taiwan, Thailand, S. Japan.

E c o l. In moist, shaded places beside streams in light forest, from c. 1500-2150 m in Sri Lanka.

N o t es. Sledge (1956 and 1962), in common with nearly all other authors, referred this species to the Javan *A. nigripes* Blume, the type-specimen of the latter (L) being remarkably and confusingly similar to the present species. Apart from using the name *A. nigripes*, Sledge's accounts, like those of Hope before him., clarified the present species very well and subsequent papers by Bir and other Indian workers have only served to confuse it.

S p e c i m e n s   E x a m i n e d. NUWARA ELIYA DISTRICT: Ramboda, Nuwara Eliya, *Thwaites C.P. 3067* p.p. (PDA); Ramboda Pass, by track from summit of pass to Maturata, 1920 m., *Sledge 1314* (BM); jungles on the Maturata side of Nuwara Eliya, *Freeman 185A, 186B* (BM); Kandapola patana, *Thwaites C.P. 3067* p.p. (PDA); Kandapola Forest Reserve, Nuwara Eliya, 1950m., *Sledge 1321* (BM); stream-gully above new road to radio-station, upper part of south side of Mt. Pidurutalaga, N. of Nuwara Eliya, 2150 m., 20 Oct. 1993, *Fraser-Jenkins Field nos. 390-392*, with *Abeysiri* (K, US, PDA, NMW); Horton Plains, on path to Haldummula, Sept. 1890, *s.coll. sn.* (PDA); Horton Plains, in shady forest by stream., 2100 m., *Holttum 39221* (SING); forested stream below and N. of road, c. 3 km E. of Farr Inn.

on road towards Ohiya, Horton Plains, S. of Nuwara Eliya, c. 1950 m., 27 Oct.1993, *Fraser-Jenkins Field no. 433*, with *Abeysiri* (K, US, PDA, NMW). LOCALITY UNKNOWN: *Thwaites C.P. 3067* p.p. (K, CGE); *Gardner s.n.* (K); *Bradford s.n.* (BM); *Wall s.n.* (E); *Hutchison s.n.* (E).

6. Athyrium solenopteris (Kunze) T. Moore, Ind. Fil.: 43. 1857.

Allantodia solenopteris Kunze, Linnaea 24: 266. 1851.
Type. From S. India (B).
Athyrium ceylanense (Klotzsch) T. Moore apud T. Moore, Ind. Fil.: 181. 1860, nom. nud.
Asplenium gymnogrammoides Klotzsch ex Mett., Abhandl. Senckenb. naturf. Ges. 3: 237. 1859. Type. Sri Lanka, "Ceylan (Gärtner Farmann) Matur. jun., Hort. bot. Berol. den 18 Januar 1857. Originalexemplar" (B).
Athyrium solenopteris var. *pusillum* (Kunze) T. Moore, Ind. Fil.: 187. 1860.
Athyrium gymnogrammoides (Klotzsch ex Mett.) Bedd., Ferns S. Ind.: 52. 1864.
Athyrium devolii Ching, Sunyatsenia 3: 1, t. 1. 1935.
Athyrium giganteum Devol, Lingnan Sci. J. 21: 79, t. 5. 1945.
Athyrium palustre Seriz., J. Jap. Bot. 45: 264. 1970.

Rhizome thick, quite long, ± erect, occasionally branched, densely clothed with lanceolate, pale-brown scales at the apex. Fronds up to c. 6, in an apical tuft, arching or held erect by undergrowth, up to c. 70 cm tall. Stipe ± thin, fragile, (encasing, thickened old stipe-bases persisting), long, up to c. same length as lamina, with few scattered, lanceolate, thin, entire, pale scales, more so near its base; rhachis similar to upper stipe. Lamina bipinnate-tripinnatifid to tripinnate, ± widely lanceolate to ovate-deltate, widest, up to c. 30 cm., shortly above base, lowest pair of pinnae usually slightly shorter, or same length as the next, often abruptly caudate at apex, very fragile, delicate, pale-green, glabrous, with small, delicate costular setae above pinnule-midribs, or at least pinna-apices, though setae often insignificant, or virtually absent; pinnae, up to c. 17 pairs, ± discrete, elongated narrowly triangular, up to c. 15 cm long, 3 cm wide, shortly stalked, acute to often rather abruptly caudate apices, symmetrical, inserted at 90°, often variously backward-deflexed, pinnate and deeply bipinnatifid, even bipinnate towards bases; pinnules triangular-ovate, with obtuse to acute apices, shortly stalked, often slightly backward-deflexed, bases markedly asymmetrical, stalk present at ± basiscopic corner, ± deeply, often pinnatifidly lobed, some of lowest ones just becoming pinnately lobed at their bases, lobes on acroscopic side of pinnule being longer, wider, nearer to 90° than the more obliquely sloping ones at basiscopic side, lobes ± rectangular with rounded to acute apices, varying markedly in width. Plants with wider, more crowded lobes merging continuously into forms with much narrower lobes and, consequently, more finely dissect fronds, lobe-apices, particularly pinnule-apices with prominent, long, narrowly acute teeth, often flabellately splayed-out around apex. Sori c. 8-20 per pinnule, often borne throughout lamina, medially or sub-basally on veins, small, varying in each pinnule from ± straight to curved at apex into a hook-shape. Indusium small, narrow, thin, white, with lacerate margins, lying over top of sorus, shrivelling considerably and turning brown on ripening. Spores small, uniform., pale-brown, smooth; perispore apparently absent. Cytotype tetraploid, sexual, n = 80 (80 bivalents at meiosis) and hexaploid, sexual, n = 120 (120 bivalents at meiosis).

D i s t r. Higher altitudes in the central mountains of Sri Lanka. Quite common. A Sino-Himalayan element. Sri Lanka, S. India, the E. Himalaya, from Sikkim east to Bhutan and Meghalaya (Khasi Hills), S.W., S. and S.E. China and S. Japan.

E c o l. Among bushes in damp, often marshy hollows in semi-open forest, also on shaded, moist road-banks, from c. 1700-1950m in Sri Lanka.

N o t es. Sledge (1956 and 1962) revived Beddome's earlier treatment of two varieties in this species, var. *solenopteris* and var. *pusillum.*, though commenting that the two merge into each other. The ranges of variation in the field, however, suggest that var. *pusillum* is not of taxonomic significance, it is not recognised here. More important are the two distinct cytotypes he discovered and illustrated (Manton & Sledge 1954), a tetraploid with more finely lobed, acute and more dissect pinnules, which probably corresponds with *A. solenopteris*, and a hexaploid with coarser, more obtuse pinnules, which appears to match better *A. gymnogrammoides*, which name Sledge had included within var. *pusillum*. These two may may well turn out to be recognisable and distinct species.

The epithet solenopteris is an indeclinable noun in apposition, its gender must not be changed to agree with the generic name.

S p e c i m e n s E x a m i n e d. BADULLA DISTRICT: Namunukula, 1875m., *Sledge 1202* (BM). NUWARA ELIYA DISTRICT: Nuwara Eliya, 10 May 1906, *Matthew s.n.,* (K); Nuwara Eliya, *Freeman 183A, 184B* (BM), s.n., July 1887 (E); Nuwara Eliya, Jan. 1847, *Gardner s.n.,* in Thwaites C.P. 1346 (PDA); Nuwara Eliya, Moon Plains, damp ground in jungle, 1800m., *Sledge 714* (BM); Kandapola Forest Reserve, near Nuwara Eliya, 1920m., *Sledge 1322, 1324, 1325, 1326* (BM); Ramboda Pass, in forest by track from summit of pass to Maturata, 1920m., *Sledge 1317* (BM); on S. side of ridge, c. 1-2km along track from c. 1km below top of Ramboda Pass (top part of Top Pass on W. side) which heads N.W. towards Maturata, W. of Nuwara Eliya, in woods, c. 1700-1800m., 25 Oct. 1993, *Fraser-Jenkins Field nos. 346-349,* with *Abeysiri* (K, US, PDA, NMW); road-bank just below top of Top Pass on W. side, upper Ramboda Pass, W. of Nuwara Eliya, c. 1950m.,

23 Sept. 1993, *Fraser-Jenkins Field no. 227,* with *Jayasekara, Samarasinghe & Abeysiri* (K, US, PDA, NMW); c. ½ km to the E. of top of Top Pass, upper Ramboda Pass, W. of Nuwara Eliya, c. 2100m., 31 Oct. 1993, *Fraser-Jenkins Field no. 495,* with the *Bandara family* (K, US, PDA, NMW); Pidurutalagala, 1950m., *Sledge 735* (BM); Hakgala, by track in jungle, 1740m., *Sledge 746* (BM). LOCALITY UNKNOWN: *Thwaites C.P. 1346* (K, BM., E); *Walker s.n.* (K); *Wall s.n.* (K, E); *Hooker & Thompson 208* (BM*); Beddome s.n.* (BM); *Bradford* in Herb. *Hance s.n.* (BM); *Hutchison s.n.* (E), Nuwara Eliya, up to 2134 m., *Wall 38/194* in Herb. Barkly (BM).

Sect. Echinoathyrium

Ching & Hsieh, Bull. Bot. Res. Harbin 6(4): 132. 1986.

7. Athyrium praetermissum Sledge, Ann. Mag. Nat. Hist., ser. 12, 9: 457, t. 16. 1956.
Type. From Sri Lanka, "Knuckles Mt.,1676 m.. *Sledge 1077,* 30 Jan. 1954" (BM).

Athyrium praetermissum var. *tripinnatum* Sledge, Ann. Mag. Nat. Hist., ser. 12, 9: 453. 1956. Type. Sri Lanka, „Ramboda Pass, at summit by track to Maturata, 6300ft, *Sledge 1304,* 17 Mar. 1954" (BM).

Rhizome thick, decumbent, long, usually unbranched, densely clothed with long, linear, dark, castaneous-brown to blackish scales at the apex. Fronds up to c. 6 in apical tuft, arching, large (up to c. 75 cm tall). Stipe thick, stiff, (thickened, encasing old stipe-bases persistent), same length as, or slightly longer than lamina, bearing dense tuft of linear, exserted, castaneous-blackish to dark russet-brown, entire scales up to shortly above base and a few, scattered, thin, pale-russet, almost hair-like scales higher, but distally absent. Rhachis similar to the upper stipe. Lamina bipinnate, often tripinnatifid, occasionally just tripinnate below, ovate-deltate, widest at base, here longest pinnae, gradually tapering at apex, slightly stiffly herbaceous, mid- to dark-green, glabrous, bearing stiff costal spine above, at junction of each pinnule-midrib with pinna-costa, especially towards tips of pinnae, but without costular setae above pinnule-midribs; pinnae, up to c. 15 pairs, contiguous or a little overlapping, elongated-deltate, up to c. 20 cm long, 5 cm wide, stalks up to c. 7 mm long, narrowly acute and acuminate to caudate at apices, ± symmetrical, inserted at an oblique angle to rhachis, pinnate, becoming bipinnatifid, occasionally just bipinnate towards base in larger plants; pinnules triangular-ovate, with obtuse or narrowly obtuse, rarely acute apices, shortly stalked, slightly or obviously falcate, sloping, markedly asymmetrical at their bases, stalk being at basiscopic corner, ± deeply, sometimes pinnatifidly lobed at their bases, rarely only just pinnatisect, lobes on acroscopic side of pinnule being longer, wider and more at 90° than obliquely sloping ones at basiscopic side, lobes ± ovate-rectangular, with obtuse apices, wide, lowest acroscopic lobe longer than next, pinnules thus acroscopically basally auriculate, lobe-apices, particularly pinnule-apices, bearing ± prominent, small, acute teeth. Sori c. 6-50 per pinnule, borne throughout the lamina, basally on veins, in two rows close to midrib, small, ± short, ± straight or slightly curved, or becoming hook-shaped. Indusium ± small, slightly thick, pale-brown, margins entire or slightly undulated, lying over top of sorus until lifting, shrinking slightly and turning reddish-brown on ripening. Spores small, uniform., pale-brown, smooth, perispore apparently absent. Cytotype unknown, but probably tetraploid, sexual, n = 80, by extrapolation from its hybrids.

D i s t r. Upper-mid to higher altitudes in the central mountains of Sri Lanka. Common. A S.E. Asian element. Sri Lanka, S. India, C. and E. Nepal, Darjeeling, Sikkim., Arunachal Pradesh, Meghalaya, S.E. Tibet, S.W. and S. China, Burma, Thailand.

E c o l. Slopes in forest, often near streams, also on shaded road-banks, from c. 1650-2100 m altitude in Sri Lanka.

N o t es. Sledge's var. *tripinnatum* merely represents large, well dissect plants of the same species, but "var. *erythrorhachis*" is consistently distinct, well separated and belongs to the next species.

A. praetermissum belongs to a group of closely related but distinct species, including *A. mackinnoniorum* (Hope) C. Chr. ["mackinnonii"] in the west Himalaya, east to W. Nepal. The E. Himalayan populations differ from the Sri Lankan and S. Indian ones in frequently, though not always, having setae above the pinnules, emphasising the close relationship between Sections *Echinoathyrium* and *Strigoathyrium.*, but they nevertheless appear to represent the same species and match well in all other respects. However they have not yet been investigated cytologically.

S p e c i m e n s E x a m i n e d: MATALE DISTRICT: Knuckles Mountain, 1676 m., *Sledge 1077*(K, BM); stream on S.W. side of Wattakelle Hill, above Wattakelle village, N.E. of Kallebokka, N. of Wattegama, W. Knuckles Range, N.E. of Kandy, 1200-1500 m., 1 Oct. 1993, *Fraser-Jenkins Field no. 266*, with Bandara family (K, US, PDA, NMW). NUWARA ELIYA DISTRICT: Udapussalawa, Apr. 1854 *s.coll. s.n.*, in *Thwaites C.P. 1344* p.p. (PDA); Nuwara Eliya, *Gamble 27564* (K); Ramboda, Sept. 1847, *Gardner s.n.* in *Thwaites C.P. 1344* p.p.(PDA); Ramboda Pass, at summit by track to Maturata, 1920 m., *Sledge 1299, 1304 and 1310* (BM); on S. side of ridge, c. 1-2 km along track from c. 1 km below top of Ramboda Pass (top part of Top Pass on W. side) which heads N.W. towards Maturata, W. of Nuwara Eliya, in woods, c. 1700-1800 m., 25 Oct. 1993, *Fraser-Jenkins Field nos. 344 and 345*, with *Abeysiri* (K, US, PDA, NMW); jungles on Maturata side of Nuwara Eliya, *Freeman 185 &186* (BM); Mt. Pidurutalagala, Nuwara Eliya, 2100 m., *Ballard 1258* (K); steep bank in Eucalyptus forest above road, just above lower Army check-post at bottom of road up S. side of Pidurutalaga Mountain, on lower slopes, N. of Nuwara Eliya, c. 1950 m., 26 Oct. 1993, *Fraser-Jenkins Field no. 375*, with *Abeysiri* (K, US, PDA, NMW); Hakgala Peak, 1800 m and 1675m., *Ballard 1102 & 1399* (K); forest just above road, c. 1km up road to Nuwara Eliya west from Hakgala Botanic Garden, E. of Nuwara Eliya, c.1760 m., 22 Sept. 1993, *Fraser-Jenkins Field no. 197*, with *Jayasekara, Samarasinghe* and *Abeysiri* (K, US, PDA, NMW); Adam's Peak, 14 Feb. 1908, *Matthew s.n.* (K); forested stream below road, c. 3 km E. of Farr Inn on road towards Ohiya, Horton Plains, S. of Nuwara Eliya, c. 1950 m., 27 Oct. 1993, *Fraser-Jenkins Field no. 435* with *Abeysiri* (K, US, PDA, NMW). LOCALITY UNKNOWN: *Thwaites C.P.* 1344 p.p. (K, BM., E, PDA); *Thwaites C.P. 1345* p.p. (K, E); *Trimen s.n.* in *Thwaites C.P. 1344* p.p. & *1345* p.p. in Herb. Beddome (K); *Gardner 1068 & 1069* (PDA and K); *Walker s.n.* (K); *Hooker & Thompson 207* p.p. (BM); *Wall s.n.* (E); *Hutchison s.n. & 187* (E).

8. Athyrium wardii (Hook.) Mak., Bot. Mag. Tokyo 13: 15. 1899.

Asplenium wardii Hook., Sp. Fil. 3: 189. 1860. Type. From Tsushima Island, S. Japan.
Athyrium gymnogrammoides var. *erythrorhachis* Bedd., Suppl. Ferns S. Brit. India : 12. 1876.
Athyrium majus (Makino) Makino, Bot. Mag. Tokyo 28: 178. 1914.
Athyrium tsussimense Koidz., Flor. Symb. Or.-As.: 41. 1930.
Athyrium praetermissum var. *erythrorhachis* (Bedd.) Sledge, Ann. Mag. Nat. Hist., ser. 12, 9: 459, t. 17. 1956.

Rhizome thick, decumbent, quite long, usually unbranched, densely clothed with long, linear, dark castaneous-brown to blackish scales at the apex. Fronds up to c. 5 in apical tuft, arching, up to c. 50 cm tall. Stipe thick, stiff (encasing, thickened old stipe-bases persistent), when alive pinkish-purple, drying ± stramineous, up to 1½ times or longer than lamina, bearing dense tuft of linear, exserted, dark-castaneous to dark russet-brown, entire scales up to shortly above base and a few, scattered, thin, pale-russet, hair-like scales slightly further up, absent above first quarter of stipe. Rhachis similar to upper stipe, pink. Lamina bipinnate, not becoming tripinnatifid, markedly deltate to widely rectangular-deltate, widest at base, basal pinnae being longest, markedly abruptly tapering at apex, slightly stiffly succulent-herbaceous, ± dark-green, glabrous, bearing stiff costal spine above at junction of each pinnule-midrib with pinna-costa, especially towards pinnae-tips, but without costular setae above the pinnule-midribs; pinnae, up to 10 pairs, ± contiguous, elongate-deltate, or lower ones slightly narrowed to their bases, up to c. 15 cm long, 3 cm wide, shortly stalked, with narrowly acute and acuminate to caudate apices, ± symmetrical, inserted at 90°, lower ones often slightly backward deflexed, pinnate; pinnules ± deltate, with obtuse to narrowly obtuse, rounded apices, very shortly stalked or becoming narrowly adnate to pinna-costa, slightly falcate, sloping, markedly asymmetrical at their bases, stalk being at basiscopic corner, ± entire and unlobed, apart from a rounded, acroscopic basal auricle, usually without teeth, or with a few, insignificant, small, ± obtuse teeth around pinnule-apex. Sori c. 4-10 per pinnule, borne throughout lamina, sub-basally in two rows, small, ± short, straight or slightly curved. Indusium ± small, thin, white, margins entire or slightly undulated, lying over top of sorus until lifting, shrinking somewhat and turning pale-brown on ripening. Spores small, uniform., pale-brown, smooth, perispore apparently absent. Cytotype (Sri Lanka and Japan) tetraploid, sexual, n = 80 (80 bivalents at meiosis).

D i s t r. Higher altitudes in the central mountains of Sri Lanka. Restricted, but very locally quite common. Previously thought to be a Sino-Japanese element; its presence in Sri Lanka is unexpected, at least until its range in China is better known once the many separate names given to it there in the herbarium have been combined. Sri Lanka, S.W.?, S. and S.E. China, Korea, Japan.

E c o l. Slopes in forest, often near streams, from c. 1700-2100 m altitude in Sri Lanka.

N o t es. Although Sledge included this taxon as a variety within *A. praetermissum.*, it is obvious, both in the herbarium and especially in the field in Sri Lanka, that it is a distinct species. It is also not connected by intermediates with the latter. The Sri Lankan plants match the Japanese and Chinese ones very well and it appears that they must, surprisingly, represent the same species (see Fraser-Jenkins 1998, in prep., who first separated this taxon as *A. wardii*).

S p e c i m e n s   E x a m i n e d. BADULLA DISTRICT: Badulla, *Thwaites C.P. 1344* p.p., in Herb. Brodie (E). NUWARA ELIYA DISTRICT: Nuwara Eliya, *Freeman 177B, 181F* (BM); Nuwara Eliya, Kandapola Forest Reserve, 1920 m., *Sledge 1338* (BM); on S. side of ridge, c. 1-2 km along track from c. 1km below top of Ramboda Pass (top part of Top Pass on W. side) which heads N.W. towards Maturata, W. of Nuwara Eliya, in woods, c. 1700-1800 m., 25 Oct. 1993, *Fraser-Jenkins Field nos. 341-343*, with *Abeysiri* (K, US, PDA, NMW); Nuwara Eliya, Moon Plains, in secondary forest, 1829 m., *Sledge 719* (BM); forested slope above Hakgala Botanic Garden on N.E. side of Hakgala Mountain, E. of Nuwara Eliya, c. 1700 m., 23 Sept. 1993, *Fraser-Jenkins Field no. 207, with Jayasekara, Samarasinghe* and *Abeysiri* (K, US, PDA); Horton Plains, *Thwaites C.P. 1344* p.p., in Herb. Brodie (E); in jungle near Horton Plains, 2060 m., *Sledge 697* (BM). LOCALITY UN-KNOWN: *Thwaites C.P. 1344* p.p. (PDA); *Thwaites C.P. 1344* p.p., in Herb. Beddome (BM); 1870, *Beddome s.n.* (K); *Gardner 1067* (K, CGE, PDA); *Wall s.n.* (E).

4. DIPLAZIUM

Sw., J. Bot. (Schrad.) 1800(2): 61. 1802. Sledge, Bull. Br. Mus., nat. Hist.
(Bot.) 2(11): 293. 1962. id., Bot. J. Linn. Soc. 84: 17. 1982. Kramer et al. in
Kubitzki (ed.), Fam. Gen. Vasc. Plants 1: 133. 1990.
Type species: *Diplazium plantaginifolium* (L.) Urban (based on *Asplenium
plantaginifolium* L.).

Diplaziopsis C. Chr., Ind. Fil. 227. 1905. Sledge, Bull. Br. Mus., nat. Hist.
(Bot.) 2(11): 317. 1962. id., Bot. J. Linn. Soc. 84: 17. 1982. Kramer in
Kubitzki (ed.), Fam. Gen. Vasc. Plants 1: 135. 1990. Type species:
Diplaziopsis javanica (Blume) C. Chr. (based on *Asplenium javanicum*
Blume = *Diplazium javanicum* (Blume) Makino).
Athyrium sensu Copel., Gen. Fil. 147. 1947; sensu Holttum, Fl. Malaya 2:
541. 1954, non Roth

Medium-sized to large terrestrial or epilithic ferns. Rhizome usually short,
erect, sometimes creeping; scales ovate to linear, light to dark brown, margins
entire or toothed. Stipes sometimes muricate at base or throughout from persisting scale bases. Rhachis adaxially grooved, sparsely scaly; costae and
costules grooved above, edge of groove strongly winged, wing interrupted,
enlarged at junction of costa and rhachis. Lamina simple to quadripinnate,
usually herbaceous. Veins usually free, rarely adjacent vein groups anastomosing rather freely. Sori elongated along veins., lowest acroscopic sorus
usually double, other sori occasionally double, sometimes all sori single and
allantodioid. Indusia lateral, double sori with two discrete indusia.

A large and taxonomically insufficiently understood genus of an estimated
400 species (but most probably fewer) in the warmer regions of the world,
only rarely and locally extending into temperate regions. Closely related and
sometimes even united with *Athyrium*. Twelve species recognised in Sri Lanka
following Sledge (1962), but moving *D. subsinuatum* and *D. zeylanicum* to
Deparia and including *Diplaziopsis javanica* into *Diplazium*. Delimitation of
species by Sledge was not changed waiting for a more detailed study of the
genus in Asia which is desperately needed but not yet available.

KEY TO THE SPECIES

1 Veins free
  2 Fronds pinnate, pinnae entire to deeply pinnatifid
    3 Pinnae deeply pinnatifid, rhachis narrowly winged . . . .
. . . . **1. D. beddomei**
    3 Pinnae not deeply pinnatifid, rhachis not winged
      4 Pinnae entire to shallowly lobed

5 Lamina apex pinnatifid . . . . **3. D. sylvaticum**
5 Lamina apex formed by a conform terminal pinna . . . .
. . . . **12. D. javanicum**
  4 Pinnae pinnatifid . . . . **4. D. dilatatum**
2 Fronds bi- to tripinnate
  6 Rhizome erect
    7 Rhizome scales up to 5 mm long, entire or inconspicuously toothed
      8 Pinnules decurrent, pinna rhachis with a continuous wing . . . .
. . . . **1. D. beddomei**
      8 Pinnules decurrent, pinna rhachis without a continuous wing . . . .
. . . . **2. D. decurrens**
    7 Rhizome scales up to 1 cm long, conspicuously toothed
      9 Stipes smooth
        10 Terminal pinna segment serrate for less than ?of its length . . . .
. . . . **4. D. dilatatum**
        10 Terminal pinna segment serrate for at least ?of its length . . . . .
. . . . **5. D. travancoricum**
      9 Stipes muricate . . . . **6. D. polypodioides**
  6 Rhizome creeping
  11 Sori not allantodioid . . . . **7. D. cognatum**
  11 Sori allantodioid
    12 Pinnule lobes 3-5 mm wide, pinna rhachis and costa with broad entire
    scales, sori close to the costa . . . . **10. D. muricatum**
    12 Pinnule lobes 5-8 mm wide, pinna rhachis and costa with narrow
    toothed scales or glabrous, sori diverging from the costa . . . .
. . . . **11. D. procumbens**
1 Veins anastomosing
  13 Stipes smooth . . . . **8. D. esculentum**
  13 Stipes muricate . . . . **9. D. paradoxum**

1. Diplazium beddomei C. Chr., Ind. Fil. 228. 1905. Sledge, Bull. Br. Mus. nat. Hist. (Bot.) 2(11): 299. 1962. id., Bot. J. Linn. Soc. 84: 17. 1982.
Type: Sri Lanka, *Gardner 1063, 1247, s.n.* in *C.P. 3100* (PDA, syn, lectotype
    to be chosen).

Diplazium schkuhrii sensu Bedd., Ferns S. India 76, pl. 230. 1863/64. Id.,
    Handb. Ferns Brit. India 181. 1883, non J.Sm.
Asplenium schkuhrii sensu Thwaites, Enum. Pl. Zeyl. 385. 1864, non Mett.

Rhizome erect. Stipes up to 45 cm long, sparsely scaly; scales small, lanceolate, acute, irregularly toothed, dark, mixed with almost hair-like ones. Lamina 30-40 × 10-30 cm., ovate- or deltoid-lanceolate usually deeply pinnate-pinnatifid, sometimes bipinnate; pinnae stalked up to 1 cm., alternate, the lowermost pair often opposite and deflexed, those of pinnate fronds 5-10 × 2-3 cm., acuminate, basal segments close to rhachis, margin subentire to crenate-lobate, those of bipinnate fronds 25-30 × 5-10 cm., pinnules spaced, sessile, basiscopic margin decurrent, forming a narrow wing along pinna rhachis, apex acute, margin bluntly crenate-lobate to pinnatifid with obtuse, ± truncate, slightly crenate lobes, firmly herbaceous, main rhachis, pinna rhachis and costae glabrous or sparsely scaly at junction of pinnae with main rhachis. Veins pinnate with 1-2 pairs of simple veinlets per segment, in larger pinnules of bipinnate fronds 4-5 pairs, mostly soriferous. Sori along almost the whole veinlet, basal acroscopic ones diplazioid, the others simple.

D i s t r. Endemic to Sri Lanka.

E c o l. Widely distributed but not common in the forests of the Central and Southern Provinces from 750 to 1650 m altitude.

S p e c i m e n s   E x a m i n e d. RATNAPURA DISTRICT: Adam's Peak, March 1846, *Gardner 1247* (CGE, PDA), Adam's Peak, 1524 m., *Gardner1247* in C.P. 3100 (K); Sinharaja Forest, above Beverley Estate, Deniyaya, 792 m., *Sledge 1395* (BM., K); Adam's Peak, Nov 1881, *Wall s.n.* in Herb. Hook. (K); Deniyaya, jungle above Beverley Estate, 914 m., 12 March 1954, *Sledge 1272* (BM); Deniyaya, 550 m., 5 Feb 1954, *Schmid 1150* (BM). NUWARA ELIYA DISTRICT: Ramboda, June 1845, *Gardner 1063* (CGE, PDA); Hakgala, *Freeman 194B* (BM). KANDY DISTRICT: Raxawa, Feb 1854, *Gardner s.n.* in C.P. 3100 (PDA); Peacock Hill, Pussalawa, *Robinson s.n.* (K). WITHOUT LOCALITY: *Gardner 1063* in Herb. Hook. (K); *Gardner 1247* (BM); *Gardner s.n.* in *C.P. 3100* (PDA); *Thwaites C.P. 3100* (BM., E, K); *Thwaites C.P. 3100* in Herb. Barkly (BM); *Thwaites C.P. 3100* in Herb. Bedd., in Herb. Hook. (K); *Trimen s.n.* (C.P.3100) in Herb. Bedd. (K); *Thwaites C.P. 3951* (PDA); *Beddome s.n.* (BM); Aug and Sep 1870, *Beckett s.n.* (K); *Naylor Beckett 31* in Herb. *T. Moore* (K); *Skinner s.n.* (K); *Wall s.n.* (E, K); *Hutchison s.n.* (E); *Ferguson s.n* (US 815513).

2. Diplazium decurrens Bedd., Ferns S. India 76, pl. 229. 1864. Sledge, Bull. Br. Mus. nat. Hist. (Bot.) 2(11): 300. 1962. id., Bot. J. Linn. Soc. 84: 17. 1982.
Type: Sri Lanka, *Thwaites C.P. 3332* (BM., holo; K, iso)

Asplenium polypodioides (Blume) Mett. var. ß Thwaites, Enum. Pl. Zeyl. 385. 1864.
Type: Sri Lanka, *Gardner 1245, s.n.* in *C.P. 3332* (PDA, syn)
Diplazium polypodioides Blume var. *decurrens* (Bedd.) Bedd., Handb. Ferns Brit. India 186. 1883.
Asplenium thwaitesianum Szyszyl. in G. Beck, Itin. Princ. Coburg 2: 125. 1888. Type: Sri Lanka, *Thwaites C.P. 3332* (W, holo)

Rhizome erect. Stipes 25-70 cm long, scaly abaxially; scales up to 6 × 1-2 mm., acute, entire, dark. Lamina ovate to deltoid-ovate, bipinnate, 30-70 × 15-50 cm., with up to 12 pairs of ascending pinnae; pinnae up to 50 × 15-20 cm., usually smaller, lower ones stalked up to 3 cm; pinnules 2-6 (up to 12) cm long, basal ones often smaller than succeeding ones, shortly stalked or sessile, with pinnatifid to entire margins, middle and distal ones broadly adnate, basiscopic margin decurrent, herbaceous, glabrous on both surfaces. Veins pinnate with 1-2 pairs of veinlets in smaller pinnules and 4-6 pairs in larger ones, mostly soriferous. Sori curved, along almost the whole veinlets, at least basal acroscopic ones diplazioid.

D i s t r. Endemic to Sri Lanka.

E c o l. In the central forests from 600 to 1500 m altitude.

S p e c i m e n s   E x a m i n e d. KEGALLE DISTRICT: Gallebodde Rock, in forest, 1219 m., 27 Jan 1954, *Sledge 1056* (BM). KANDY DISTRICT: Corbet's Gap, in secondary jungle, 1219 m., 9 Dec 1950, *Sledge 542* (BM); Near Patragala Plains, March 1846, *Gardner 1245* (CGE, PDA); Ambagamuwa, Nov. 1854, *Thwaites C.P. 3332* (PDA); Upper slopes of Gongala Hill, in forest, 11 March 1954, *Sledge 1260* (BM); Above Hoolankande, by path through forest, 4500 ft., 20 Jan 1954, *Sledge 1016, 1023* (BM). Forest above Le Vallon Estate, 1524 m., 9 Feb 1954, *Sledge 1126* (BM). RATNAPURA DISTRICT: Adam's Peak, *Gardner s.n.* in C.P. 3332 (PDA); Adam's Peak, 14 Feb 1908, *Matthew* (K); Adam's Peak, 5 Jan 1951, *Manton 839* (BM); Deniyaya, forest above Beverley Estate, 792 m., 4 Apr 1954, *Sledge 1404* (BM). NUWARA ELIYA DISTRICT: Pidurutalagala, 2370 m., *v. Fridau s.n.* (GZU); Between Adam's Peak and Nuwara Eliya, *Gardner 1246* (CGE, K). WITHOUT LOCALITY: *Thwaites C.P. 3332* (BM., E; W, photo BM); *Thwaites C.P. 3332* in Herb. Barkly (BM); *Thwaites C.P. 3332* in Herb. Bedd. (BM., holotype): *Thwaites C.P.3332* (K); *Beddome s.n.* (K); Southern Prov., 600-1200 m., *Wall s.n.* (BM); Central & Southern Prov., 610-1219 m., *Wall 38/246* in Herb. Barkly (BM); *Walker s.n.* (K); *Ferguson s.n* (US 815499, 815509, 816418).

3. Diplazium sylvaticum (Bory) Sw., Syn. Fil. 92. 1806. Bedd., Ferns S. India 53, pl. 161. 1863/64. Id., Handb. Ferns Brit. India 177. 1883. Sledge, Bull. Br. Mus. nat. Hist. (Bot.) 2(11): 301. 1962. id., Bot. J. Linn. Soc. 84: 17. 1982.

Callipteris sylvatica Bory, Voy. Iles Afrique 1: 282. 1804. Type: Mascarene Islands, Mauritius, *Bory s.n.* (P, holo).

Asplenium sylvaticum (Bory) C. Presl, Rel. Haenk. 1: 42. 1825. Thwaites, Enum. Pl. Zeyl. 385. 1864.

Diplazium firmum Fée, Mém. Fam. Foug. 10: 30, pl. 38, fig. 2. 1865. Type: Sri Lanka, *Thwaites C.P. 1349* (RB, holo; BM., E, K, PDA, iso).

Rhizome short, erect. Stipe up to 40 cm long, usually shorter, scaly at least near base; scales up to 15 × 2 mm., very dark, margin toothed, those in upper part lighter and smaller; lamina simply pinnate, 50-80 × 25 40 cm., herbaceous; pinnae numerous, 12-20 pairs below pinnatifid apex, middle and basal ones stalked, up to 20 × 2 cm., base truncate to broadly cuneate, margin subentire to crenate-serrate or lobed to ¹/₃ towards costa, apex acuminate, serrate. Veins in pinnate groups with 3-4 pairs of unbranched veinlets. Sori 3-6 mm long, basal acroscopic ones diplazioid, others simple, straight, usually neither touching costa nor margin.

D i s t r. Mauritius, India, Sri Lanka, eastwards to S. China and the Philippines, southwards to Java.

E c o l. In forests from 600 to 1200 m altitude.

S p e c i m e n s   E x a m i n e d. MATALE DISTRICT: Matale, Dec 1845, *Gardner s.n. in C.P. 1349* (PDA); N side of Brae Gap by path to Hoolankande, in forest, 1067 m., 4 March 1954, *Sledge 1234* (BM); Gongala Hill, in forest, 11 March 1954, *Sledge 1286* (BM). KANDY DISTRICT: Kandy, Lady Horton's Walk, *Chevalier s.n.* (BM); Corbet's Gap, in forest, 1219 m., 22 Jan 1954, *Sledge 1037* (BM). RATNAPURA DISTRICT: Adam's Peak, 5 Jan 1951, *Manton P. 384* (BM). WITHOUT LOCALITY: *Gardner 34* in Herb. Hill (BM), in Herb. T.Moore (K); 1847, *Gardner 1059* (K); *Gardner s.n.* in C.P. 1349 (K); *Thwaites C.P. 1349* (RB, holotype of *Diplazium firmum* Fée); 1863, *Thwaites C.P. 1349* (BM., E, K); *Thwaites C.P. 1349* in Herb. Barkly (BM); July 1866, *Thwaites 3892* (CGE, K, PDA); *Thwaites C.P. 1349* in Herb. Hook. (K); *Robinson s.n.* (K); Aug 1870, *Naylor Beckett s.n.* (E, K); *s.coll. s.n.* in Herb. T.Moore (K); *Macrae 882* (BM); *Robinson C 77* (K); Central Prov., 610-219 m., *Wall 38/217* in Herb. Barkly (BM); *Wall s.n.* (E); *Wight 1951* (E); *Ferguson s.n.* (US 815515); *Hancock 53* (US 1277203).

4. Diplazium dilatatum Blume, Enum. Pl. Javae 2: 194. 1828. Bedd., Ferns S. India 53, pl. 162. 1863/64. Sledge, Bull. Br. Mus. nat. Hist. (Bot.) 2(11): 303. 1962. id., Bot. J. Linn. Soc. 84: 17. 1982.

Type: Indonesia, Java, Burangrang, *Blume s.n.* (L, holo, photos BM., US).

Asplenium latifolium D.Don, Prodr. Fl. Nepal. 8. 1825, nom. illeg., non Bory 1803. Type: Nepal, *Wallich* (K).

Diplazium latifolium T. Moore, Ind. Fil. 141, 331. 1860, nom. nov. for *Asplenium latifolium* D. Don. Bedd., Handb. Ferns Brit. India 187 p.p. 1883.

Asplenium dilatatum (Blume) Hook., Sp. Fil. 3: 258. 1860. Thwaites, Enum. Pl. Zeyl. 385. 1864.

Rhizome erect. Fronds tufted. Stipe up to 60 cm long, smooth, basally scaly; scales up to 15 × 1 mm., dark brown with black toothed margins. Lamina 60-90 × 30-45 cm., pinnate-pinnatifid to bipinnate with 10-15 pairs of pinnae; pinnae of bipinnate fronds up to 35 × 15 cm., basal pinnules shortly stalked to sessile, becoming adnate towards acuminate, pinnatifid apex; pinnules up to 8 × 2 cm., oblong, base truncate to broadly cuneate, margin entire to coarsely serrate, apex acute to acuminate. Costae with scattered, very narrow, brown scales. Veins of each pinnule forming at least 10 pairs of veinlet groups, basal acroscopic veinlet of each group fertile. Sori extending from costa half-way to margin, sometimes shorter sori on other veinlets. Lower and middle pinnae of pinnate-pinnatifid fronds up to 35 × 3-5 cm., distinctly stalked, progressively less deeply lobed towards acuminate, serrate apex, lobes rounded, subentire. Costae sparsely scaly; distal pinnae shallowly lobed and serrate to subentire below pinnatifid frond apex. Veins forming 5-10 pairs of veinlets, often many forked, mostly soriferous. Sori up to 1 cm long, extending from costa almost to margin, those of basal acroscopic veinlet diplazioid, others simple.

D i s t r. India, Sri Lanka, China, Indochina, Burma southwards throughout Malesia to N. Australia.

E c o l. Common in forests from 900 to 1650 m altitude.

S p e c i m e n s   E x a m i n e d. MATALE DISTRICT: Forested stream-gulley beside and below road c. 1 km W. of top of pass between Rattota and Laggala, N.E. of Matale, c.1200 m., 15 Sept 1993, *Fraser-Jenkins* with *Jayasekara & Bandara, Field No.134* (K). KANDY DISTRICT: Kintyre Estate, 4 June 1932, *Simpson 9744* (BM); Hoolankande, in forest, 1350 m., 20 Jan 1954, *Sledge 1002, 1003, 1004, 1006, 1024* (BM); Hantane, Nov 1854, *Thwaites C.P. 1350* (PDA); Hantane, 1158 m., 8 Dec 1950, *Ballard 1045* (K); Near Kandy, jungle at Oodawella, 1219 m., 8 Dec 1950, *Sledge 529, 536* (BM); Corbet's Gap, in secondary jungle, 1330 m., 9 Dec 1950, *Sledge 570* (BM); Ambagamuwa, jungle, in stream., 19 Jan 1954, *T.G.Walker T135* (BM). BADULLA DISTRICT: Badulla, *Freeman 193A, 195C* (BM); Namunukula, *Freeman 201A* (BM); Namunukula, jungle on Tonacombe Estate, 1372 m., 23 Feb 1954, *Sledge 1177, 1178, 1181* (BM).RATNAPURA DISTRICT: Adam's Peak, *Moon s.n.* (BM). NUWARA ELIYA DISTRICT: Forest on Ramboda Pass, June 1845, *Gardner 1058* (CGE); Nuwara Eliya, 1847, *Gardner s.n.* in C.P. 1350 (PDA); Forest between Adam's Peak and Nuwara Eliya, March 1846, *Thwaites C.P. 1248* (CGE); Hakgala, gulley near Botanical Garden, ground fern in wet jungle, 1760 m., 30 Dec 1950, *Ballard 1347* (K); *Gardner 1058* in Herb. Hook. (K); Forest stream gully c. 2 km W. of Hakgala Botanic Garden, on lower N.W. side of Hakgala Mountain, E. of Nuwara Eliya, 1650m., 23 Sept 1993, *Fraser-Jenkins* with *Jayasekara, Samarasinghe & Abeysiri Field No. 218* (K); Nuwara Eliya, *Freeman 202B* (BM); Rambodda

Pass, *Freeman 196D* (BM); Nuwara Eliya, 1950 m., 11 March 1954, *Schmid 1518* (BM., G); Nuwara Eliya, *Chevalier s.n.* (BM). WITHOUT LOCAL-ITY: *Gardner 35, 36* in Herb. Hill (BM); *Gardner 1058* in Herb. J.Smith (BM); *Thwaites C.P. 1248* (K); *Thwaites C.P. 1350* p.p. (BM); *Thwaites C.P. 1353* (E); *Thwaites C.P. 1353* in Herb. Barkly (BM); *Macrae 895* (BM); Central Prov., 914-1524 m., *Wall 38/249* in Herb. Barkly (BM); *Ferguson s.n.* (US 815496, 815508).

5. Diplazium travancoricum Bedd., Handb. Ferns Brit. India 188. 1883. Sledge, Bull. Br. Mus. nat. Hist. (Bot.) 2(11): 305. 1962. id., Bot. J. Linn. Soc. 84: 17. 1982.
Type: India, Athraymallay, Tinnevelly, *Beddome s.n.* (K, holo).

Asplenium travancoricum (Bedd.) Baker, Ann. Bot. 5: 310. 1891.

Rhizome erect, fronds tufted. Stipe up to 80 cm., scaly; scales narrowly lanceolate, acuminate, 10 × 1 mm., dark brown, margins toothed. Lamina 70-110 × 35-55 cm., bipinnate with 10-15 pairs of pinnae below the pinna-like apex; largest pinnae 30-45 × 10-15 cm., stalked, lanceolate, somewhat attenuate at base, upper half or third forming a long, broad, shallowly incised or serrate apex with acuminate or caudate tip, pinnules subsessile, becoming adnate and decurrent, largest 5-8 × 2 cm., oblong-lanceolate, entire below acuminate, serrate tip, basal ones smaller than adjoining ones; upper pinnae almost entire except for acuminate, serrate apex; frond apex pinna-like, 10-20 cm long, basally lobed, distally serrate. Veins 15-25 pairs per pinnule, forked up to 3 times in larger pinnules, mostly soriferous. Sori up to 1 cm long, usually several pairs diplazioid.

D i s t r. S. India, Sri Lanka.

E c o l. Rare in montane forest.

S p e c i m e n s   E x a m i n e d. BADULLA DISTRICT: Yelumali, Namunukula, 12 March 1907, *Silva* (PDA); Forest on Namunukula, 1750 m., 24 Feb 1954, *Sledge 1197, Walker T. 581* (BM). NUWARA ELIYA DISTRICT: Ramboda Pass, by stream., 1753 m., 28 Dec 1950, *Sledge 761, 762* (BM).WITHOUT LOCALITY: *Gardner 1059* (E); *Robinson 82 a* (K).

6. Diplazium polypodioides Blume, Enum. Pl. Javae 2: 194. 1828. Bedd., Ferns S. India 54, pl. 163. 1864. id., Ferns Brit. India pl. 293. 1868. id., Handb. Ferns Brit. India 184. 1883. Sledge, Bull. Br. Mus. nat. Hist. (Bot.) 2(11): 306. 1962. id., Bot. J. Linn. Soc. 84: 17. 1982.
Type: Indonesia, Java, Parang Mts., *Blume s.n.* (L, holo).

Diplazium asperum Blume Enum. Pl. Javae 2: 195. 1828. Bedd., Handb.
Ferns Brit. India 184. 1883. Type: Indonesia, Java, Mt. Tjerimai, *Blume*
s.n. (L, holo).
Asplenium polypodioides (Blume) Mett., Fil. Hort. Bot. Lips. 78. 1856.
Thwaites, Enum. Pl. Zeyl. 385. 1864, non Sw. 1800.
Diplazium polypodioides Blume var. *brachylobum* Sledge, Bull. Brit. Mus.
nat. Hist., Bot. 2(11): 307. 1962. id., Bot. J. Linn. Soc. 84: 17. 1982.
Type: Sri Lanka, Corbet's Gap, *Sledge P.78* (BM., holo).

Rhizome stout, erect. Stipes up to 1 m long, muricate, scaly near base;
scales narrow, elongate, brown, with black-toothed margin. Lamina bipinnate
to commonly bipinnate-tripinnatifid, 60-120 × 35-70 cm., largest pinnae
30-60 cm long, stalked, with up to 15 pairs of pinnules below pinnatifid apex,
pinnules up to 12 × 3 cm., often smaller, shortly stalked or sessile, base
truncate, margin usually lobed ¾ towards costa, sometimes only shallowly
lobed, apex acuminate, lobes oblong, slightly oblique, acroscopic margin
± straight, basiscopic margin rounded. Veins 6-10 pairs in deeply lobed pinnules, forked in larger lobes, mostly soriferous. Rhachises and costae glabrescent or with scattered minute, toothed scales, slightly muricate like stipe. Sori
straight, 2-4 mm long, stretching from costa ½ to ¾ way to margin, the basal
acroscopic ones diplazioid, others simple.

D i s t r. S. India, Sri Lanka, Peninsular Malaysia, Philippines.

E c o l. Common in forests from 600 to 1350 m altitude.

N o t e. Sledge (1962) has separated specimens from montane forest with
shallowly lobed pinnules as var. *brachylobum* which seems to be connected to
typical *D. polypodioides* by a whole range of intermediate forms and, therefore, is not recognised here.

S p e c i m e n s E x a m i n e d. MATALE DISTRICT: Forested stream-
gulley beside and below road c. 1 km W. of top of pass between Rattota and
Laggala, N.E. of Matale, c.1200 m., 15 Sept 1993, *Fraser-Jenkins* with
Jayasekara & Bandara, Field No.135 (K). KEGALLE DISTRICT: Gallebodde,
near stream in jungle, 610 m., 26 Jan 1954, *Sledge 1048* (BM). KANDY
DISTRICT: Hantane, 1847, *Gardner s.n.* in C.P. 1352 (PDA); Corbet's Gap,
in secondary jungle, c. 1220 m., 9 Dec 1950, *Sledge P. 78* (BM., holotype of
var. *brachylobum* Sledge); Stream gully in forest above and S.E. of "Hunas
Falls Hotel" W. side of Hunnasgiriya mountain, S.S.E. of Elkaduwa, N.E. of
Kandy, 1200-1300 m., 25 Aug 1993, *Fraser-Jenkins* with *Jayasekara,*
Samarasinghe & Abeysiri Field No. 44 (K); Central Province, E of Madugoda,
streamside bordering road through jungle, 762 m., 9 Jan 1954, *Sledge 950*
(BM); Hunnasgiriya, streamside in jungle, 884 m., 16 Jan 1954, *Sledge 972*
(BM); Hoolankande, by path through forest, 1372 m., 20 Jan 1954, *Sledge*
1009 (BM); Hoolankande, edge of jungle, c. 1370 m., 20 Jan 1954, *T.G.*

Walker T141 (BM). BADULLA DISTRICT: Badulla, *Freeman 191A, 192B* (BM); Namunakula, jungle at Tonacombe Estate, 1310 m., 21 Feb 1954, *Sledge 1159* (BM). RATNAPURA DISTRICT: Adam's Peak, *Moon s.n.* (BM); Deniyaya, jungle above Beverley Estate, 914 m., 12 March 1954, *Sledge 1276* (BM., K). NUWARA ELIYA DISTRICT: Nuwara Eliya, 1847, *Gardner s.n.* in C.P. 1353 (PDA). GALLE DISTRICT: Hiniduma, moist roadsides, partially shaded, common, c. 25 m., 6 Dec 1976, *Faden & Faden 76/501* (K). DISTRICT UNKNOWN: WITHOUT LOCALITY: *Gardner 37* in Herb. Hill (BM), in Herb. T.Moore (K); *Gardner 1060* (BM); *Gardner 1061, 1062* (K, PDA); 1847, *Gardner 1062* in Herb. J.Smith (BM), in Herb. Hook. (K); *Gardner s.n.* in C.P. 1352 (K); *Gardner s.n.* (C.P. 1352) in Herb. Hook. (K); *Thwaites C.P. 1350* (BM*); Thwaites C.P.1352* (BM., K); *Thwaites C.P. 1352* in Herb. Barkly (BM), in Herb. Hook. (K); *Thwaites C.P. 1353* p.p. (BM); *Thwaites C.P. 3098* (BM., E, PDA); Central Prov., *Naylor Beckett 30* (BM., E); *Robinson 80, 81* (K); *Mrs. Walker* in Herb. Hook. (K); *Skinner s.n.* (K); Nov 1879, *Wall s.n.* (K); *Wight 1922* (E); *Ferguson s.n.* (US 815507).

7. Diplazium cognatum (Hieron.) Sledge, Bull. Br. Mus. nat. Hist. (Bot.) 2(11): 308. 1962. id., Bot. J. Linn. Soc. 84: 17. 1982.

Asplenium australe sensu Hook., Sp. Fil. 3: 232. 1860 p.p. sensu Thwaites, Enum. Pl. Zeyl. 385. 1864, non Brack.

Diplazium assimile sensu Bedd., Ferns Brit. India pl. 294 p.p. (as to C.P. 1347 only). 1868, non (Endl.) Bedd.

Athyrium assimile sensu Bedd., Suppl. Ferns S. India Brit. India 12. 1876, non (Endl.) C. Presl

Diplazium umbrosum var. *assimile* sensu Bedd., Handb. Ferns Brit. India 190 p.p. (as to Sri Lankan specimens only). 1883, non (Endl.) Bedd.

Athyrium cognatum Mett. [in sched.] ex Hieron., Hedwigia 59: 321. 1917.
    Type: Sri Lanka, *Thwaites C.P. 1347* (B, holo; K, PDA, iso).

Rhizome creeping. Stipes up to 60 cm long, scaly at base; scales narrowly linear, entire, brown, up to 5 mm long. Fronds deltoid, tripinnate-quadripinnatifid, glabrous, membranous; lower pinnae stalked, up to 30-40 × 15-20 cm; secondary pinnae 10 × 3-4 cm gradually acuminate; tertiary pinnae very shortly stalked basally, adnate and decurrent distally, apex blunt, divided ½ to ⅔ towards costa into c. 5 segments on each side, segments small, slightly falcate, entire or with 1 or 2 teeth on acroscopic margin, rarely on basiscopic margin as well. Veins pinnate in segments, anterior one soriferous. Sori oblique, diplazioid, extending from costa, other veins also soriferous in larger segments.

D i s t r. Endemic to Sri Lanka.

E c o l. Rare in the forests of the interior from 600 to 1500 m altitude.

S p e c i m e n s   E x a m i n e d. MATALE DISTRICT: Pittawella, Matale, 610 m., *Wall 38* in Herb. Barkly (BM). BADULLA DISTRICT: Haputale, Apr & May 1856, *Thwaites C.P. 1347* (PDA). NUWARA ELIYA DISTRICT: Dimboola, June 1848, *Thwaites C.P. 1347* (PDA). WITHOUT LOCALITY: *Thwaites C.P. 1347* in Herb. Barkly (BM); *Thwaites C.P. 1347* in Herb. Bedd., in Herb. Hook., in Herb. T.Moore (all K); *Thwaites C.P. 1347* (E, K); Jan 1872, *Wall s.n.* (K); *Beddome s.n.* (K, Ferns Brit. India pl. 294); *Ferguson s.n.* (US 815519); *Hutchison s.n.* (E).

Sect. Anisogonium

(C. Presl) Sledge, Bull. Br. Mus. nat. Hist. (Bot.) 2(11): 309. 1962.

Anisogonium C. Presl, Tent. Pterid. 115. 1836.
Type species: *Anisogonium fraxinifolium* (C. Presl) C.Presl (based on *Diplazium fraxinifolium* C. Presl)

8. Diplazium esculentum (Retz.) Sw., J. Bot. (Schrad.) 1801(1): 312. 1803. Sledge, Bull. Br. Mus. nat. Hist. (Bot.) 2(11): 310. 1962. id., Bot. J. Linn. Soc. 84: 17. 1982.

Hemionitis esculenta Retz., Observ. Bot. 6: 38. 1791. Type: India orientalis, *Koenig s.n.* (LD, holo).
Asplenium esculentum (Retz.) C. Presl, Rel. Haenk. 1: 45. 1825. Hook., Sp. Fil. 3: 268. 1860. Thwaites, Enum. Pl. Zeyl. 385. 1864.
Anisogonium esculentum (Retz.) C. Presl, Tent. Pterid. 116. 1836. Bedd., Handb. Ferns Brit. India 192, fig. 94. 1883
Callipteris esculenta (Retz.) J. Sm. ex Houlston & T. Moore, Gard. Mag. Bot. [3]: 265. 1851. Bedd., Ferns S. India 54, pl. 164. 1863/64.

Rhizome erect. Stipes 30-60 cm long, scaly at base, otherwise smooth; scales 10 × 1 mm., toothed. Lamina bipinnate, up to 100 × 50 cm., herbaceous; lower pinnae 40-50 × 15 cm., pinnules 5-10 × 1-2 cm., lowermost ones shortly stalked, others sessile, base truncate to broadly cuneate, often with basal auricles on one or both sides, margin crenate to shallowly lobed, lobes toothed, apex acuminate, serrate; veins pinnate, 6-10 pairs per lobe, lower 2-3 pairs of adjacent groups anastomosing, forming an irregular excurrent vein to sinus between lobes. Rhachis glabrescent, costae on lower surface with scattered minute, ovate, toothed scales. Sori mostly on all lateral veins, extending along nearly entire length;basal acroscopic ones diplazioid.

D i s t r. India, Sri Lanka, eastwards to S. China and Taiwan, throughout Malesia to Samoa.

E c o l. Common on river banks and in wet open places below 900 m altitude.

S p e c i m e n s E x a m i n e d. COLOMBO DISTRICT: Kelani River, near Colombo, *Chevalier s.n.* (BM). KEGALLE DISTRICT: Mawanella, 27 Jan 1954, *Schmid 1061* (BM., G). KANDY DISTRICT: Kandy, *Robinson 68* (K). BADULLA DISTRICT: Badulla, *Freeman 203A, 204B, 205C* (BM). RATNAPURA DISTRICT: Near Ratnapura, wet bushy ground by road, 30 m., 13 March 1954, *Sledge 1288* (BM., PDA); Ratnapura, Apr 1855, *Thwaites C.P. 3270* (PDA). GALLE DISTRICT: Near Galle, in open forest, Apr 1844, *Gardner 1058* (CGE); Galle, *Thwaites C.P. 3270* (PDA). WITHOUT LOCALITY: *Thwaites C.P. 3270* (BM., E, K); *Thwaites C.P. 3270* in Herb. Hook. (K); Aug 1870, *Naylor Beckett s.n.* (K); *Robinson 83, 84* (K); Western & Central Prov., below 914 m., *Wall 38/274* in Herb. Barkly (BM); 1837, *Wight 133* (E); *Ferguson s.n.* (US 815483, 815484).

9. Diplazium paradoxum Fée, Mém. Fam. Foug. 5: 214. 1852. Sledge, Bull. Br. Mus. nat. Hist. (Bot.) 2(11): 310. 1962. id., Bot. J. Linn. Soc. 84: 17. 1982. Type: Sri Lanka, *Gardner 36* (?P, holo)

Asplenium dilatatum sensu Thwaites, Enum. Pl. Zeyl. 385. 1864 (as to *C.P. 1350* p.p. and *Gardner 1060*), non (Blume) Hook.
Asplenium smithianum Baker in Hook. & Baker, Syn. Fil. 245. 1867. Type: Sri Lanka, *Gardner 1351* (K, holo; K, iso)
Callipteris smithiana (Baker) Bedd., Ferns Brit. India pl. 332. 1870.
Anisogonium smithianum (Baker) Bedd., Handb. Ferns Brit. India 192. 1883.

Rhizome oblique, decumbent. Stipes 30 cm long, muricate, furfuraceous throughout, scaly at base; scales 10 mm long, narrow, brown, margin toothed but not black. Lamina bipinnate, broadly ovate to deltoid-ovate, up to 60-70 cm long, firmly herbaceous; lower pinnae pinnate, middle ones pinnatifid becoming serrate only close to pinnatifid frond apex; largest pinnae up to 45 × 15 cm., pinnules oblong, 5-10 × 1-2 cm., sessile, base broadly cuneate to subtruncate, margin usually entire, in large fronds serrate-dentate to shallowly pinnatifid, apex serrate, rather abruptly acuminate; both surfaces glabrous. Rhachis and pinna rhachises of lower pinnae thinly furfuraceous and often sparsely muricate. Veins 2-6 pairs in each group, basal acroscopic one (sometimes two) fusing with basiscopic one of next group about half-way or more between costa and margin. Sori elongate, up to 7 mm long; basal acroscopic ones diplazioid.

D i s t r. Endemic to Sri Lanka.

E c o l. Uncommon in the forests of the Central Province.

S p e c i m e n s E x a m i n e d. MATALE DISTRICT: Matale East, Apr. 1869, *Naylor Beckett s.n.* in *C.P. 3990* (PDA). KANDY DISTRICT: Hantane Range, in forest, Aug 1844, *Gardner 1060* (CGE, PDA); Hantane,

1850, *Thwaites C.P. 1350* (PDA); Hantane, 2 Feb 1954, *Sledge s.n.* (PDA); Oodawella, 914 m., *Wall 38/275* in Herb. Barkly (BM); Lagalla, 900 m., *Naylor Beckett 33* (K). WITHOUT LOCALITY: *Gardner 1058* (BM); *Gardner 1060* in Herb. J. Smith (BM); *Gardner s.n.* in C.P. 1351 (K, holo- and isotype of *Asplenium smithianum* Baker); *Thwaites C.P. 1359* (E); *Thwaites C.P.3990* (BM); *Thwaites C.P.3990* in Herb. Bedd. (K); *Beddome s.n.* (BM., K); Central Prov., *Naylor Beckett 34* (BM); *Robinson C 82* (K); Nov 1881, *Wall s.n.* (E, K); 1829, *Macrae s.n.* in Herb. Hook. (K); *s.coll. s.n.* in Herb. Hook. (K); 1899, *Anderson s.n.* (E); *Wight 1921* (E); *Ferguson s.n.* (US 815514).

Subgen. Pseudallantodia
(C.B. Clarke) Sledge, Bull. Br. Mus. nat. Hist. (Bot.) 2(11): 312. 1962.

Asplenium subgen. *Pseudallantodia* C.B. Clarke, Trans. Linn. Soc. London, Ser. 2, Bot. 1: 495. 1880.

Type species: *Asplenium procerum* (Hook. & Baker) Wall. ex C.B. Clarke (based on *Asplenium umbrosum* var. *procerum* Hook. & Baker) [= *Diplazium muricatum* (Mett.) Alderw.].

10. Diplazium muricatum (Mett.) Alderw., Malayan Ferns 829. 1909. Sledge, Bull. Br. Mus. nat. Hist. (Bot.) 2(11): 312. 1962. id., Bot. J. Linn. Soc. 84: 17. 1982.

Athyrium gymnogrammoides sensu Bedd., Ferns S. India 52, pl. 156. 1864 (as to *C.P. 1344*). id., Handb. Ferns S. India 168. 1883, non (Klotzsch ex Mett.) Bedd.
Athyrium australe sensu Bedd., Ferns S. India 52, pl. 158. 1864 p.p., non Brack.
Asplenium gymnogrammoides sensu Thwaites, Enum. Pl. Zeyl. 385. 1864, non Klotzsch ex Mett.
Asplenium muricatum Mett., Ann. Mus. Bot. Lugd.-Bat. 2: 239. 1866. Type, Indonesia, Java, *Zippel s.n.* (L, holo)
Diplazium umbrosum var. *australe* sensu Bedd., Handb. Ferns Brit. India 189 p.p. 1883, non (R.Br.) Bedd.

Rhizome creeping. Stipes up to 1 m long, usually somewhat muricate, sparsely scaly when young; scales thin, ovate, brown, deciduous. Lamina broadly ovate, bipinnate-tripinnatifid to tripinnate-quadripinnatifid, 40-90 cm long, pinnate throughout, firmly herbaceous; largest pinnae in tripinnate-quadripinnatifid lamina up to 60 × 30 cm., secondary pinnae up to 18 × 6 cm in tripinnate one, oblong-lanceolate, apex acuminate, tertiary pinnae up to 3 cm long, acroscopic base truncate, basiscopic base cuneate, apex acute, margin lobed ¹/₃ to ½ to costa; lobes 3 mm wide at base, falcate-rounded with toothed edges; secondary pinnae in smaller bipinnate laminae scarcely exceeding 3 × 1 cm., lobed almost to costa, lobes 4-5 mm wide at base, oblong, obtuse, margin usually serrate. Rhachis and costae on lower surface with scattered ovate-acute, brown, entire scales. Veins forked. Sori allantodioid, short, oblong to subquadrate, 1-3 mm long, straight, forming two rows close to costa, sometimes with additional sori in ultimate segments; basal acroscopic sori often diplazioid, others simple. Indusium thin, membranaceous.

D i s t r. S. India, Sri Lanka, Taiwan, Himalayas, Burma, Thailand, Java.

E c o l. In montane forests above 1800 m altitude.

S p e c i m e n s   E x a m i n e d. BADULLA DISTRICT: Namunukula, 1920 m., 24 Feb 1954, *Sledge 1189* (BM). NUWARA ELIYA DISTRICT: Nuwara Eliya, in woods, Sep 1844, *Gardner 1066* (CGE); Nuwara Eliya, *Freeman 176A, 180E, 199A, 200B* (BM); Horton Plains, in jungle, 2100 m., 30 Dec 1950, *Sledge 786* (BM., PDA); Ramboda Pass, by track to Maturata, 1920 m., 17 March 1954, *Sledge 1313* (BM); Udapussalawa, Apr 1854, *s.coll. 1069* in *C.P. 1344* p.p. (PDA). WITHOUT LOCALITY: *Gardner 1066* (K); *Gardner 1066* in Herb. Hook. (K); *Thwaites C.P. 1344* (E); *Walker s.n.* (K).

11. Diplazium procumbens Holttum, Gard. Bull. Sing. 11: 95, fig. 4. 1940. Sledge, Bull. Br. Mus. nat. Hist. (Bot.) 2(11): 315. 1962. id., Bot. J. Linn. Soc. 84: 17. 1982.
Type: Malaysia, Pahang, Fraser's Hill, *Holttum S.F.N. 36503* (SING, holo).

Asplenium polypodioides sensu Thwaites, Enum. Pl. Zeyl. 385. 1864 (as to *C.P. 1352* p.p.)., non (Blume) Mett.
Asplenium dilatatum sensu Thwaites, Enum. Pl. Zeyl. 385. 1864 (as to *C.P. 1350* p.p.), non (Blume) Hook.
Athyrium procumbens (Holtt.) Holtt., Fl. Malaya 2: 572. 1954.

Rhizome creeping, black. Stipes up to 75 cm long, black and slightly muricate at base, sparsely scaly when young; scales small, brown, toothed. Lamina broadly ovate to deltoid-ovate, bipinnate, up to 70 × 60 cm; largest pinnae 45 × 20 cm., pinnules shortly stalked basally, becoming sessile to adnate towards apex, 6-12 × 1.5-3 cm., base truncate, apex acuminate, pinnatifid ¹/₃ to ²/₃ towards costa; lobes 5-8 mm wide at base, broadly truncate-rounded, margin entire to slightly toothed towards apex. Rhachis and costae with scattered, elongate, narrow, brown, sparsely toothed scales, particularly at junction of pinnule and pinna rhachis. Veins 4-7 pairs per lobe, simple, in larger lobes forked. Sori allantodioid, narrow, 3-6 mm long, spreading from costa along veins for ¹/₃ to ½ of their length; basal acroscopic ones diplazioid. Indusium very thin, whitish.

D i s t r. Sri Lanka, Peninsular Malaysia (Malaya).

E c o l. Rare in montane forest from 1350 to 1950 m altitude.

S p e c i m e n s   E x a m i n e d. KANDY DISTRICT: Pallagalla, Oct 1853, *Thwaites C.P. 1350, 1352* (PDA); Hoolankande, in forest, 1372 m., 20 Jan 1954, *Sledge 1005* (BM). BADULLA DISTRICT: Namunukula, shady forest,1980 m., 24 Feb 1954, *Sledge 1192* (BM). NUWARA ELIYA DISTRICT: Hakgala, 1800 m., 23 Dec 1950, *Holttum S.F.N. 39173* (SING). WITHOUT LOCALITY: *Thwaites C.P. 1353* (E); *s.coll. s.n.* (PDA); *Trimen* (C.P. 3100) in Herb. Bedd. (K); *Thwaites C.P. 3098* (BM); 1899, *Anderson s.n.* (E); *Palliser s.n.* (US 684019).

12. Diplazium javanicum (Blume) Makino, Bot. Mag. Tokyo 20: 85. 1906.

Asplenium javanicum Blume, Enum. Pl. Javae 2: 175. 1828. Type: Indonesia, Java, Mt. Tjerimai, *Blume s.n.* (L, holo, photo BM).
Asplenium reticulatum Wall., Num. List 188. 1828, nom. nud.
Allantodia brunoniana Wall., Pl. Asiat. Rar. 1: 44. 1830. Bedd., Ferns S. India 52, pl. 159. 1863/64. Thwaites, Enum. Pl. Zeyl. 385. 1864. Type: Nepal, Sheopuri, 1821, *Wallich 188* in Herb. Hook. (K, holo).
Allantodia javanica (Blume) Trevis., Nuov. Giorn. Bot. Ital. 7: 159. 1875. Bedd., Handb. Ferns Brit. India 195, fig. 97. 1883.
Diplaziopsis javanica (Blume) C. Chr., Ind. Fil. 227. 1905. Sledge, Bull. Br. Mus. nat. Hist. (Bot.) 2(11): 317. 1962. id., Bot. J. Linn. Soc. 84: 17, 27. 1982.

Medium-sized terrestrial ferns. Rhizome short, erect; scales entire, brown. Stipes up to 30 cm long, scaly at base, glabrous above. Lamina up to 60 cm long, up to 30 cm wide, simply pinnate with up to 12 pairs of pinnae and a conform terminal one; pinnae 10-17 × 2.5-4 cm., sessile, oblong, base truncate, margins entire to slightly crenulate towards caudate apex, glabrous on both surfaces, herbaceous. Veins forked near costa, anastomosing about half-way to margin, forming several rows of ± elongate areoles towards margin. Sori linear-oblong, allantodioid, confined to anterior branch. Indusium very thin, closely enveloping young sporangia, opening irregularly or asplenioid at maturity.

D i s t r. Widespread in S. and E. Asia, throughout Malesia eastwards to Samoa, Fiji and Tahiti.

E c o l. Terrestrial in forest from 600 to 1200 m altitude.

S p e c i m e n s   E x a m i n e d. KANDY DISTRICT: Hantane, Jan 1854, *Thwaites C.P. 2543* (PDA); Hantane Range, Oct 1844, *Gardner 1057* (PDA); Nillumalle, Madulkelle, Oct 1887, *s.coll.* (PDA); Palagalla, Oct 1853,

Thwaites C.P. 2543 (PDA). RATNAPURA DISTRICT: forest above Beverley Estate, 914 m., 12 March 1954, *Sledge 1275* (BM); Beverley Estate (part of Sinharaja Forest), c. 915 m., 17 March 1954, *T.G.Walker T784* (BM). NUWARA ELIYA DISTRICT: Nuwara Eliya, *Freeman 206A* (BM). WITHOUT LOCALITY: 1847, *Gardner 1057* in Herb. J. Smith (BM), in Herb. Hook. (K); *Thwaites C.P. 2543* (BM., CGE); *Thwaites C.P. 2543* in Herb. Barkly (BM); Central Prov., 900 – 1500 m., *Naylor Beckett 79* (BM); *Robinson C 84* (K); *Wall s.n.* in Herb. Gamble (K); Central Prov., 610-1219 m., *Wall 39/1* in Herb. Barkly (BM); *s.coll.* 12 in Herb Hance (BM); *s.coll.* [ex Herb. Hook.] in Herb. J.Smith (BM).

NEW NAMES PUBLISHED
IN THIS VOLUME

New or recently published names used in this volume

ASPLENIACEAE

Asplenium bipinnatum (Sledge) Philcox, stat. nov.

DRYOPTERIDACEAE

Four groups are now treated as subfamilies of the Dryopteridaceae:
subfamily Peranematoideae (C. Presl) Fraser-Jenk., comb. nov. (Basionym.
  Filicineae tribus Peranemataceae ["*Peranemaceae*"] C. Presl, Tent.
  Pterid.: 64. 1836;
subfamily Dryopteridoideae;
subfamily Polystichoideae Fraser-Jenk., subfam. nov. (segmentibus laminae
  acutibus plerumque dentis uniter terminantibus. Holotype *Polystichum*
  Roth)
subfamily Tectarioideae B. K. Nayar.

GRAMMITIDACEAE

Ctenopteris perplexa Parris, sp. nov.   Fern Gaz. 16(4): 201-202.2001
Ctenopteris epaleata Parris, sp. nov.   Fern Gaz. 16 (5):239-244.
Grammitis sledgei Parris, sp. nov.   "   "   "   "   "   "
Prosaptia ceylanica Parris, sp. nov.   "   "   "   "   "   "

HYMENOPHYLLACEAE

Crepidomanes campanulatum (Roxb.) Jayasekara, comb.nov.
Mecodium gardneri (Bosch) Jayasekara, comb.nov.

POLYPODIACEAE

Selliguea montana (Sledge) Hovenkamp, comb.nov.

Endemic Species of the Flora of Sri Lanka (in Bold)

Species restricted to Sri Lanka and Southern India are also listed and
marked with an asterisk (*)
(For authorities check index)

Amauropelta hakgalensis
Asplenium disjunctum, A. longipes, A. bipinnatum*, A. decrescens*, A.
   serricula*
Asplenium zenkerianum*
Athyrium hohenackerianum*
Bolbitis subcrenata
Calymmodon glabrescens
Cheilanthes thwaitesii, C. bullosa*
Christella meeboldii
Coniogramme serra(?)
Crepidomanes intramarginale*
Ctenitis thwaitesii
Ctenopteris epaleata, C. thwaitesii, C. perplexa*
Cyathea hookeri, C. sinuata, C.walkerae, C.crinita*
Davallia denticulata
Deparia polyrhizos
Didymoglossum wallii, D. exiguum*
Diplazium beddomei, D. cognatum, D. decurrens, D. paradoxum,
Diplazium travancoricum*
Dryopsis obtusiloba
Dryopteris macrochlamys, D. approximata*, D. deparioides (*?) D. sledgei*
Elaphoglossum ceylanicum
Grammitis beddomeana, G.medialis, G. sledgei, G. wallii, G. zeylanica
Grammitis attenuata*
Huperzia ceylanica*, H. vernicosa*
Idiopteris hookeriana*
Lindsaea caudata, L. schizophylla, L. venusta*
Meringium macroglossum
Microlepia majuscula
Osmunda collina
Polystichum anomalum, P. walkerae, P. harpophyllum*
Pronephrium gardneri, P. thwaitesii*
Prosaptia ceylanica
Pseudocyclosorus tylodes*
Pteridrys zeylanica
Pteris gongalensis, P. praetermissa, P. reptans, P. confusa(*?), P.
   multiaurita*, P. x otaria*,

Pteris quadriaurita*
Pyrrosia gardneri, P. **pannosa,** P. ceylanica*, P. heterophylla*
Selaginella calostachya, S. latifolia, S. cochleata(*?), S. crassipes*
Selaginella integerrima*, S. praetermissa*
Tectaria thwaitesii, T. pteropus-minor*, T. trimenii*
Trigonospora angustifrons, T. **glandulosa,** T. **obtusiloba,** T. **zeylanica**
Vittaria microlepis*
Xiphopteris cornigera
see also: Sledge W.A., 1982. An annotated check-list of the Pteridophyta of
     Ceylon. *Bot. J. Linn. Soc.* **84**: 1-30

SELECTIVE GLOSSARY OF TERMS

Compiled by Monika Shaffer-Fehre

Abaxial	side away from axis in a lamina or pinna; the underside.
Acicular	of thin, needle-shaped (often white) hairs; characteristic of (though not confined to) the Thelypteridaceae
Acicular ray	rays are branches of stellate hairs, a.-rays are usually straight and needle-like, shape species-specific, they are characteristic of some taxa in the Polypodiaceae
Acroscopic	the side (pinna-half) towards the apex (cf. basiscopic)
Acrostichoid	sorus covers entire abaxial ( ) side of lamina: e.g. *Acrostichum., Elaphoglossum*
Acumen	tapering tip (here of scales)
Acuminate	tapering gradually / abruptly to narrow point (scales / pinnae)
Adaxial	disposed towards axis i.e. in a lamina or pinna: top surface (abaxial)
Adpressed	(of scales) lying close along surface (appressed)
Adventitious	arising in an unusual or abnormal position
Adnate	united to the structure of a different kind e.g. pinna-segment to a costa
Aerophore a)'lateral line' b) swollen...	a) in a continual pale line, breathing pores often laterally on stipe; b) swollen patch (dark-coloured when dry) at pinna-base: e.g. *Pseudocyclosorus tylodes,* Thelypteridaceae
Allantodioid	sausage-shaped
Amphiphloic	having phloem on both sides of the xylem
Anadromous	at least lower acroscopic pinnae inserted closer to rhachis than basal pinnule; also of veins: the first set(s) of veins in each segment of the frond originates from acroscopic side of midrib (catadromous)
Analogous	the same as / bearing resemblance
Anastomosing	reticulate: of veins that repeatedly unite to form a network connivent = converging, not joined, but here appropriate;

characteristic of some taxa of the Thelypteridaceae (formerly *Nephrodium*) in which lateral veinlets form a ± herring-bone pattern, joining in an excurrent ( ) vein that goes to sinus ( ); see also goniopterid anastomoses ( )

Angustate	narrowed
Annulus	ring of thickened cells occupying ± circumference of sporangium (facilitates explosive dehiscence of sporangium) (cf. indurated); position can be taxonomic criterion
Antheridium	multicellular, sperm-producing organ of spermatogenous tissue, surrounded by sterile jacket
Antrorse	pointing towards the apex; (opposite: retrorse)
Apomixis	=apogamy: reproduction without fusion of gametes, sporophyte generated from vegetative cells of gametophyte. see in references in Introduction: Lovis, J. 1977
Appressed	closely flattened against support ('adpressed')
Archegonium	multicellular, egg-producing gametangium with a jacket of sterile cells
Areole	small area of lamina encircled by uniting veins
Aristate	having an awn-like bristle terminating pinna / pinnule
Articulated	= jointed
Ascending	of rhizomes: turning upward from a creeping position (cf. decumbent)
Auriculate	base of pinna or pinnule with ear-like lobe
Auricle	very short triangular pinna at base of lamina in some taxa e.g. *Christella papilio,* Thelypteridaceae
Baculate	of a spore: having pillar-like processes, always longer than broad, higher than 1 μ
Basiscopic	side pointing to base: lower half of pinna (acroscopic)
Bundle sheath	differentiated tissue, often thick-walled, surrounding vascular bundle (in assimilating tissues like lamina)
Caduceous	of scales or indusia that are shed early
Campteroid	(venation) here only a single series of areoles parallel to costae, other veins free
Capitate hairs	hairs with enlarged, often bulbous tip; often glandular
Carinate	keeled, (of leaf shape) e.g. *Selaginella,* Selaginellaceae
Catadromous	basal basiscopic pinnules inserted closer to rhachis than acroscopic ones; the first set of veins in each segment of frond originates from basiscopic side of midrib
Catenate	chain-like; pertaining to hairs (a single line of cells) most cells collapsed, but alternating at right angle to one another

Caudate   abruptly ending in a long, slender tail-like tip or append-
                age

Caudex   erect rhizome ( )(resembling trunk; of varying sizes)

Ceraceus   waxy (of glands that may be white, yellow, orange) see
                farina

Chartaceus   papery

Ciliate   fringed with hairs

Circinate   (=circinnate) circinate vernation: e.g. lamina inrolled in
                spiral, with tip of leaf innermost; characteristic of true fern
                (crozier)

Clathrate   (isotoechous) of scales with a latticed appearance due to
                thickened radial walls; (periclinal) surface walls transpar-
                ent, often iridescent: e.g. Vittariaceae, most Aspleniaceae

Clavate   club-shaped (of trichomes)

Clustered   densely crowded

Coenosorus   compound sorus of several contiguous or coalesced sori

Collection set A collection (usually a duplicate), bearing the collector's
                name and unique number and consisting of one or more
                sheets :i.e. rhizome, stipe, several sheets for a large frond;
                all hold valuable information.

Commissure   vascular strand / vein below receptacle of ± marginal sori

Concolorous   of one colour throughout

Cone   Strobilus ( ) made up of sporophylls (*Selaginella*)

Confluent   running into each other, combining into one

Contiguous   touching, adjoining, neighbouring

Cordate   heart-shaped

Coriaceus   leathery

Costa   the midrib of a pinna (also pinna-rhachis)

Costal arch   transverse section of costa resembling an inverted "V"

Costule   mid-vein of pinnule

Crenate   margin notched with regular blunt or rounded teeth

Crispaceous   crisped, irregularly waved and twisted, curled

Cristate   crested

Crozier   the coiled, circinate leaf bud in true ferns

Ctenitoid   of hairs: articulated with dark red septae (Dryopteridaceae)

Cucullate   hooded

Cuneate   wedge-shaped

Cyanobacterium blue-green "alga", a prokaryote, (symbiont) of *Azolla*

Decrescent   decreasing, tapering gradually or abruptly

Decompound	pinna several times subdivided; some Thelypteridaceae
Decumbent	lying prostrate, only tip growing upwards (tip ascending)
Decurrent	with a (decreasing) wing continuing downward from point of insertion
Deflexed	turned abruptly downward (of some basal pinnae e.g. *Christella parasitica*), Thelypteridaceae
Deltate	triangular
Dichotomous	forking regularly; the two ± equal branches forking again repeatedly e.g. *Psilotum.*, Psilotaceae
Dictyostele	vascular tissues making up stele form discrete cylinders
Didymosorous condition	closely contiguous sori along costa, either side of deep sinus e.g. *Christella hispidula, C. parasitica*, Thelypteridaceae
Dimidiate	½ of organ much smaller e.g. basiscopic lamina reduced : (*Adiantum*)
Dimorphous	of two forms: e.g. distinctly different fertile and sterile fronds (*Leptochilus, Blechnum colensoi, Anemia*)
Discrete	clearly separated
Dorsiventral	of a plant part in which upper and lower sides are structurally different
Echinate	covered in spine-shaped structures
Eglandular	without glands
Endospore	internally formed, thick-walled spore of bacteria / blue-green algae; &: innermost layer of spore coat (peri-, ex-ospore)
Entire	margin without indentations of any kind
Epilithic	growing on stone, boulder or rock / rockface
Epiphytic	growing on other plants, but not parasitic in nature
Erecto-patent	spreading at an angle of about 45°
Erose	with a ragged edge or margin
Excurrent vein	veinlets of neighbouring pinnules join forming between them excurrent vein leading to sinus, e.g. some Thelypteridaceae
Exindusiate	indusium absent (indusiate)
Exospore	the outer layer of the spore wall
Exserted	projecting beyond envelope, e.g. indusium; here straight or curved bristle-like sporangiophore ( ) Hymenophyllaceae
Facultative	applied to organisms that can adopt an alternative mode of existence; e.g. terrestrial in soil, but able to survive rooted in a stream = facultative rheophyte ( )

Falcate	sickle-shaped; simultaneously curved and tapering, scythe-shaped
False indusium	see pseudo-indusium
False veins	vein-like structures, not connected to vascular system (among Hymenophyllaceae)
Farina	a meal-like powder on some frond surfaces (ceraceus)
Fibrils	firm, thread-like fibres; hair-like scales (Dryopteridaceae)
3-fid, 2-fid	3y-, 2y segments still connected by wing of lamina decurrent along their costa / rhachis support
Flabellate	broadly wedge-shaped, fan-shaped
Free	veins not anastomosing
Frond	Equivalent to angiosperm leaf: unit of stipe ( ) & lamina ( )
Fovea, Foveolate	pit containing the sporangium (at the leaf base of *Isoetes*) minutely pitted
Fossulate	surface with small depressions, small furrows
Fugacious	ephemeral, fleeting, transient
Fusiform	narrow, acuminate pinnae e.g. *Trigonospora angustifrons*
Gametophyte	thallus bearing reproductive organs antheridia ( ) / archegonia ( ) (alternation of generations in ferns, gametophyte generation:=n)
Glabrescent	process of losing indument, becoming glabrous
Glabrous	without hairs
Glaucous	covered with white or pale bloom
Glochid	barb, (glochidate provided with barbs) : e.g. in *Azolla*
Goniopterid (anastomoses)	goni, gonia = angled, angular- veinlets angled against one another, like those forming excurrent vein :Thelypteridaceae
Granulose	appearing as to be covered with minute grains
Gymnogrammoid	sori spreading in ± irregular lines along (anastomosing) veins; they are not covered by indusium (acrostichoid)
Haplostele	cylindrical protostele ( ) with smooth margin in transection
Hastate	Spear-shaped, of leaf blade with narrow centre with 2 basal lobes spreading at right angle
Heterosporous	pteridophyte with 2 sizes of spores; (reverse: homosporous)
Heterostelic	( ) polystelic
Hirsute	covered with fairly coarse and stiff, long erect / ascending hairs (cf. Villous)
Homologous	corresponding in general structure and descent
Hyaline	translucent, almost like clear glass
Hydathode	the enlarged tip of a vein, mostly guttating water

Imbricate	similar to, closely overlapping, as shingles of a roof (scales)
Indefinite growth	e.g. elongated tip of frond bends down, roots and forms new fern: e.g. *Adiantum caudatum*; fronds in which rhachis keeps elongating indefinitely e.g. *Lygodium.*, Schizaeaceae
Index	index, wherever used, is the length / width proportion of a 2-dimensional shape; it is used for scales, lamina and pinnae, after giving measurements of individual length and width
Indument (plural noun)	surface cover by trichomes, hairs, scales, one or several kinds: mono-, di-, hetero- morphic indument
Indurated	becoming hardened: rib-like cells of annulus ( ) in sporangia ( )
Indusium (see also pseudo-[false]-indusium)	membranous flap of tissue covering (or sometimes subtending or enclosing) sporangia on sorus ( ) of many ferns; absence / presence, form of attachment and shape can be taxonomically important; indusium present = indusiate
Inflexed	turned inward
Inframarginal	position of sorus: slightly set back from the margin (also intramarginal)
Inframarginal vein	connects free vein endings; one to few mm inside of, and parallel to, margin (see commissure)
Involucre	protecting cover; e.g. indusium of some Hymenophyllaceae
Isosporous	spores of same shape and size
Isotoechous	see clathrate ( )
Isodromous	first set of veins in pinnule opposite; (ana- catadromous)
Laciniate	'flap', with taper-pointed incisions
Lamina	expanded portion of a frond; referred to also as 'blade'
Leaf gap	gap in stelar cylinder through which vascular strands of fronds emerge
Litobrochioid	(of venation) complete reticulum in which areoles lack free included veins
Marginal	position of sorus: attached to or on the edge (above commissural vein)
Massula (ae)	aggregates of microspores enclosed in hardened mucilage *Azolla,*Azollaceae; *Salvinia,* Salviniaceae
Megaspore	the larger spores of a heterosporous fern *Azolla, Marsilea, Salvinia*
Medial	position of sori ½ -way between midrib (costa) and margin
Melleous	with qualities of honey: -coloured, flavoured etc.(of spores)

Microsporangium	in a heterosporous pteridophyte, containing smaller spores
Monoecious	producing both microspores and megaspores on one individual e.g. *Selaginella latifolia,* Selaginellaceae
Monolete	a single scar 'fissure' in exine of spore (formed by simultaneous division)
Monopodial	non-branching; (reverse: sympodial)
Monotypic	taxon represented by only one subordinate taxon e.g. a genus with only a single species
Mucronate	hair-pointed teeth
Ochreous	yellow, ochre
Over-topping	one frond (usually the fertile one) always significantly longer than the sterile one e.g. *Trigonospora obtusiloba*
Palmate	frond in which lobes or divisions radiate from a single point; a compound blade without a rhachis
Papillose-echinate	bearing spine-shaped papillae
Paraphyses	sterile structures resembling variously shaped ± branched hairs, terminal cell mostly glandular
Patent	spreading ( ) from axis at c. 90°
Pectinate	comb-like
Peltate	fixed at the centre, umbrella-shaped
Pericytic	of stomata: guard cell surrounded by ring/doughnut-shaped neighbouring cell; cells NOT connected by anticlinal wall; occurs in e.g. *Anemia, Pyrrosia, Monogramma*
Perispore	outer covering of spore (possibly derived from tapetum) sometimes called epispore
Petiolule	the leaf stalk is a petiole, the stalks of the leaflets (pinnules) in a pinnate lamina are petiolules
Phyllopodium	an outgrowth of rhizome that supports a stipe; characteristic of Polypodiaceae
Pinna	the primary division of a compound frond
Pinnate	pertaining to fern frond whose primary divisions extend fully to the rhachis, usually attached by costa only; equivalent to '1-pinnate'; the terms '2-pinnate' '3-pinnate' refer to blades with further degrees of division
Pinnate venation	veinlets arising from next larger vein in a ± herring bone pattern; anstomoses here absent
Pinnatifid	pinnately divided; the sinuses extending half-way or more to rhachis, costa or costule; [2-pinnatifid, 2-pinnate] hence a bi-pinnatifid frond is not bi-pinnate: where the 3 main axes are completely free of lamina

Pinnatisect	pinnately divided, the divisions attached by their full width to rhachis, costa or costule; intergrades with above term
Pinnule	secondary or tertiary division (separate no. of laminar wings) of a compound frond making the latter 2- or 3-pinnate
Pinnulet	the smallest division of a compound, fully pinnate frond – referred to as 'ultimate segment' by some authors
Plastic species	those exhibiting wide degree of variation within species and its ability to respond to (e.g. environmental) conditions
Ploidy	number of sets of chromosomes (haploid=n, diploid=2n, 3, 4, polyploid) = ploidy levels
Polymorphic genera	in genetics, the existence of 2 or more forms that are genetically distinct from one another, but contained within the same interbreeding population; the polymorphism may be transient or may persist for many generations, when it is said to be balanced.
Polystelic	containing more than one or two steles (within rhizome)
Prothallus	(adj. prothallial) tissue generated by germinating spore (here antheridia, archegonia) see also gametophyte ( ) = prothallus
Protostele	a solid core of xylem surrounded by phloem
Pseudo-indusium (false indusium)	modified, recurved margin of lamina in some ferns which covers and protects the sori (at least when young) = false indusium; characteristic of e.g. *Adiantum*, Adiantaceae
Pseudo-peltate	only appearing to be peltate (centrally attached)
Puberulous	densely covered in minute hairs
Pubescent	downy
Pulvinus	cushion-like swelling at base of stipe or petiole; its changes in turgidity result in movements e.g. *Angiopteris*, Marattiaceae (more frequent in angiosperms)
Pustular	with low projections resembling blister or pimple
Quoad	with respect to: quoad plantam typicam = as regards the type
Rank	" fronds in 2 ranks", lines of said organ(s)
Receptacle	specialized tissue below sorus bearing sporangia
Receptacular paraphyses	one group of glandular hairs / paraphyses growing out from the receptacle (and not on or attached to sporangium)
Reflexed (recurved)	abruptly bent backward, downward: of pinnae and margin (e.g. false indusium of *Adiantum*)
Remote	stipes, pinnae attached distinctly distant from neighbouring. similar element on common axis

Reticulate	(of venation) a network characteristic of e.g. *Tectaria*
Rhachis	main axis of fern-lamina, beginning at top of stipe (stalk of that lamina) (further divisions with qualifying prefixes: 'pinna-rhachis') (the spelling: 'rachis' is also correct)
Rheophyte	plant growing in flood zone of permanent water source
Rhizoid	very thin 'rootlet' of gametophyte or rhizome
Rhizome	stem of pteridophytes; varying forms: small, corm-like e.g. *Ophioglossum* short to long-creeping , climbing, scandent
Rugose	wrinkled
Ruminate	very uneven surface, copiously folded
Rupestrial	cliff-, rock-dwelling
Scandent	loosely scrambling, loosely climbing
Scarious	of thin, dry membranous texture and not green
Segment	in a pinna the individual portions between two sinuses
Septate	divided by partitions 'septum/a' e.g. septate hair has cross walls between cells; (reverse: aseptate)
Serrate	having sharp teeth pointing forward
Sessile	without a stalk (of parts spanning all sizes: pinna, sporangium or gland)
Setose	covered with bristles
Simple	of frond: undivided, not compound
Simple	of margin (entire)
Sinus	the recess / notch between two lobes; can have distinct (noteworthy!) membrane at lowest point (Thelypteridaceae)
Sinusate /sinusoid	having a wavy margin in one plane
Siphonostele	a vascular cylinder of xylem and phloem around a central pith or hollow (*Salvinia*)
Solenostele	tubular stele: phloem borders xylem in- and externally 'amphiphloic'
Sorus	assembly of sporangia on receptacle ( ), may or may not be covered by indusium( )
Soriferous	of fertile frond covered in sori / sporangia (= fertile)
Sorophore	elongate, gelatinous strand holding sori (*Marsilea*)
Sori	plural noun: collection of sporangia on receptacle ( )
Spiciform	resembling a spike
Sporangium	in modern ferns capsule containing (usually 64) spores
Sporangiophore	a spike-like support for sporangia, exserted ( ) beyond bell-shaped indusium (Hymenophyllaceae)

Spore	a one-celled, asexual reproductive body
Sporeling	the young fern plant (sporophyte)
Sporipherous	lamina bearing spores
Sporocarp	a hard, bony capsule containing sporangia, as in *Marsilea*
Sporophyll	Specialized leaf-like organ subtending one or more sporangia (sporophylls make up cone/strobilus of *Selaginella*)
Sporophyte	the conspicuous pteridophyte plant (alternation of generations in ferns –sporophyte generation = 2n)
Spreading hairs	hairs spreading at various angles (see also patent)
Squamule	small scale
Squarrose	rough with scales, tips of bracts projecting outwards, usually at 90°
Stele	the vascular tissue of axes (stem, rhizome, caudex)
Sterile	frond without sori ( )
Stipe	support of lamina in ferns, equivalent to the stalk or petiole in angiosperm leaves
Stipitate	with a stipe, as opposed to 'stipe absent'
Stramineous	straw-coloured (ochreous = yellow ochre)
Strigose	covered in strigae : short, straight, rigid, appressed bristle-like hairs
Strobilus	cone-shaped fertile portion of axis, consisting of imbricate sporophylls e.g. Lycopodiaceae and Selaginellaceae
Subulate	awl-shaped, tapering gradually from narrow or moderately broad base to a fine point
Sulcate	stipe with a (mostly adaxial) groove; also of spore (& pollen grain) with narrow, elongated pore
Supramedial	position of sori: close to and along commissural vein
Symbiosis	living together of dissimilar organisms:mutual or individual benefit e.g. cyanobacterium living in symbiosis with *Azolla*
Synangium	Sporangia fused to form an organ with two to several compartments; sessile ( ) e.g. *Marattia* Marattiaceae
Tapetum	matrix surrounding archesporial cell
Terete	round in transverse section
Ternately pinnate	with three secondary pinnae arising from tip of common petiole e.g. *Pteridium longipes* Dennstaedtiaceae
Terrestrial	on soil-, ground-dwelling
Trilete	tri-radiate scar on a spore (arising by successive divisions); 3 germinating furrows; (triquetrous)

Triploid	Having three times the basic haploid chromosome number of its taxon. (ploidy). A triploid plant is often the product of hybridisation between a diploid and a tetraploid.
Truncate	ending abruptly as if cut straight accross
Tufted	bunched, caespitosus
Uncinate	barbed, hooked, hook-shaped
Uniseriate	one line of "subject" (uni-, bi- multi-seriate line of cells etc.)
Veinlet	branch of a vein
Veins free	i.e. veins not anastomosing (may or may not reach margin)
Vernation	arrangement of unexpanded fronds in a bud
Verrucose	warty, wrinkled (bullulate)
Villous	bearing long, weak / soft hairs (cf. hirsute)
Xerophytic	ability to endure scarcity of water for prolonged periods

REFERENCES CONSULTED FOR GLOSSARY

Burrows, J.E. 1990. South African Ferns and Fern Allies.xiii, 359 p. Sandton, Frandsen Publishers.

Cody, W.J. & Britton, D.M., 1989. Ferns and fern allies of Canada. iv, 430 p. Ottawa, Research Branch Agriculture, Canada, Publication 1829/E.

Johns, R.J., 1991. An introduction to the ferns and fern allies. Student manual. Royal Botanic Gardens, Kew. Unpublished.

McCarthy, P.M., (vol. ed.) 1998. Flora of Australia,48 xxii, 766 p. Ferns, Gymnosperms and Allied Groups. Canberra, CSIRO.

Proctor, G.R., 1985. Ferns of Jamaica. viii, 631 p. British Museum (Nat. Hist.) London, The Natural History Museum.

Stearn, W. T., 1992. Botanical Latin. 4th edn., xiv, 546 p. Newton Abbot David & Charles.

INDEX

---

1 Note: Names adopted in the Revised Handbook to the Flora of Ceylon are printed in roman type. Synonyms and all other names mentioned in the text are in italics.
Author abbreviations follow AUTHORS OF SCIENTIFIC NAMES IN PTERIDOPHYTA compiled by R.E.G. Pichi Sermolli Publ.Royal Botanic Gardens. Kew. 1996.

Pneumatopteris Nakai 457, 462
  callosa (Blume) Nakai 462
  truncata (Poir.) Holtt. 462
Poecilopteris hookeriana T. Moore 240
Polybotrya sect. *Egenolfia* (Schott.) Diels 239
    appendiculata (Willd.) J. Sm. 242
      var. *asplenifolia* (Bory) Bedd. 242
    asplenifolia (Bory) C. Presl 241
POLYPODIACEAE Bercht. & J. Presl 311
Polypodium L. 311
  subg. *Drynaria* Bory 312
  subg. *Lepisorus* C. Chr. 339
  subg. *Pleopeltis* sect. *Lepisorus* C. Chr. 339
  acrostichoides auct. 323
  (*Phlebodium*) *aureum* L. 348
  adnascens Sw. 320
  anomalum Hook. & Arn. 137
  artistatum Forst.f. 129
  attenuatum Willd. 183
  auriculatum L. 286 287, 288
  beddomeanum Alderw. 178
  bergianum Schltdl. 510
  contiguum (G. Forst.) J. Sm. 209
  cordifolium L. 286
  cornigerum Baker 202
  corticolum C. Chr. 191
  crenatum Forssk. 535
  cucullatum sensu Bedd. 211
  decorum Brack. 193
  dentatum Forssk. 477
  dichotomum Thunb. ex Murray 173
  dilatatum Wall. 326
  distans D. Don 498
  dubium Roxb. 75
  ellipticum Thunb. 348
  europhyllum C. Chr. 326
  evectum G. Forst. 273
  excavatum Roxb. 332
  gardneri Mett. 317
  glandulosum Hook. 190, 191
  granulosum sensu Thwaites 492
  harpophyllum Zenker ex Kunze 134
  hirsutulum G. Forst. 290

insigne Blume 326
lanceolatum L. 342
lepidotum Willd. ex Schltdl. 343
leptophyllum L. 364
lineare Burm.f. 174
lingua auctt. 322
lingulatum Sw. 331
linnei Bory 314
macrocarpum Willd. 342
mediale Baker 185
membranaceum D. Don 327
membranifolium R.Br. 328
mollicomum Nees & Blume 195
moultonii Copel. 193
nigrescens Blume 328
nudum Kunze 340
obliquatum Blume 204
paludosum sensu Bedd. 499
palustre Burm.f. 46
pannosum Mett. 322
parasiticum L. 474
parasiticum Mett. 183
pedunculatum Salomon 334
phymatodes L. 332
porosum Mett. 323
porosum Wall. 323
proliferum (Retz.) Roxb. 489
pteroides Retz. 159, 470
pteropus Blume 329
punctatum Sw. 331
punctulatum Poir. 331
pyrolaefolium Goldm. 34
pyrrhorhachis Kunze 498
quercifolium L. 313
repandulum Mett. 200
rugulosum sensu Bedd. 78
scolopendrium Burm. 332
sparsisorum Desv. 314
speluncae L. 76
subfalcatum var. *β glabrum* Bedd. 200
subpinnatum C.B. Clarke 545
subtriphyllum Hook. & Arn. 165
subtripinnatum C.B. Clarke 545
subtruncatum Bory 503
tenericaule Hook. 466
thwaitesii Bedd. 194